Optical Compressive Imaging

SERIES IN OPTICS AND OPTOELECTRONICS

Series Editors: **E Roy Pike**, Kings College, London, UK
Robert G W Brown, University of California, Irvine, USA

Recent titles in the series

Optical Compressive Imaging
Adrian Stern

The Limits of Resolution
Geoffrey de Villiers and E. Roy Pike

Polarized Light and the Mueller Matrix Approach
José J Gil and Razvigor Ossikovski

Light—The Physics of the Photon
Ole Keller

Advanced Biophotonics: Tissue Optical Sectioning
Ruikang K Wang and Valery V Tuchin (Eds.)

Handbook of Silicon Photonics
Laurent Vivien and Lorenzo Pavesi (Eds.)

Microlenses: Properties, Fabrication and Liquid Lenses
Hongrui Jiang and Xuefeng Zeng

Laser-Based Measurements for Time and Frequency Domain Applications: A Handbook
Pasquale Maddaloni, Marco Bellini, and Paolo De Natale

Handbook of 3D Machine Vision: Optical Metrology and Imaging
Song Zhang (Ed.)

Handbook of Optical Dimensional Metrology
Kevin Harding (Ed.)

Biomimetics in Photonics
Olaf Karthaus (Ed.)

Optical Properties of Photonic Structures: Interplay of Order and Disorder
Mikhail F Limonov and Richard De La Rue (Eds.)

Nitride Phosphors and Solid-State Lighting
Rong-Jun Xie, Yuan Qiang Li, Naoto Hirosaki, and Hajime Yamamoto

Molded Optics: Design and Manufacture
Michael Schaub, Jim Schwiegerling, Eric Fest, R Hamilton Shepard, and Alan Symmons

An Introduction to Quantum Optics: Photon and Biphoton Physics
Yanhua Shih

Optical Compressive Imaging

Edited by

Adrian Stern

Department of Electrical and Computer Engineering
Ben-Gurion University of the Negev
Beer-Sheva, Israel

CRC Press is an imprint of the
Taylor & Francis Group, an **informa** business
A TAYLOR & FRANCIS BOOK

CRC Press
Taylor & Francis Group
6000 Broken Sound Parkway NW, Suite 300
Boca Raton, FL 33487-2742

First issued in paperback 2020

ISBN-13: 978-1-4987-0806-7 (hbk)
ISBN-13: 978-0-367-78268-9 (pbk)

Library of Congress Cataloging-in-Publication Data

Names: Stern, Adrian, author.
Title: Optical compressive imaging / Adrian Stern.
Description: Boca Raton : CRC Press, 2017. | Series: Series in optics and optoelectronics | Includes bibliographical references and index.
Identifiers: LCCN 2016038159 | ISBN 9781498708067 (acid-free paper)
Subjects: LCSH: Compressed sensing (Telecommunication) | Optical data processing.
Classification: LCC TK5102.9 .S754 2017 | DDC 006.6--dc23
LC record available at https://lccn.loc.gov/2016038159

Visit the Taylor & Francis Web site at
http://www.taylorandfrancis.com

and the CRC Press Web site at
http://www.crcpress.com

To my beloved wife, Sharon; wonderful kids: Elad, Ma'ayan, and Idan; my mother, Nora; and in the memory of my father, Iuliu.

Contents

Section II Compressive Imaging Systems

Section III Multidimensional Optical Compressive Sensing

Section IV Compressive Sensing Microscopy

Section V Phase Retrieval

Preface

> He will, besides, endeavor to find such an hypothesis which would require the least complicated motion and the least number of spheres: he will therefore prefer an hypothesis which would explain all the phenomena of the stars by means of three spheres to an hypothesis which would require four spheres.
>
> **Moses Maimonides (c. 1190)***

Ever since the seminal works of Donoho and Candes et al. were published in 2006, compressive (or compressed) sensing (CS) has attracted a great deal of attention in a variety of areas, including applied mathematics, computer science, and engineering—in fact, in almost every field that involves data sensing. Compressive sensing has influenced many scientists and engineers to reconsider the way information is captured, represented, processed, and stored. Maybe what has made CS so interesting and attractive is the fact that it has broken the sensing paradigm that, in practice, dominated for nearly a century—namely, what has followed in the wake of the celebrated Shannon–Nyquist sampling theory, which provides the sampling condition for band limited signals. Since every physical signal and system has a finite bandwidth, this theory is universal. Compressive sensing builds upon another feature that is virtually universal to all signals and systems, namely, sparsity. Sparsity expresses the idea that discrete physical signals depend on a number of degrees of freedom that is much smaller than the number of samples. For continuous signals, this means that the "information rate" signal may be much smaller than implied by its bandwidth. Sparsity can be considered as one of the manifestations of the *parsimony principle* (also known as Ockham's razor) that presumes that simpler theories, explanations, and representations are preferable to more complex or complicated ones. The quote mentioned earlier from Moses Maimonides (c. 1190 in 1904) writings is an example of the utility of this principle, which is used to refute Ptolemy's proposal of an epicyclical model for the revolution of planets around the Earth. Compressive sensing exploits the fact that most signals are sparse or compressible in the sense that

* Maimonides, M., *The Guide for the Perplexed*, New York: E.P. Dutton, 1904, Part 2, Chapter 11, c. 1190, translated by M. Friedlander (1903).

they have concise representations under appropriate transforms. By exploiting the sparsity, together with appropriate sensing schemes and reconstruction algorithms, CS enables reduction of the number of samples, compared to that required by traditional Shannon–Nyquist sampling theorem. The possibility of reduction of the number of measurements has obviously generated a lot of interest in a world in which the capture of sensed data and images is increasing continuously. This holds true particularly for optical sensing and imaging, for which the typical volume of data is both huge and highly compressible.

Since CS can substantially reduce the number of measurements needed to accurately capture a scene, it has the potential of making a huge impact on optical system design and performance. In general, CS has potential value whenever the marginal cost (in terms of size, weight, power, price, etc.) of a measurement is high. It offers reduced cost owing to both the reduction in the number of photosensors needed and the concomitant reduction in sensor size and weight; it also offers a reduction of the acquisition time and increase of the frame rates. It may also offer improved detector-noise-limited measurement fidelity, because the same total number of photons can be measured using fewer photodetectors. It also offers the possibility of breaking resolution limits.

Given the high level of interest in CS and recognizing the potential impact of CS in optical applications, this book aims to introduce the reader to the evolving field of optical compressive sensing. This book intends to teach, enlighten, and provide the reader with an introduction to CS in the optics realm. It is intended to serve as a reference as well as an introduction to researchers and industry professionals who may wish to apply CS in optical sensing and imaging problems. The target audience includes optics practitioners and optical system designers, electrical and optical engineers, mathematicians (applied and theoretical), and signal processing professionals that seek to better understand the practical needs in optical systems. It is appropriate for both academic researchers (faculty, graduate students, and postdoctoral researchers) and industry professionals.

During the last decade, a large corpus of research devoted to CS research has been published. In fact, since the idea of CS was first proposed about a decade ago, there have been thousands of papers written on this topic. The growth in publications is geometric, with many new theoretical and practical results being published. The first few years of CS research focused largely on basic sensing designs, mathematical analysis of random measurements, reconstruction guarantees, and algorithms for signal reconstruction. Many excellent tutorials and books were written providing a theoretical introduction to CS (Eldar and Kutyniok 2012; Foucart and Rauhut 2013). However, despite the remarkable advances that have been made within the theory of CS, its application in practical settings has generally lagged behind. More recently, there has been increased attention devoted to the particular details of effectively applying CS in various application areas. This book is the first to focus on the application of CS in optical sensing and imaging. Its purpose is to support the growing community of active researchers in this area, and it benefits from the contributions of internationally recognized authors, each bringing insights from the perspective of a variety of optical applications.

It should be mentioned that both the theory of CS and most of its optical application have numerous antecedents. This is, in fact, common to almost any scientific discovery and engineering breakthrough, but should, nevertheless, not undermine their value and potential impact. Compressive sensing theory evolved as a branch of sparse signal processing and has made such a strong impact that many today identify the entire field of sparse and redundant representations with it. In this book, we will try to focus on CS only. In the same way as CS has its antecedents, optical systems designed to capture data parsimoniously were proposed long before the emergence of CS theory. However, the novelty derives from the fact that the CS theory provided clear guidelines for implementing systems that capture data more efficiently. Readers are introduced to these guidelines in the first section of this book. First, you will be exposed to the basic CS notions necessary for optical implementations. The section gradually leads through more advanced theory and applications. The second section is devoted to compressive imaging designs. Various design schemes and examples for THz compressive imaging are presented. The field of THz imaging is a prominent example of the case in which CS can help to overcome the high price of sensors. The third section is devoted to the application of CS for multidimensional optical sensing. Multidimensional optical sensing is appealing for CS application because of the typically high redundancy within such signals and the high acquisition effort. In the fourth section, the utility of CS for microscopy is demonstrated. It is shown how CS helps to achieve high spatial and temporal resolution with STORM and various on-chip microscopy designs. Section five deviates from the *classical* linear CS model and addresses the phase retrieval problem. It introduces new theoretical tools for this well-studied problem in optics. The section aims to demonstrate the power of these new tools and invite its optical application.

I believe that the ideas and concepts presented in this book will help to harness the CS model in the realm of optical applications. We hope that it will foster further improvements and create many successful collaborations based on a deeper understanding of concepts at the intersection of optics and signal/image processing.

I thank the authors for their outstanding contributions and the T&F editors and staff for their support.

References

Eldar, Y. C. and G. Kutyniok, eds., *Compressed Sensing: Theory and Applications.* Cambridge University Press, Cambridge, UK, 2012.

Foucart, S. and H. Rauhut, *A Mathematical Introduction to Compressive Sensing*, Vol. 1, no. 3. Birkhäuser, Boston, MA, 2013.

Maimonides, M., *The Guide for the Perplexed*, New York: E.P. Dutton, 1904, Part 2, Chapter 11, c. 1190, translated by M. Friedländer (1903).

Editor

Adrian Stern, PhD, is an associate professor and head of the Department of Electro-Optical Engineering at Ben-Gurion University of the Negev, Israel. He earned all his degrees in electrical and computer engineering at Ben-Gurion University of the Negev. He was a postdoctoral fellow at the University of Connecticut, Storrs, and also served as a senior research and algorithm specialist for GE Molecular Imaging, Israel. In 2014–2015, during his sabbatical leave, he was a visitor scholar and professor at Massachusetts Institute of Technology (MIT). His current research interests include compressive imaging, 3D imaging, computational imaging, and phase-space optics.

Stern has published over 150 technical articles in leading peer-reviewed journals and conference proceedings, more than one quarter of them being invited papers. He is an elected fellow of SPIE and a member of the IEEE and Optical Society of America (OSA). He acted as editor for several journals, including the *Optics Express* journal for six years.

Editor

Contributors

Ravindra A. Athale
Electro-Optic and Infrared Imaging Sensors
Office of Naval Research
Arlington, Virginia

Isaac Y. August
Department of Electro-Optical Engineering
Ben-Gurion University of the Negev
Beer-Sheva, Israel

Pere Clemente
Central Service for Scientific Instrumentation
University Jaume I
Castelló, Spain

Vicente Durán
Institute of New Imaging Technologies
University Jaume I
Castelló, Spain

Yonina C. Eldar
Department of Electrical Engineering
Technion—Israel Institute of Technology
Haifa, Israel

Albert Fannjiang
Department of Mathematics
University of California, Davis
Davis, California

Mercedes Fernández-Alonso
Institute of New Imaging Technologies
University Jaume I
Castelló, Spain

Lu Gan
Department of Electronic and Computer Engineering
Brunel University
Uxbridge, United Kingdom

Babak Hassibi
Department of Electrical Engineering
California Institute of Technology
Pasadena, California

Ryoichi Horisaki
Graduate School of Information Science and Technology
Osaka University
Osaka, Japan

Bo Huang
Department of Pharmaceutical Chemistry
and
Department of Biochemistry and Biophysics
University of California, San Francisco
San Francisco, California

Esther Irles
Institute of New Imaging Technologies
University Jaume I
Castelló, Spain

Kishore Jaganathan
Department of Electrical Engineering
California Institute of Technology
Pasadena, California

Jun Ke
School of Optoelectronics
Beijing Institute of Technology
Beijing, People's Republic of China

Jesús Lancis
Institute of New Imaging Technologies
University Jaume I
Castelló, Spain

Mark A. Neifeld
College of Optical Sciences
Department of Electrical and Computer Engineering
The University of Arizona
Tucson, Arizona

Jonathan M. Nichols
Optical Sciences Division
Naval Research Laboratory
Arlington, Virginia

Aydogan Ozcan
Department of Electrical Engineering and
Department of Bioengineering
University of California, Los Angeles
Los Angeles, California

Yair Rivenson
Department of Electrical Engineering
University of California, Los Angeles
Los Angeles, California

Ikbal Sencan
Athinoula A. Martinos Center for Biomedical Imaging
Department of Radiology
Massachusetts General Hospital (MGH)
Harvard University
Charlestown, Massachusetts

Hao Shen
Shanghai Advanced Research Institute
Chinese Academy of Sciences
Shanghai, People's Republic of China

Yao-Chun Shen
Department of Electrical Engineering and Electronics
University of Liverpool
Liverpool, United Kingdom

Fernando Soldevila
Institute of New Imaging Technologies
University Jaume I
Castelló, Spain

Adrian Stern
Electro-Optical Engineering Unit
Ben-Gurion University of the Negev
Beer-Sheva, Israel

Enrique Tajahuerce
Institute of New Imaging Technologies
University Jaume I
Castelló, Spain

Lei Zhu
Department of Modern Physics
School of Physical Sciences
University of Science and Technology of China
Hefei, Anhui, People's Republic of China

I

The Theory of Compressive Sensing and Its Applications in Optics

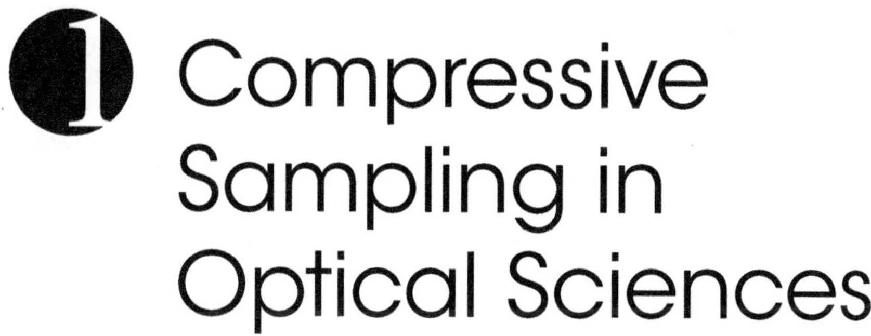

1 Compressive Sampling in Optical Sciences

Jonathan M. Nichols and Ravindra A. Athale

Contents

The field of compressive sampling (CS) is now roughly a decade old and has enjoyed a great deal of attention and measurable success in overcoming the temporal or spatial sampling constraints in modern data acquisition devices. Many of these successes have come in the Optical Sciences in diverse application areas such as imaging, spectroscopy, and photonic links, to name a few.

In the chapters to follow, the reader will be introduced to many of these success stories and obtain a good feel for what CS can and cannot be expected to accomplish in the coming years.

First, however, it makes sense to review the fundamentals of CS. In doing so, we revisit the basic questions: What are we trying to accomplish with CS? What is the mathematical framework by which CS operates? What are the conditions needed for successful implementation? And what are the main hurdles to implementation?

A very large number of papers on the mathematics of CS and the associated estimation algorithms have already been written. Our goal here is therefore not to give a comprehensive overview, but rather to cover the basics and point the reader toward those references we feel do a good job of explaining the field.

A Historical Perspective

Before launching directly into the topic of "compressed sensing," aka "compressive sampling" (CS), it is worthwhile to first think about what it is we are trying to accomplish when we sense the world around us. Sensing is a fundamental operation that serves as an interface between the physical world and our digital storage and communications infrastructure, that is, our information systems. On the front end, sensing involves process of transduction where signals in the physical world (electromagnetic waves, acoustic waves, temperature, pressure, etc.) are converted into an intermediate representation (usually electrical but also mechanical, optical, or chemical in some instances). This is followed by sampling/quantization in the sensed domain (e.g., spatial, temporal, spectral). Transduction followed by sampling/quantization results in measurements, which we can store, transmit, and perhaps later use to reconstruct (if possible) or display the analog quantity being sensed. For example, perhaps we want to replicate an image on a screen or playback an audio signal.

Now, most of the physical processes and parameters of interest are continuous while the measurements are usually discrete. This mapping from a continuum to a discrete space is intrinsically compressive as it takes an infinite set of points (an analog signal) and reduces it to a finite set of points (our measurement). In other words, the fact that we are capturing a discretized version of an analog process implies our sampling is compressive in nature.

Of course, we often exploit our prior knowledge about the information bandwidth of the signal and use this finite set of points to *infer* the signal values anywhere, that is, recover the continuous waveform. This is accomplished via Nyquist sampling, which according to the well-known Shannon–Whittaker–Kotelnikov theorem allows, in principle, lossless recovery of the information contained in the original analog signal.

Measurements are also intrinsically compressive when there is a dimensionality mismatch between the domain being sensed and the measurement space. This situation arises naturally in imaging systems when a 3D scene radiance map is projected on a 2D measurement space. In this case, the inverse problem involves inferring 3D information from multiple, engineered 2D measurements combined with prior knowledge and postprocessing. This has been a very active area of research for many decades.

Finally, any measurement instrument (sensing device) is bound to have finite bandwidth (spatial or temporal) or a broadened impulse response, which also results in compressive measurements. The research areas of superresolution or signal recovery have been extremely active area for many decades, and they also carry the same guiding principles of making multiple nonredundant measurements and utilizing prior knowledge in postmeasurement processing algorithms to recover (decompress) information that was contained in the compressed measurements.

The foregoing discussion is meant to convey that the notion of data compression is intrinsic in the process of sensing. It can therefore be convincingly argued that *all* sensing is compressive and that the optics community has been performing compressed sensing for a very long time. It is also a true statement that the use of discrete measurements of limited dimensionality and/or bandwidth have long been used to attempt to infer or recover the information of interest.

Indeed, designing sensing systems such that this intrinsic compression does not lead to loss of information has been an active research area from the beginning of the information age. Similarly, the strategy of combining multiple measurements and prior knowledge in postprocessing algorithms to recover or infer information that was implicitly embedded in the compressed measurements also has a long and rich history.

So what do we mean when we refer to "compressed sensing"? What is new in the current wave of research is the underlying theory that determines conditions under which the desired information can be inferred and with what accuracy. As in the past, the new theory provides us a principled means of incorporating *a priori* information about the data of interest. This new theory also suggests that rather than measure the data directly one instead measure linear *projections* of the desired data (a topic to be discussed in the "CS Theory: Mathematical Description" section). Typically, this means modulating the analog data in some predefined manner prior to discretely sampling. In the field of optics, recent manufacturing advances have facilitated an unprecedented ability to manipulate light, hence we have a means of experimentally implementing what the mathematics suggests. Some of these devices will be featured in the chapters that follow.

Another important distinction between historical sensing approaches and "today's" CS is that current techniques are increasingly used in those domains where compression is *not* intrinsically driven by the physics of the sensing system but is rather aimed at overcoming the spatial, temporal, or spectral resolution of the sampling device. From this perspective, the goal of CS is to reduce the burden on the physical sampling, storage, and communication subsystems at the cost of increasing the subsequent processing complexity (i.e., solving the inverse problem). This strategy is particularly attractive where the sensing is performed on platforms with constraints on size, weight, and power but where the resulting data exploitation is performed on workstations, which do not have the same constraints.

In short, while the field we have come to know as CS relies on different mathematics and improved hardware, the aims are more or less the same as historical sampling approaches. Discretely sample the "analog world" and then use those samples to accurately estimate the information of interest. In what follows, we review the basic principles of today's CS, present some applications of the theory, and discuss some of the hurdles to implementation.

CS Theory: Heuristic Description

The modern theory of CS gives us prescriptions for design and solving linear, underdetermined (more unknown values than data) inverse problems where typically we are trying to estimate data values at a higher temporal or spatial resolution than that provided by our sampling device. The more we know about the data, the better our estimates will be. The relevant theory behind the approach is now a decade old, having been set forth in References 1 and 2. However, before discussing these prescriptions and the requirements on *a priori* information, consider an intuitive example.

Assume there exists a wall with some unknown fraction painted white and the other black (see Figure 1.1a). The white side reflects a uniform exitance of 1 mJ/s/μm^2 while the black side reflects no light. Assume further that we would like to take a high-resolution picture of this wall with an $N \times N$ pixel camera where each pixel is 1 μm^2 in area. However, due to cost constraints, the only camera we have at our disposal consists of just two pixels, each measuring 10 μm × 5 μm = 50 μm^2 in area (exactly 50× the area of the detectors we wish we had).

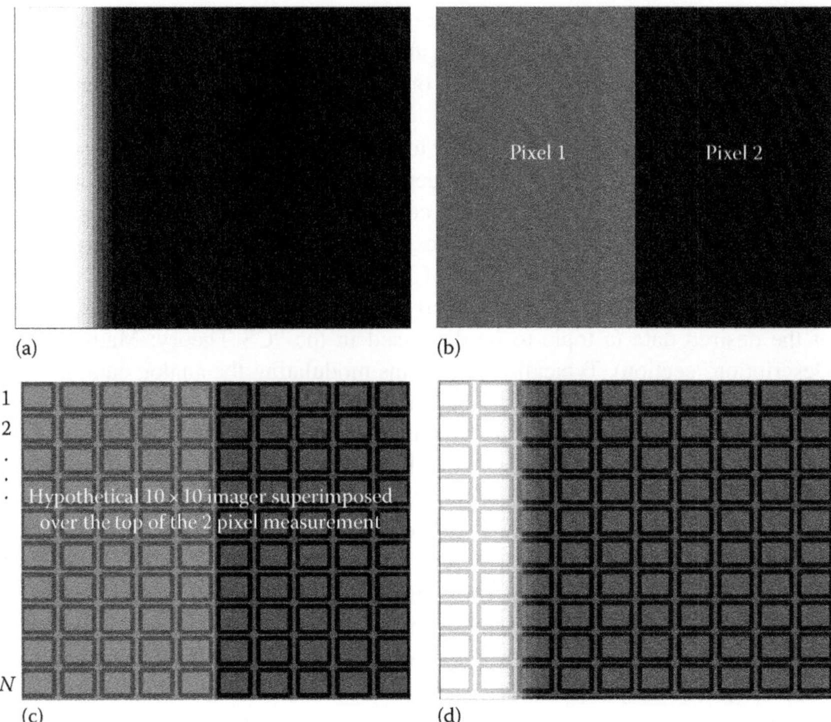

Figure 1.1 (a) True scene to be imaged. (b) Measurements collected by the 2 pixel imager. Pixel 2 sees no light; however, pixel 1 records a mixture of high and low intensity. (c) Superposition of the desired 100 pixel imager on top of the 2 pixel measurement. Based on the grayscale value of the 2 pixel measurement, we can reliably estimate the fraction of the 100 pixels that contain light and dark pixels. (d) Recovered 10 × 10 pixel image using only the 2 pixel measurement.

While we obviously cannot make the N^2 pixel measurements directly, we can make an accurate estimate for each of the N^2 values as follows.

First, imagine our hypothetical $N \times N$ pixel array overlaid on top of our 2 pixel detector as shown in Figure 1.1c. Denote C_1 as the (unknown) number of the N sensor columns that would be white if we had the larger sensor. Similarly, denote C_2 as the number of white pixel columns (again unknown) captured by the second measurement. Our large, cheap pixels simply sum up the impinging radiation. Hence, the energy captured by pixel 1 in 1 second is $g_1 = N \times C_1$ mJ (number of hypothetical 1 μm^2 pixels being irradiated) while pixel 2 records and $g_2 = N \times C_2$ mJ.

Given the measurements g_1, g_2, we can trivially solve for C_1 and C_2. Moreover, once C_1 and C_2 are known, we have sufficient information to estimate what the $N \times N$ image of our wall would look like using only two measurements! We simply create an $N \times N$ image in which the first $C_1 + C_2$ columns are white.

Now, what allowed us to solve for arbitrary "N," despite having only two measurements, was the fact that we knew an incredible amount of information about the piece of data we were interested in (our wall). Specifically, we knew our wall consisted of a transition from high intensity to zero intensity. Moreover, we knew that this "step" occurred at the same horizontal location in the image, regardless of the vertical position. That is to say, we only need to determine a single row of the image and then simply make all $N = 10$ rows identical. The only thing we did not know was where the low–high transition occurred. However, we were able to figure out this transition point by some simple reasoning and solving two equations for two unknowns. It is worth mentioning at this point that we have implicitly assumed that the transition will occur at exactly one of the N horizontal locations. What if the true transition had occurred precisely between two of our "high res" columns? This question lies at the heart of one of the big challenges to CS and will be addressed, at least in part, in the "Hurdles to Implementation" section.

CS Theory: Mathematical Description

The process we have just gone through can be generalized to more complex data types. However, doing so requires us to place our heuristic reasoning into a more solid mathematical framework. The first step is to come up with a mathematical description for our data, that is, formulate a *data model.*

In the aforementioned example, our data model was simply the family of N, 1D step functions where the step occurred at each of the possible "N" values along the horizontal axis of our hypothetical sensor array. If we were to describe this family of functions mathematically (again, for the $N = 10$ case), it would simply be

$$\Psi = \begin{bmatrix} 1 & 1 & 1 & 1 & 1 & 1 & 1 & 1 & 1 & 1 \\ 0 & 1 & 1 & 1 & 1 & 1 & 1 & 1 & 1 & 1 \\ 0 & 0 & 1 & 1 & 1 & 1 & 1 & 1 & 1 & 1 \\ 0 & 0 & 0 & 1 & 1 & 1 & 1 & 1 & 1 & 1 \\ 0 & 0 & 0 & 0 & 1 & 1 & 1 & 1 & 1 & 1 \\ 0 & 0 & 0 & 0 & 0 & 1 & 1 & 1 & 1 & 1 \\ 0 & 0 & 0 & 0 & 0 & 0 & 1 & 1 & 1 & 1 \\ 0 & 0 & 0 & 0 & 0 & 0 & 0 & 1 & 1 & 1 \\ 0 & 0 & 0 & 0 & 0 & 0 & 0 & 0 & 1 & 1 \\ 0 & 0 & 0 & 0 & 0 & 0 & 0 & 0 & 0 & 1 \end{bmatrix} \tag{1.1}$$

where each column represents one possible intensity distribution in the horizontal direction in Figure 1.1.

Thereby capturing each of the possible "N" transition points from high to low intensity. There are no units on these entries as they simply describe a high-to-low transition. Recall that each row of our example image (see Figure 1.1) was accurately described by the second column of this matrix. A wall with a different transition location would select a different column as the best match.

The selection of a particular column can be mathematically described by the product $\boldsymbol{\Psi\alpha}$, where "$\boldsymbol{\alpha}$" is an N-element vector with a single 1 placed at the position of the column we wish to select and the rest of the elements zero. The units of the signal are determined by $\boldsymbol{\alpha}$, which in this case is 1 mW/μm^2. Thus, the vector $\boldsymbol{\alpha}$ is the unknown we wish to solve for.

Note also that while in our example we knew that $\boldsymbol{\alpha}$ had a single nonzero entry, the more general model $\mathbf{f} = \boldsymbol{\Psi\alpha}$ states that the data, $\mathbf{f}$, is accurately described as a linear combination of the columns of $\boldsymbol{\Psi}$.

A second assumption we made was that the detector acts to sum the impinging intensity. This means that the measurement g_1 sums the intensity over the area associated with the first $N/2=5$ columns of our desired detector (an area of $N \times N/2$ μm²) while g_2 does the same for the second $N/2$ columns. Thus, letting

$$\Phi = \begin{bmatrix} N & N & N & N & N & 0 & 0 & 0 & 0 & 0 \\ 0 & 0 & 0 & 0 & 0 & N & N & N & N & N \end{bmatrix} \tag{1.2}$$

we arrive at a mathematical description relating our measurements, **g**, to the piece of data we hope to recover, **f**,

$$\mathbf{g} = \Phi\Psi\alpha + \mathbf{n} = \Phi\mathbf{f} + \mathbf{n} \tag{1.3}$$

where we have added the vector **n** to denote an additive noise process that inevitably contaminates any real measurement. Note this is just a linear system of equations where we have 2 equations (measurements) and the 10 unknown values in the vector $\boldsymbol{\alpha}$. We know from undergraduate mathematics that this system does not admit a unique solution in general. However, we have just shown that if enough is known about the data we seek to estimate a unique solution is possible. This is the essence of CS. With a discrete model for the data, $\boldsymbol{\Psi}$, and for the sampling hardware, $\boldsymbol{\Phi}$, unique solutions for the N elements in $\boldsymbol{\alpha}$ are possible given only $M < N$ measurements **g**. Overviews of this basic approach to sampling, and subsequently recovering, data can be found in References [3,4]. In the next section, we briefly review how one can solve for $\boldsymbol{\alpha}$ given **g** and discuss the conditions under which such a solution is expected to be accurate.

CS Solutions

Numerous references have discussed how to solve Equation 1.3 (see, e.g., [5,6]) and under what conditions such a solution is possible (see, e.g., [1,7]).

Recall that we are trying to *estimate* $\boldsymbol{\alpha}$ given **g** (indeed, if we could sample **f** directly, we would not require CS in the first place). To accomplish this, we form the estimate $\hat{\alpha}$ (hat denotes estimate), which then yields $\hat{\mathbf{f}} = \Psi\hat{\alpha}$. In other words, we are really estimating the coefficients of our signal model so that, when multiplied by the model matrix, we obtain an estimate of the data of interest at the desired spatial or temporal resolution.

While different approaches to solving (1.3) exist, one of the most frequently used in practice is the so-called l_1 minimization, which forms

$$\hat{\alpha} = \arg\min \|\mathbf{g} - \mathbf{A}\alpha\|_2^2 + \tau\|\alpha\|_1 \tag{1.4}$$

where $\mathbf{A} = \Phi\Psi$ is the so-called $M \times N$ *sampling matrix*. The first term is the usual minimization of discrepancy between data and model (see Equation 1.3) while the second term minimizes the sum of the elements in $\boldsymbol{\alpha}$, that is, the so-called l_1 norm.* The degree to which we penalize the sum of the elements is dictated by

* The pth norm of a vector $\boldsymbol{\alpha} \in \mathbb{C}^N$ is defined as $l_p = \left(\sum_{i=1}^{N} |\alpha_i|^p\right)^{1/p}$.

the positive constant τ. The idea is to find solutions that accurately describe the data, but with as few nonzero values as possible.

This is a familiar concept in science, sometimes referred to as the principle of parsimony. The idea of finding models that (1) fit the data but (2) possess a minimum number of terms is a common theme in model selection, for example, when using the Akaike information criteria. Of course, penalizing the sum of the model parameters is not the same as penalizing the number of nonzero terms. Strictly speaking, this would be accomplished by minimizing the l_0 norm (penalize the number of nonzero coefficients) as a regularizer [8]. Indeed, a number of "greedy" algorithms are available for solving (1.4), but with l_0 replacing l_1. The reason for using l_1 as a surrogate is that it renders (1.4) convex, hence amenable to algorithmic approaches to optimization that are in many ways preferable to the greedy approach. It should also be mentioned that parsimony can be enforced in other ways. For example, some have taken a Bayesian estimation philosophy in which choice of a prior distribution guides the solution toward one with few nonzero terms (see, e.g., [9]). For the remainder of this introduction, however, we seek a solution to the problem (1.4).

Considerable literature has been devoted to the estimator (1.4) and fast solvers are freely available (mainly in MATLAB®) from many of the authors. A comprehensive listing of references and solvers is maintained at http://dsp.rice.edu/cs. One of the more popular algorithms is the Gradient Projections for Sparse Reconstruction (GPSR) given by Figueiredo et al. [5] and the more general SpaRSA algorithms [10]. For its simplicity and ease of use, GPSR is used in the examples of subsequent sections as the algorithm of choice.

The use of l_1 regularization predates CS (see, e.g., [11]); however, what CS brings to the table is a set of conditions on both $\boldsymbol{\alpha}$ and $\mathbf{A}$ that guarantee an accurate estimate. First, the vector $\boldsymbol{\alpha}$ must be *sparse*, meaning it possesses a small number K of nonzero values. In our toy example, $\boldsymbol{\alpha}$ only had a single ($K=1$) nonzero value, made possible by the fact that we knew that our signal was accurately described by only a single column in $\boldsymbol{\Psi}$. The key ratio in determining the "solvability" of Equation 1.3 will therefore, not surprisingly, be K/M, that is, the number of nonzero values we seek to estimate relative to the number of acquired data.

Secondly, our measurements must be linear combinations (i.e., projections) of the values in $\boldsymbol{\alpha}$, as determined by the sampling matrix $\mathbf{A}$. In other words, the measurement process needs to be accurately modeled by the matrix $\boldsymbol{\Phi}$ that projects the signal $\mathbf{f}=\boldsymbol{\Psi\alpha}$ onto a set of measurements $\mathbf{g}$. Perhaps the most important (and much researched) question is therefore what constitutes a "good" sampling matrix?

It turns out that the error in the estimate can be bounded for a given sampling matrix $\mathbf{A}$. To begin, we state that a given $\mathbf{A}$ matrix satisfies the restricted isometry property (see References 1 and 12) of order K and parameter $\delta_K \in [0,1]$ if

$$(1-\delta_K)\|\alpha\|_2^2 \leq \|A\alpha\|_2^2 \leq (1+\delta_K)\|\alpha\|_2^2 \tag{1.5}$$

holds for all sparse vectors $\boldsymbol{\alpha}$ having no more than K nonzero entries. Such matrices are denoted RIP(K, δ_K). This is a Parseval-like inequality and speaks to the degree to which the energy in the compressed samples is captured in the signal model coefficients $\boldsymbol{\alpha}$.

If **A** satisfies RIP($2K$, δ_{2K}) with $\delta_{2K} < \sqrt{2} - 1$, and the noise vector in (1.3) is such that $\|\mathbf{n}\|_2 \leq \varepsilon$, then the estimator (1.4) obeys the bound

$$\|\alpha - \hat{\alpha}\|_2 \leq C_{1,K}\varepsilon + C_{2,K}\frac{\|\alpha - \alpha_K\|_1}{\sqrt{K}} \tag{1.6}$$

where $C_{1,K}$ and $C_{2,K}$ are constants that depend on K but not on N or M. The term $\boldsymbol{\alpha}_K$ is the best K-sparse approximation of $\boldsymbol{\alpha}$; that is, $\boldsymbol{\alpha}_K$ is the approximation obtained by keeping the K largest entries of $\boldsymbol{\alpha}$ and setting the others to zero. In other words, the error grows linearly with the noise level and shrinks according to how well the data can be approximated with K coefficients [4].

It turns out that sampling matrices populated with independent, randomly chosen entries tend to satisfy the RIP property; hence, many CS strategies involve the random modulation of a signal prior to capture (the random modulation is captured by the hardware model $\mathbf{\Phi}$, which, in turn, creates a "random" **A** matrix). Certainly, other nonrandom sampling matrices can be (and have been) used in designing a CS system [13].

Of course, the aforementioned development does not provide much in the way of a practical guide for building systems. Defining the RIP property for a given sampling strategy and assessing the error bounds is clearly a nontrivial proposition [12]. An alternative is to look at the *mutual coherence* among the columns of **A**. Defining the Gram matrix $\mathbf{G} \equiv \mathbf{A}^{\mathrm{T}}\mathbf{A}$, we can define

$$\mu(\mathbf{A}) \equiv \max_{\substack{1 \leq i,j \leq N \\ 1 \neq j}} |G_{ij}| \tag{1.7}$$

which takes the maximum inner product among all N^2 possible elements of the matrix **G**. Unlike the RIP property, coherence is fairly straightforward (if not memory intensive) to compute for a given instantiation of **A**. It can be shown that if $K \leq (\mu(\mathbf{A})^{-1} + 1)/4$, then the estimation error associated with (1.3) is given by [14]

$$\|\alpha - \hat{\alpha}\|_2^2 \leq \frac{4\varepsilon^2}{1 - \mu(\mathbf{A})(4K - 1)} \tag{1.8}$$

The bounds given by (1.6) and (1.8) are useful in the sense of predicting when a given system architecture (as embodied in **A**) is likely to produce good estimates (recovery) of the data from the compressed samples. While underdetermined linear systems have certainly been solved prior to CS (see again [11]), theoretical guarantees such as these have increased our understanding of such problems.

It can therefore be stated that a "good" CS system is one that minimizes coherence among the columns of **A** while at the same time provides an accurate K-term model of the data. The former property can be at least partially controlled by selecting the physical hardware (i.e., choosing $\mathbf{\Phi}$) that gives us "good" compressed samples **g**. The latter requirement, a good K-term approximation, is predicated on our ability to model the data. In our toy example, the hardware is simply a detector that sums impinging intensity and can therefore be modeled by the simple matrix (1.2). However, in subsequent examples, we will show how random modulation of a piece of data prior to sampling produces a $\mathbf{\Phi}$ matrix with desirable properties.

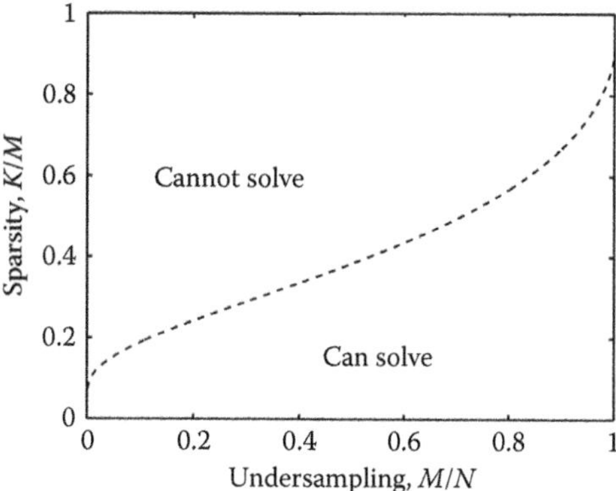

Figure 1.2 Problem space associated with compressive sampling. Below the curve solutions are generally possible while above the curve they are not. The key problem parameters are the degree of undersampling (M/N) and the sparsity of our data model relative to the number of acquired data (how good our data mode is). (Adapted from Donoho, D.L. and Tanner, J., *IEEE Trans. Inform. Theory*, 56(4), 2002, 2010.)

While concepts such as RIP and mutual coherence are useful and can be used in design, we have found yet another theory to be of more value when deciding when we are and are not likely to successfully recover our data. Begin by defining the key ratios M/N, that is, how few samples we collect relative to the number of data we want to estimate, and K/M, which is how sparsely we can model the data relative to the number of compressed samples.

Given these two parameters, the work of Donoho and Tanner [7] tells us when we can and cannot estimate $\boldsymbol{\alpha}$, provided we have chosen $\mathbf{A}$ correctly (presumably in accordance with the RIP or low coherence criteria). Figure 1.2 shows a curve dividing the problem space into two regions.

Above this curve, we cannot hope to reliably estimate our signal from the compressed samples; below the curve we can. This curve accurately describes the performance of a large number of CS systems and we have found it to be an excellent guidepost in predicting the performance of actual CS hardware (see, e.g., [15]). What the curve implies is not surprising: if we can sparsely model a piece of data, we can undersample by a large amount and still recover at a high resolution. The converse is also true.

In short, we choose $\boldsymbol{\Phi}$ (our hardware model) to minimize coherence of the sampling matrix and $\boldsymbol{\Psi}$ (our signal model) to minimize K and then measure $\mathbf{g}$. If the aforementioned conditions are met, a number of software tools are available for accurately estimating $\boldsymbol{\alpha}$ via (1.4), even when the M elements of $\mathbf{g}$ are much less than the N desired values in $\boldsymbol{\alpha}$. Some brief examples of optical CS systems are provided next.

CS Examples in Optical Sciences

Imaging Architectures

In order to facilitate the estimation of $\boldsymbol{\alpha}$, the most common CS architectures modulate the data at fast temporal or spatial scales prior to sampling, thus giving rise to rectangular $\boldsymbol{\Phi}$ matrices with more columns than rows. It is therefore

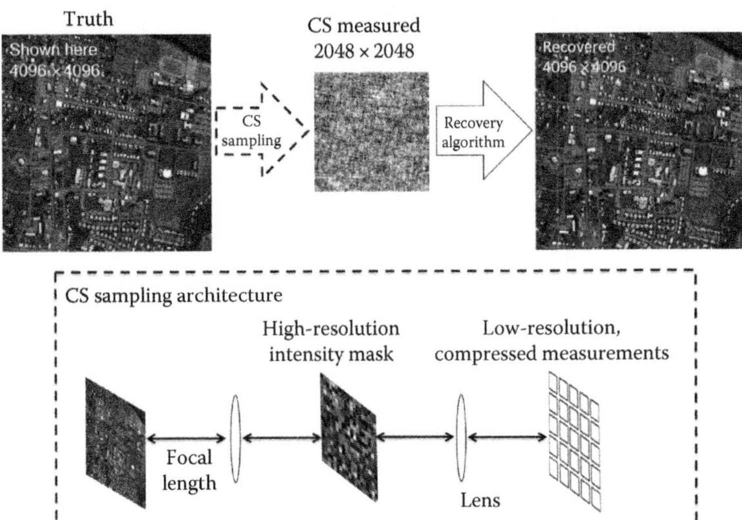

Figure 1.3 One possible compressive imaging architecture. The scene intensity is randomly modulated by a mask (coded aperture) in the Fourier plane before being imaged onto a low-resolution focal plane. The low-resolution measurements appear to possess little relationship to the true image. However, based on our knowledge that there is structure in the scene (roads, buildings, etc.) combined with our knowledge of the intensity mask allows us to accurately estimate (recover) the full 4096 × 4096 image from the 2048 × 2048 compressed measurements.

the hardware model that causes the problem to be underdetermined. Moreover, as we have discussed, random modulation gives rise to $\mathbf{\Phi}$ matrices with the needed properties. In the context of imaging, this can be accomplished by modulating the incoming light with a coded aperture, located at a specific point in the lens assembly (often the Fourier plane) [16]. However, although the entries are chosen randomly from some probability distribution, they are values that we know and can therefore be used to specify $\mathbf{\Phi}$ in the recovery algorithm. A typical CS architecture for such an imaging device is shown in the lower portion of Figure 1.3.

Given the position of the coded aperture in the lens assembly, it can be shown that this setup is effectively modulating the Fourier transform of the image. This is the so-called "random convolution" architecture described in Reference 16.

As an example, consider the sailboat image in Figure 1.4. Directly sampling this 256 × 256 image would yield the image in the lower left corner of the figure. Instead, we can use the 128 × 128 compressed measurements to estimate the full-resolution image. The estimated image is shown in the lower right position of the figure.

The estimate is good, although we can see some mild artifacts. Such artifacts are common in imagery recovered from CS devices and are currently a topic of research. As we will discuss in the next section, these artifacts are at least partially due to the fact that it is very difficult to come up with a K-sparse model for an image that allows recovery for even a modest amount of undersampling. As we have already seen (Figure 1.2), the ability to accurately recover is strongly influenced by how small we can make K relative to how much we wish to undersample.

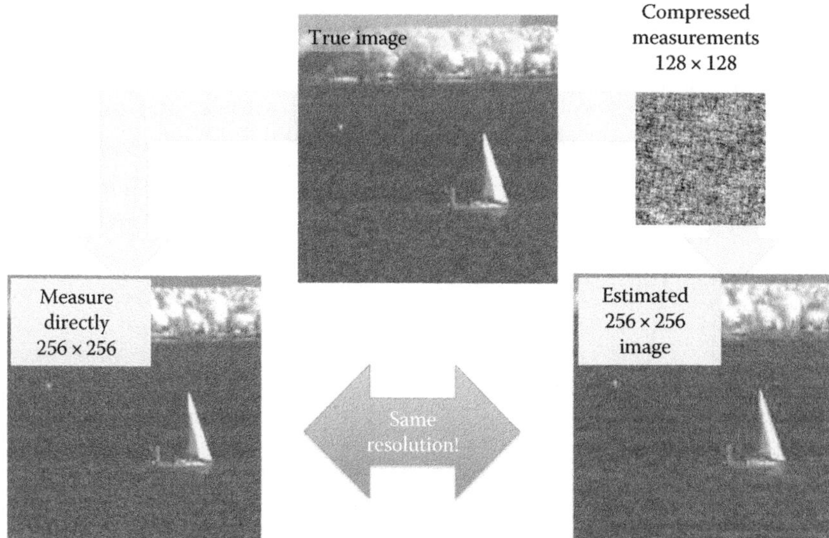

Figure 1.4 Second example of CS process described in Figure 1.3. A sailboat is imaged with a short-wave infrared camera using a 256 × 256 pixel camera. The 128 × 128 compressed measurements are also shown. Based on these measurements, the full 256 × 256 image can be accurately recovered using the GPSR estimation algorithm.

Nonetheless, researchers have successfully produced CS imaging architectures [17], the most famous being the single-pixel camera [18]. Other imaging architectures that use CS principles to improve either power consumption [19] or performance of a particular imaging task have also been developed [20].

Optical ADC

There are many reasons for moving to optical architectures for certain communication devices including increased bandwidth, low loss, and low size and weight of the components. While an optical fiber can support instantaneous bandwidths into the terahertz, current analog-to-digital converters (ADCs) are limited to the tens of giga-samples per second (GS/s). CS provides at least one means of sampling in the tens of GS/s yet recovering information at sampling intervals consistent with a TS/s digitizer.

In an effort to explore this possibility, the system given in Figure 1.5 was built and tested in the laboratory in recovering both single multitone signals. In keeping with the aforementioned CS theory, the incoming signal is first modulated with a pseudorandom bit sequence prior to filtering and then sampling at the low rate. This sequence of operations is accurately modeled by a matrix $\mathbf{\Phi}$ that, when multiplied by the signal model, gives a sampling matrix $\mathbf{A}$ with low coherence among the columns. Details of the experiment, including the coherence calculations for this system, can be found in Reference 21.

The signal model $\mathbf{\Psi}$ for this problem is simply the Fourier basis, containing sines and cosine functions at a discrete set of frequencies. That is to say, for a single tone (with arbitrary phase), we would expect $K=2$, that is, the signal can be represented by the sum of a single sine and single cosine vector.

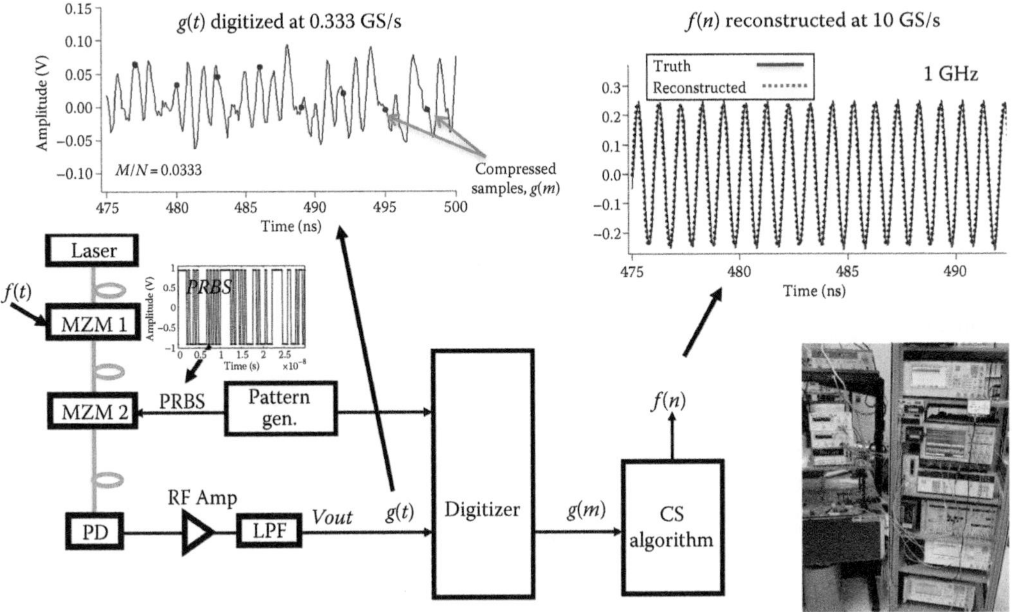

Figure 1.5 A compressive photonic ADC. The compressed measurements are collected at a low sampling rate from a randomly modulated filtered waveform $g(t)$. These samples, along with a Fourier basis (signal model) and the $l-1$ minimization algorithm, are sufficient to accurately reconstruct a 1 GHz tone, despite the fact that this signal is above the Nyquist sampling limit associated with the compressed samples.

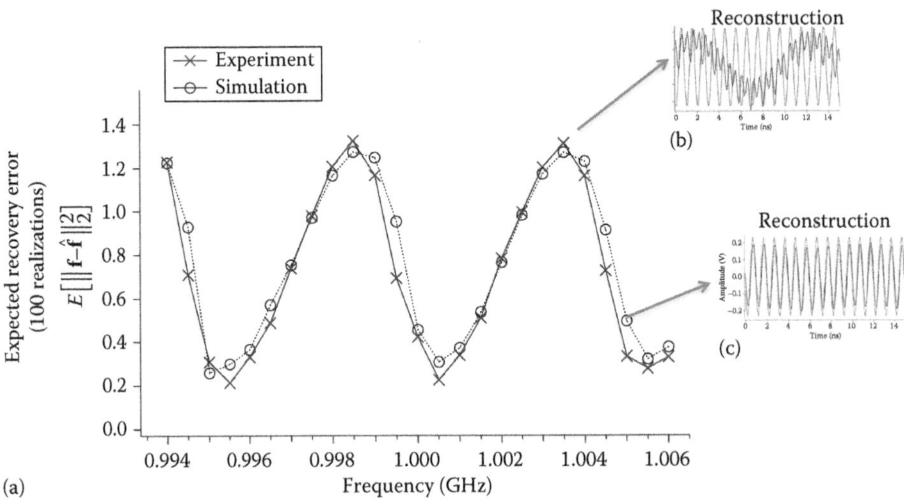

Figure 1.6 (a) Expected recovery error as a function of signal frequency. The discrete signal model was defined for frequencies 0.995, 1.000, and 1.005 GHz resulting in significantly smaller recovery errors (i.e., signal and model frequencies are the same). The recovery error is largest when the mismatch between signal frequency and signal model frequencies is greatest. Sample time-domain recovered signals are shown for both accurate (c) and inaccurate (b) signal models.

However, what we did not mention was that the frequency we sought to recover (1 GHz) was *exactly* one of the frequencies in the Fourier basis, that is, we chose the discrete Fourier model $\mathbf{\Psi}$ to include a vector at 1 GHz. We know from signal processing that if the frequency of interest is not exactly represented in the Fourier model, a much larger number of coefficients K are required to describe the signal, a phenomenon commonly referred to as spectral "leakage."

Thus, if we repeat the experiment in Figure 1.5 but with tones that vary in frequency with respect to the model, we see the reconstruction error fluctuates considerably (see Figure 1.6). This is a direct consequence of the varying sparsity K required to accurately model a waveform when there is uncertainty in a model parameter (in this case, frequency).

This is a particularly troubling result as we seldom know the form of the data model so precisely, much less know the parameters! We have two choices, either reduce the degree of undersampling (thereby reducing the benefits of CS) or find alternative strategies to (1.4) for improving estimator accuracy due to errors in the signal or data model.

Hurdles to Implementation

As in most areas of science, success depends on the quality of our model. As we have seen, we need to model both hardware $\mathbf{\Phi}$ and data $\mathbf{\Psi}$. Not surprisingly, successful implementation depends on the ability to which these two matrices predict what is observed, that is, the compressed samples $\mathbf{g}$.

The last example clearly illustrated one of the biggest hurdles to implementing CS, namely, finding a good (sparse + accurate) data model $\mathbf{\Psi}$. Even though we knew exactly the form of the model (sines and cosines), we did not know the precise frequencies (parameters) of the model. This is the same problem we mentioned at the end of our heuristic example in the "CS Theory: Mathematical Description" section. In that example, we also knew the functional form of our signal model (high-to-low intensity transition) but did not know the model parameter (point of transition).

One solution is to find a more flexible signal model and/or estimator. For example, what if both the Fourier coefficients and the frequencies were left as unknowns in (1.4)? Can algorithms be developed that estimate both quantities? It turns out that at least for some types of data, the answer is yes. The alternating convex search (ACS) algorithm, for example, does precisely that by updating first the Fourier coefficients, and then the frequencies of the nonzero vectors in $\mathbf{\Psi}$, until a convergence criteria is met [22]. The result is that the recovery error is no longer dependent on the signal coinciding with a signal model vector.

Revisiting the CS photonic link described earlier, we implemented the ACS estimator documented in Reference 22. The results are shown in Figure 1.7 and clearly show that variations in signal frequency no longer cause the previously observed fluctuations in the error since frequency is being accurately estimated along with the coefficient values.

This more flexible estimator is extremely useful when it comes to estimating *multiple* tones at arbitrary frequencies. Adding tones increases K by two (sine and cosine required for arbitrary phase); however, for every frequency not included in the signal model, the increase in K is likely an order of magnitude larger.

Figure 1.8 shows both frequency and time-domain reconstructions of a four-tone signal using traditional CS estimators (specifically the GPSR estimator mentioned earlier) and one that estimates the entire signal model (frequency and amplitude).

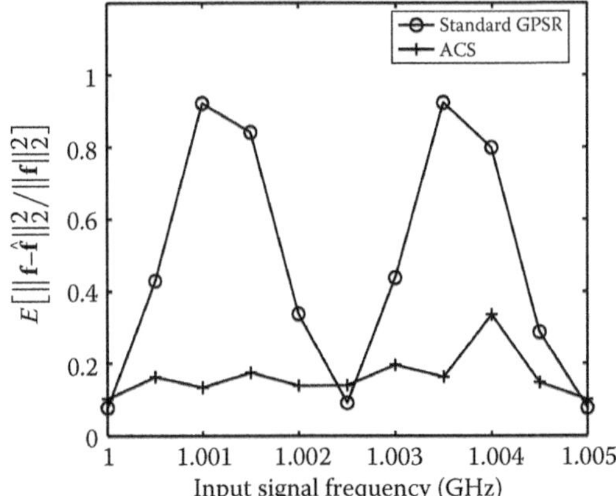

Figure 1.7 Experimental implementation of a recovery algorithm which estimates both signal frequencies and coefficient magnitudes [22]. The result is that the recovery error no longer requires the signal to coincide with a signal model vector.

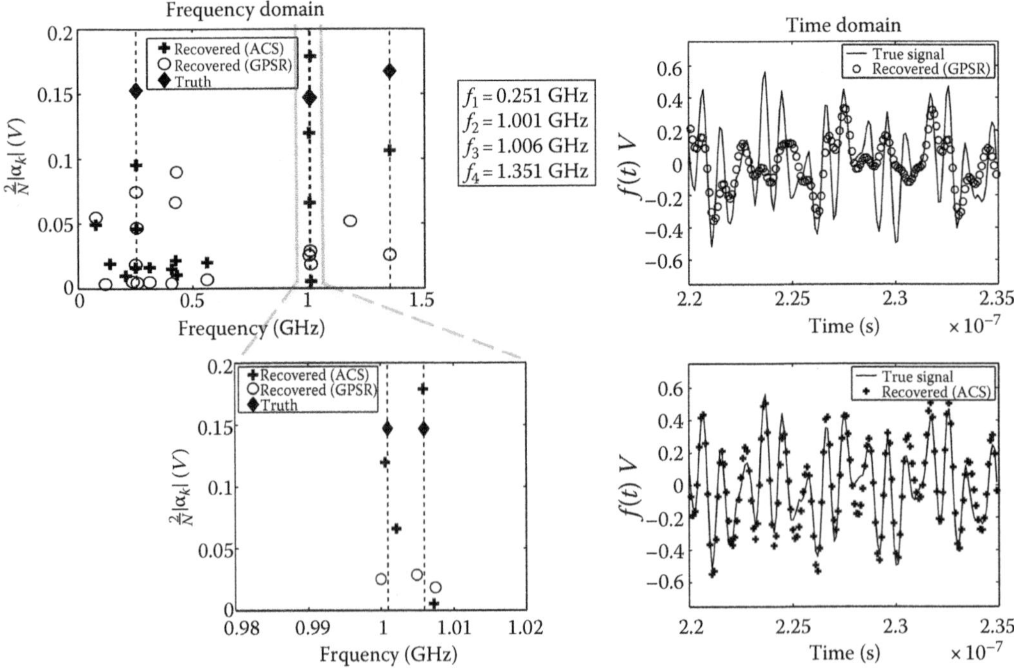

Figure 1.8 Improvement realized in a four-tone experiment by using the ACS estimator. Two of the tones are very closely spaced relative to the signal bandwidth; all four tones are not included in the set of signal model frequencies. The standard estimator struggles for the aforementioned reasons while ACS offers noticeable improvement. This is particularly obvious in the time-domain reconstructions of the signal appearing at the right of the figure.

There is clearly a benefit to the more flexible model and associated estimator, the main cost being the computational time required to find the additional model parameters (frequencies). This cost turns out to be quite modest in this case (see again [22]). Other strategies for mitigating this general problem in the context of CS waveform sampling are discussed in References 23 and 24 and in Chapter 3.

The second issue primary obstacle is developing an accurate hardware model. Here, we have at least one advantage, namely, that since we design the CS system, we control all aspects of the hardware construction and can calibrate the system. In other words, if there are discrepancies between what the hardware is doing (as manifest in our compressed samples $\mathbf{g}$) and what we think it is doing (our model predicted samples), we can make adjustments to $\mathbf{\Phi}$ to better match the data. We do not require the hardware model to be sparse either, only that it be accurate. Accuracy without sparsity is a much easier task! Nonetheless, hardware model calibration is a big part of building a working CS system as has been stressed in the literature (see, e.g., [25]) and will be highlighted in the examples of Chapters, 4, 6, 9, and 12.

Summary

Our aims in this chapter were to provide a general overview of CS in optics, give a flavor for the basic principles of operation, and provide a few examples to illustrate both the benefits and challenges. The remaining chapters will cover in much greater detail some of the system architectures that researchers are using to overcome both spatial and temporal sampling limitations of conventional optical devices. These range from imagers that can resolve details of a scene at angular resolution better than the pixel-level field of view, to spectroscopic methods that superresolve the wavelengths of the sensed light, and to video recording devices that temporally compress in order to estimate high-speed motions that exceed the rate of frame capture. It will become apparent that there exists a great deal of design freedom in compressive architectures and that much remains to be explored.

References

1. E. J. Candes and T. Tao, Decoding by linear programming, *IEEE Trans. Inf. Theory* **51**(12), 4203–4215, 2005.
2. D. L. Donoho, Compressed sensing, *IEEE Trans. Inf. Theory* **52**(4), 1289–1306, 2006.
3. R. Baraniuk, Compressive sensing, *IEEE Signal Process. Mag.* **24**(4), 118–121, 2007.
4. R. M. Willett, R. F. Marcia, and J. M. Nichols, Compressed sensing for practical optical imaging systems: A tutorial, *Optical Engineering* **50**(7), 072601, 2011.
5. M. A. T. Figueiredo, R. D. Nowak, and S. J. Wright, Gradient projection for sparse reconstruction: Application to compressed sensing and other inverse problems, *IEEE J. Sel. Top. Signal Process.* **1**(4), 586–597, 2007.
6. J. M. Bioucas-Dias and M. A. T. Figueiredo, A new TwIST: Two-step iterative shrinkage/thresholding algorithms for image restoration, *IEEE Trans. Image Process.* **16**(12), 2992–3004, 2007.
7. D. L. Donoho and J. Tanner, Exponential bounds implying construction of compressed sensing matrices, error-correcting codes, and neighborly polytopes by random sampling, *IEEE Trans. Inf. Theory* **56**(4), 2002–2016, 2010.

8. A. M. Bruckstein, D. L. Donoho, and M. Elad, From sparse solutions of systems of equations to sparse modeling of signals and images, *SIAM Rev.* **51**(1), 34–81, 2009.
9. L. He, H. Chen, and L. Carin, Tree-structured compressive sensing with variational Bayesian analysis, *IEEE Signal Process. Lett.* **17**(3), 233–236, 2010.
10. S. J. Wright, R. D. Nowak, and M. A. T. Figueiredo, Sparse reconstruction by separable approximation, *IEEE Trans. Signal Process.* **57**(7), 2479–2493, 2009.
11. B. K. Natarajan, Sparse approximate solutions to linear systems, *SIAM J. Comput.* **24**(2), 227–234, 1995.
12. R. G. Baraniuk, M. Davenport, R. A. DeVore, and M. B. Wakin, A simple proof of the restricted isometry property for random matrices, *Constr. Approx.* **28**(3), 253–263, 2008.
13. A. Ashok and M. A. Neifeld, Compressive imaging: Hybrid measurement basis design, *J. Opt. Soc. Am. A* **28**(6), 1041–1050, 2011.
14. D. L. Donoho, M. Elad, and V. N. Temlyakov, Stable recovery of sparse overcomplete representations in the presence of noise, *IEEE Trans. Inf. Theory* **52**(1), 6–18, 2006.
15. F. Bucholtz and J. M. Nichols, Compressive sampling demystified, *Opt. Photon. News* 46–49, October 2014.
16. J. Romberg, Compressive sampling by random convolution, *SIAM J. Imag. Sci.* **2**(4), 1098–1128, 2009.
17. R. F. Marcia, R. M. Willett, and Z. T. Harmany, Compressive optical imaging: Architectures and algorithms, in *Optical and Digital Image Processing: Fundamentals and Applications*, eds. G. Cristobal, P. Schelkens, and H. Thienpont. Wiley-VCH, New York, pp. 485–505, 2011.
18. M. Duarte, M. Davenport, D. Takhar, J. Laska, T. Sun, K. Kelly, and R. Baraniuk, Single-pixel imaging via compressive sampling, *IEEE Signal Process. Mag.* **25**(2), 83–91, March 2008.
19. Y. Oike and A. El Gamal, A 256 × 256 CMOS image sensor with delta-sigma-based single-shot compressed sensing, in *IEEE International Solid-State Circuits Conference (ISSCC) Digest of Technical Papers*, San Francisco, CA, pp. 386–387, February 2012.
20. A. Mahalanobis and B. Muise, Object specific image reconstruction using a compressive sensing architecture for application in surveillance systems, *IEEE Trans. Aerosp. Electron. Syst.* **45**(3), 1167–1180, 2009.
21. J. M. Nichols and F. Bucholtz, Beating Nyquist with light: A compressively sampled photonic link, *Opt. Express* **19**(8), 7339–7348, 2011.
22. J. M. Nichols, C. V. McLaughlin, and F. Bucholtz, A solution to basis mismatch in an experimental, compressively sampled photonic link, *Opt. Express* **23**(14), 18052–18059, 2015.
23. G. C. Valley, G. A. Sefler, and T. J. Shaw, Sensing RF signals with the optical wideband converter, in *Proceedings of SPIE, Broadband Access Communication Technologies VII*, San Francisco, CA, February 2013, vol. 8645.
24. G. Tang, B. N. Bhaskar, P. Shah, and B. Recht, Compressive sensing off the grid, in *IEEE 50th Annual Allerton Conference on Communication, Control, and Computing (Allerton)*, Monticello, IL, pp. 778–785, 2012.
25. M. E. Gehm and D. J. Brady, Compressive sensing in the EO/IR, *Appl. Opt.* **54**(8), C14–C22, 2015.

Quick Dictionary of Compressive Sensing Terms for the Optical Engineer

Adrian Stern

Contents

Introduction

Readers with classical optics education may not find the compressive sensing (CS) notions introduced in Chapter 1 so natural or corresponding well with the familiar optical science and engineering lexicon. This is partly because CS theory was born and has evolved mainly in fields such as information theory, computational mathematics, and signal processing, and, therefore, its formalism, notions, and concepts do not necessarily overlap those found in typical optics textbooks. This chapter aims to bridge the gap between CS formalism and that of standard optics textbooks. Readers who found the mathematics of Chapter 1 sufficiently clear, or do not feel the need for a better intuitive understanding, may skip this chapter and proceed to the following chapters.

In this chapter, we try to facilitate the introduction of the optics-trained reader into the realm of CS. We do this by reviewing some of the basic CS notions and examine them using optics language with basic optical system examples. We hope that this will help the reader to gain some intuition about the underlining CS notions. Unfortunately, sometimes, intuition may not go hand in hand with rigor and, therefore, in some such cases we chose to sacrifice the rigor and completeness for the sake of clarity.

In the "Interpretation of the Compressive Sensing Model" section, we revisit the CS system mathematical model and explain it in the context of a simple imaging system. We review the roles of the system matrix and its adjoint and explain these in terms of point spread functions (PSFs) and optical transfer functions (OTFs). In the "How Appropriate Is a Sensing Matrix for CS?" section, we review the "restricted isometry property" and provide some intuition into its meaning using transfer function terminology. Then we turn to the CS notion of mutual coherence, explain it, and distinguish it from the similar term found in statistical optics. Finally, in the "Two-Point Imaging Resolution Revisited Using CS Tools" section, we employ the CS notions that we have just related to optics to break the classical Rayleigh resolution limit of an imaging system.

Interpretation of the Compressive Sensing Model

The common mathematical description of the CS process is as given in (1.3):

$$\mathbf{g} = \mathbf{\Phi f} + \mathbf{n}, \tag{2.1}$$

where

- $\mathbf{g}$ and $\mathbf{f}$ are m and n dimensional vectors, respectively
- $\mathbf{\Phi}$ is a matrix with m rows and n columns
- $\mathbf{n}$ is the additive noise vector

Now, let us take a conventional incoherent imaging system as a case study, as shown in Figure 2.1a. We discretize the object plane to have n points and the image plane to have m pixels. The vectors $\mathbf{g}$ and $\mathbf{f}$ are the vectorized representation of the irradiance in the image and object planes, $g(x',y')$ and $f(x,y)$, respectively. The conversion from the 2D distributions $g(x',y')$ and $f(x,y)$ to the vectors $\mathbf{g}$ and $\mathbf{f}$, respectively, can be done by lexicographical reordering. With these notations, the algebraic sensing model is as in Equation 2.1. The imaging system matrix $\mathbf{\Phi}$ defines the object-to-image mapping while $\mathbf{n}$ represents additive noise such as thermal (Johnson) or "read noise." The output vector $\mathbf{g}$ for a given impulse f_i is defined by the i'th column of $\mathbf{\Phi}$. Thus, we see that the sensing matrix is built of n columns, each representing the (discrete) PSF corresponding to an impulse at location i in the object plane. In the geometrical approximation of the imaging system in Figure 2.1a, each object point is mapped to a single image point. In such a case, the system matrix is the identity matrix (Figure 2.1b). More generally, conventional imaging systems can be approximated as shift invariant systems (Goodman 1996; Brady 2009), meaning that the PSF shape is location independent; that is, the PSF exhibits the same distribution, which is only shifted with respect to the location of the impulse. For such systems, the matrix $\mathbf{\Phi}$ has the form of a Toeplitz matrix which in practical cases is conveniently approximated by a circulant matrix in the case of 1D imaging system or a block-circulnat matrix in the case of 2D imaging systems (Barrett and Myers 2003).

In CS setups, m is much smaller than n, meaning that the number of pixels in $\mathbf{g}$ is smaller than the number of object points in $\mathbf{f}$. In this case, the forward model (2.1) for the system in Figure 2.1a is ill-posed, and we cannot assure the reconstruction of all objects from the measurements $\mathbf{g}$. Therefore, for CS, we will need to properly encode the input image by one of the methods described in Chapter 5. For example, the imaging system in Figure 2.1a can be converted to performing

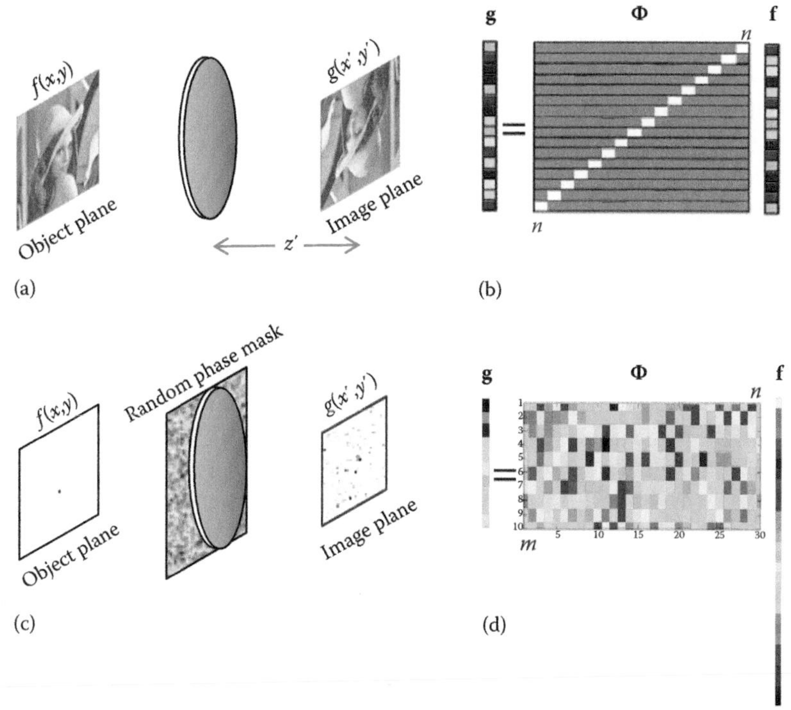

Figure 2.1 (a) Single-lens imaging system and (b) its sensing model (b). (c) By introducing a random phase transparency, (d) the object is encoded, as described figuratively by the model.

CS by introducing a random phase mask, as shown in Figure 2.1c (Stern and Javidi 2007). The respective sensing model is shown in Figure 2.1d.

Let us turn to $\mathbf{\Phi}^T$, the transpose of $\mathbf{\Phi}$, or, more generally, the adjoint (Hermitian conjugate) of $\mathbf{\Phi}$. The operator $\mathbf{\Phi}^T$ plays an important role in the analysis of the sensing process and is used in most CS reconstruction algorithms. With reference to our examples in Figure 2.1, we notice that the i'th row of $\mathbf{\Phi}^T$ is the (complex conjugate of the) i'th column of $\mathbf{\Phi}$, which is the PSF of a point source at location i in the object plane. Alternatively, the j'th column in $\mathbf{\Phi}^T$ is the j'th row in $\mathbf{\Phi}$ and thus represents the weight of all object points contributing to pixel g_j. In other words, the columns of $\mathbf{\Phi}^T$ represent the PSF in the object plane generated by an impulse in the image plane. Thus, the optical meaning of $\mathbf{\Phi}^T$ can be understood by swapping the roles of the object and image planes. In accordance with the Helmholtz reciprocity principle, the matrix $\mathbf{\Phi}^T$ represents the process of back projection of the light from right to left in the optical system in Figure 2.1. Note that in the *geometrical approximation* of the *ideal system* with the model in Figure 2.1a, $\mathbf{\Phi}^T$ is also the inverse of $\mathbf{\Phi}$, that is $\mathbf{\Phi}^T = \mathbf{\Phi}^{-1}$. This means that if we capture the image $g(x',y')$ of an object $f(x,y)$ with the system in Figure 2.1a and then swap the roles of the source and the camera, that is, we generate the optical distribution $g(x',y')$ and measure the image in the original object plane, then we would capture the exact replica of the original object $f(x,y)$.

How Appropriate Is a Sensing Matrix for CS?

CS stands on three pillars: sparsity of the signal, sparse signal-promoting algorithms, and appropriate sensing mechanisms. What is a good sampling matrix?

In the "CS Solutions" section in the previous chapter, the *restricted isometry property* (RIP) and *mutual coherence* were introduced to specify the quality of the CS sensing matrix. In the following, we shall revisit these two notions and try to put them in the context of optics.

Restricted Isometry Property

The RIP was introduced in (1.5):

$$(1-\delta_K)\|\mathbf{f}\|_2^2 \leq \|\boldsymbol{\Phi}\mathbf{f}\|_2^2 \leq (1+\delta_K)\|\mathbf{f}\|_2^2. \tag{2.2}$$

This property states that we require the distance between the measurement vectors, $\mathbf{g}_1$ and $\mathbf{g}_2$, of any two sparse signals, $\mathbf{f}_1$ and $\mathbf{f}_2$, to be proportional to the distance between the original signal vectors. (Here, we assume $\mathbf{f}$ is sparse; otherwise (2.2) should be rewritten with $\mathbf{f}$ replaced by its sparse decomposition $\boldsymbol{\alpha}$, and $\boldsymbol{\Phi}$ replaced by the overall sensing matrix $\boldsymbol{\Phi\Psi}$ defined in Chapter 1.) Such a property allows us to guarantee that, for low enough noise, two sparse vectors that are far apart from each other cannot lead to the same (noisy) measurement vector.

An equivalent expression to (2.2) can be written as the condition:

$$\left\|\boldsymbol{\Phi}_s^{\mathrm{T}}\boldsymbol{\Phi}_s - \mathbf{I}\right\|_2 \leq \delta_K, \tag{2.3}$$

to hold true for any full rank matrix $\boldsymbol{\Phi}_s$, that is obtained by restricting $\boldsymbol{\Phi}$ to have a number of columns S smaller than or equal to the sparsity of $\mathbf{f}$, $S \leq k$. Equation 2.3 implies that all eigenvalues of $\boldsymbol{\Phi}_s^{\mathrm{T}}\boldsymbol{\Phi}_s$ are approximately equal and, hence, $\boldsymbol{\Phi}_s^{\mathrm{T}}\boldsymbol{\Phi}_s$ is far from being noninvertible, thus providing robustness of the sensing process. A similar requirement is commonly used in imaging system analysis. Let us take, for example, an incoherent imaging system (e.g., Figure 2.1a) and carry out our analysis in the Fourier domain. We denote by $\hat{\mathbf{f}}$ and $\hat{\mathbf{g}}$ the Fourier transform of $\mathbf{f}$ and $\mathbf{g}$ so that $\mathbf{f} = \Im^{-1}\hat{\mathbf{f}}$ and $\hat{\mathbf{g}} = \Im\mathbf{g}$, where $\Im$ denotes the Fourier transform operator. Substituting in (2.1), we obtain

$$\hat{\mathbf{g}} = \Im\boldsymbol{\Phi}\Im^{-1}\hat{\mathbf{f}} + \hat{\mathbf{n}} = \mathbf{H}\hat{\mathbf{f}} + \hat{\mathbf{n}}. \tag{2.4}$$

The single-lens imaging system in Figure 2.1a can be approximated to be shift invariant. The systems matrix, $\boldsymbol{\Phi}$, of a shift invariant system is commonly approximated by a circulant matrix for which $\mathbf{H} = \Im\boldsymbol{\Phi}\Im^{-1}$ and is a diagonal matrix with the eigenvalues of $\boldsymbol{\Phi}$ on the diagonal. Thus, the Fourier transform of the output, $\hat{\mathbf{g}}$, is related to that of the input, $\hat{\mathbf{f}}$, by a simple elementwise multiplication with the diagonal elements of $\mathbf{H}$ that can be recognized as the samples of the *OTF*, $H(\omega_i)$ to be high above the noise level. Following this observation, we recognize (2.3) as the well-known OTF design guideline, requiring that square absolute of the OTF, $|H(\omega)|^2$. Equivalently, we require the square of the modulation transfer function (MTF) to be as close to unity as possible. The larger the MTF values are, the more robust the system is to noise, meaning that the signal can be reconstructed from the noisy measurements. Thus, the requirement of approximately uniform eigenvalues translates in our example to the requirement of an approximately uniform OTF, which is a common system design guideline.

Coherence

The mutual coherence was presented in the "CS Solutions" section as an alternative CS sensing matrix quality measure. The mutual coherence expression in (1.7) can be equivalently written as

$$\mu(\mathbf{\Phi}) \triangleq \max_{i \neq j} |\mu_{i,j}| = \max_{i \neq j} \frac{|\langle \phi_i, \phi_j \rangle|}{\|\phi_i\|_2 \|\phi_j\|_2}, \tag{2.5}$$

stating that $\mu(\mathbf{\Phi})$ is the largest absolute inner product between any two normalized columns, ϕ_i, ϕ_j, of $\mathbf{\Phi}$. In the examples of the "Interpretation of the Compressive Sensing Model" section, μ is the largest cross-correlation between the PSFs of all possible pairs of impulses in the object plane. As explained in the "CS Solutions" section, lower μ values guarantee better CS reconstruction. It can be shown that μ has values between $\sqrt{(n-m)/m(n-1)}$ and 1. Note that for $n \gg m$, the lower bound is approximately inversely proportional to the number of measurements, $\mu(\mathbf{\Phi}) \geq \sqrt{1/m}$.

The mutual coherence defined earlier should not be confused with the mutual coherence term used in statistical optics. In statistical optics, the *mutual coherence function* defines the cross-correlation of the field $u(\mathbf{r}, t)$ at pairs of positions, $\mathbf{r}_1$ and $\mathbf{r}_2$, and time instants separated by an interval τ:

$$\Gamma_{1,2}(\tau) \triangleq \langle u(\mathbf{r}_1, t+\tau), u^*(\mathbf{r}_2, t) \rangle. \tag{2.6}$$

Obviously, the two terms in (2.5) and (2.6) are not the same and do not measure the same quantities. Therefore, for example, a "coherent optical system" may or may not have a low $\mu(\mathbf{\Phi})$. To avoid confusion, in the following, we shall refer to $\mu(\mathbf{\Phi})$ in (2.5) as the *mutual parameter.*

With some effort, in some very particular cases, the mutual coherence and coherence parameters can be related. Such a relation can be found if we consider the *second-order* statistics of a *back*-projected light *field*. For instance, one can show that a CS mutual parameter μ is the maximum of the mutual intensity $\Gamma_{1,2}(0)$ in the object plane after back-propagating a totally incoherent pure field from the image plane back to the object plane.

Two-Point Imaging Resolution Revisited Using CS Tools

As an exercise using the CS notions and their optical meaning explained in the previous sections, we shall derive a new two-point imaging resolution limit. We shall do this by evaluating the coherence parameter μ of a coherent imaging system and using a reconstruction guarantee that depends on μ.

Let us consider an imaging system such as that shown in Figure 2.1a with coherent illumination, so that $f(x)$ and $g(x')$ are the input and output fields, respectively. In the following, we perform a 1D analysis for the sake of simplicity of notations. Under appropriate conditions (Goodman 1996), the system can be approximated to be shift invariant such that

$$g(x') = \int f(x_r) h(x' - x_r) dx, \tag{2.7}$$

where x_r is the reduced coordinate obtained by scaling x by the lateral magnification of the system M_L, $x_r = M_L x$, and $h(x)$ is the coherent impulse response (aka the coherent PSF), which is related by a Fourier transform to the *coherent transfer function* (*CTF*) (Brady 2009), aka the *amplitude transfer function*, $\hat{h}(u)$ (Goodman 1996). By sampling the image plane and the reduced object plane with an interval Δ, we can approximate (2.7) by

$$g[l] \approx \sum_{l} f[s]h[l-s], \tag{2.8}$$

where

$g[l] = g(l\Delta) \quad l = 1,\ldots,m$
$f[s] = f(sM\Delta) \quad s = 1,\ldots,n$
$h[l] = \Delta h(l\Delta)$

In relation to our model described in the "Interpretation of the Compressive Sensing Model" section, we identify the i and j columns of the sensing matrix as the shifted version of $h[l]$:

$$\phi_i = h[l-i], \quad \phi_j = h[l-j]. \tag{2.9}$$

The coherence parameter (2.5) is given, up to a normalization factor, by the inner product:

$$\langle \phi_i, \phi_j \rangle = \sum_{l=1}^{n} h[l-i]h^*[l-j] = \sum_{l=1}^{n} h[l-j-p]h^*[l-j] \equiv h[p] * h^*[p], \tag{2.10}$$

where

$p \triangleq i-j$
$*$ is the convolution operator

By using the convolution property of Fourier pairs

$$h[p] * h^*[p] = \Im_d^{-1}\left\{\hat{h}_d(e^{j\Omega})\hat{h}_d^*(e^{-j\Omega})\right\}, \tag{2.11}$$

where

$j = \sqrt{-1}$
$\Im_d^{-1}$ is the discrete parameter ("time") Fourier transform (DTFT) operator evaluated at frequencies Ω [rad/samples], related to the spatial frequency $u = \Omega/2\pi\Delta$ (1/m)

$\hat{h}_d$ is the DTFT of $h[l]$; $\hat{h}_d(e^{j\Omega}) = \Im_d\{h[l]\} \triangleq \sum_l h[l]e^{-j\Omega l}$. $\hat{h}_d$ can be related to the continuous parameter ("time") Fourier transform $\hat{h}(u)$ by

$$\hat{h}_d(e^{j\Omega}) = \sum_k \hat{h}\left(\frac{\Omega}{2\pi\Delta} + k2\pi\right). \tag{2.12}$$

Thus, (2.10) through (2.12) provide a means for calculating the *coherence parameter* (2.5) through the *transfer function* $\hat{h}(u)$.

Now let us apply the aforementioned result for a coherent imaging system. Assuming that the system produces images close to the center of the image field, the CTF can be approximated by the pupil function P (Goodman 1996):

$$\hat{h}(u) = P(-\lambda z'u), \tag{2.13}$$

where

λ is the wavelength

z' is the imaging distance (Figure 2.1a)

u is the spatial frequency

Substituting in (2.10) through (2.12), we obtain

$$\langle \phi_i, \phi_j \rangle = \Im_d^{-1}\left\{ \sum_k P\left(-\lambda z'\left(\frac{\Omega}{2\pi\Delta} + k2\pi\right)\right) \sum_k P^*\left(-\lambda z'\left(-\frac{\Omega}{2\pi\Delta} + k2\pi\right)\right)\right\}. \tag{2.14}$$

Now let us consider the common case of an imaging system with a circular aperture with diameter D. Then the pupil function is

$$P(r) = circ\left(\frac{r}{D}\right), \tag{2.15}$$

where r is the radial coordinate. Its respective coherent PSF is expressed in terms of the *jinc* (aka "besinc") function (Goodman 1996):

$$h(r) = \frac{D^2}{\lambda^2 z'^2} jinc\left(\frac{r}{\lambda z'/D}\right), \tag{2.16}$$

where $jinc(\rho) \triangleq J_1(\pi\rho)/\pi\rho$. The appropriate CTF, $\hat{h}(u)$, and the DTFT $\hat{h}_d$ are plotted in Figure 2.2 for the case where the sampling interval Δ is sufficiently small to avoid aliasing; that is, the replicas in (2.12) and in Figure 2.2c do not overlap. Note that in this case $\hat{h}_d(e^{j\Omega})\hat{h}_d^*(e^{-j\Omega}) = \hat{h}_d(e^{j\Omega})$ and (2.14) reduces to

$$\langle \phi_i, \phi_j \rangle = \Im_d^{-1}\left\{ \sum_k P\left(-\lambda z'\left(\frac{-\Omega}{2\pi\Delta} + k2\pi\right)\right)\right\} = h[p\Delta] = \frac{D^2}{\lambda^2 z'^2} jinc\left(\frac{D}{\lambda z'}(i-j)\Delta\right). \tag{2.17}$$

If we take sufficient samples m so that the energy leakage is negligible, then $\|\boldsymbol{\Phi}_j\|_2 = \|h[l-j]\|_2 \approx D/\lambda z'$ and the coherence parameter (2.5) is

$$\mu(\Phi) \triangleq \max_{i \neq j} |\mu_{i,j}| = \max_{p \triangleq i-j \neq 0} \left| jinc\left(\frac{D}{\lambda z'} p\Delta\right)\right|. \tag{2.18}$$

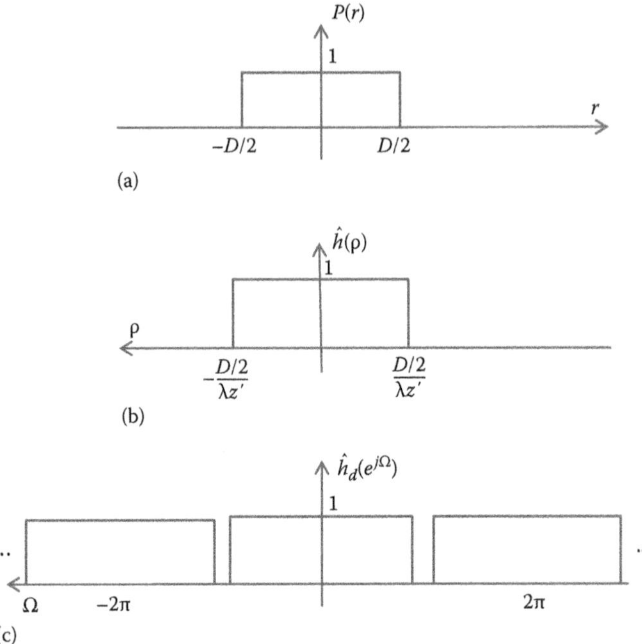

Figure 2.2 (a) Radial cross section of the pupil function of a circular lens, (b) the CTF along the spatial frequency in the radial direction ρ, and (c) the respective DTFT $\hat{h}_d(e^{j\Omega})$, appropriate to the sampled PSF. Note that $\hat{h}_d(e^{j\Omega})\hat{h}_d^*(e^{-j\Omega}) = \hat{h}_d(e^{j\Omega})$.

For small enough Δ, the expression in (2.18) monotonically decreases with *p*, so the maximum of (2.18) is achieved for $p = 1$ (that is for adjacent PSFs):

$$\mu(\mathbf{\Phi};\Delta) = \left| jinc\left(\frac{D}{\lambda z'}\Delta\right)\right|, \tag{2.19}$$

which is plotted in Figure 2.3.

Now that we have evaluated the coherence parameter for our imaging system, we can use it together with one of the many reconstruction guarantees based on μ (Eldar and Kutyniok 2012). For instance, in the case of negligible noise, a unique reconstruction solution of $\mathbf{g} = \mathbf{\Phi f}$ is guaranteed if the sparsity of **f** is $K \le \frac{1}{2}(1 + 1/\mu)$ (Donoho and Elad 2003, 2197–2202). For two object points $K = 2$, this implies that $\mu \le 1/3$. Figure 2.3 plots the coherence parameter (2.19) for our system. It can be seen that $\mu \le 1/3$ is fulfilled for

$$\Delta \ge 0.85\frac{\lambda z'}{D}. \tag{2.20}$$

Equation 2.18 implies that two points can be resolved if they are spatially separated by more than $0.85\lambda z'/D$ (i.e., the angular separation is greater than $0.85\lambda/D$ radians). This resolution limit is ~1.44 more relaxed than the classical Rayleigh resolution limit $\Delta \ge 1.22\lambda z'/D$.* In fact, since CS reconstruction

* Rayleigh derived this limit for incoherent imaging, but it is commonly used for coherent imaging systems as well.

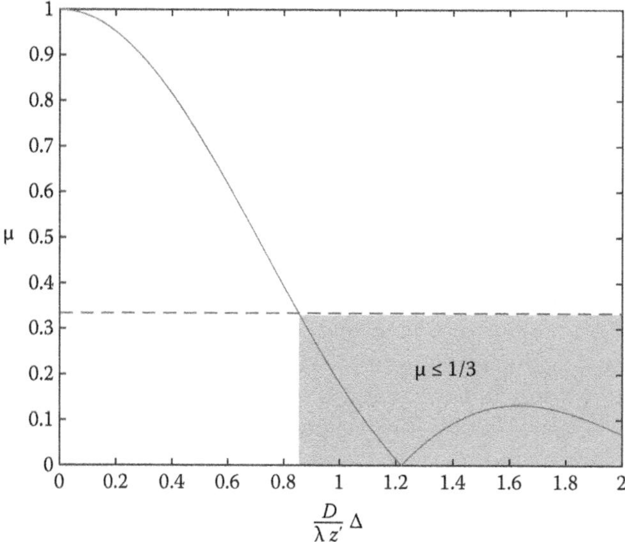

Figure 2.3 Coherence parameter μ for an imaging system with a circular aperture. For $\Delta(D/\lambda z') \geq 0.85$, the coherence parameter is smaller than 1/3 and two-point reconstruction is guaranteed.

guarantees using μ are pessimistic (Elad 2010) in practice, (2.20) is not the sharpest bound. One such case is demonstrated in compressive holography in section "Reconstruction Guarantees for 3D Object Tomography from Its 2D Hologram" of Chapter 8.

The fact that we broke the classical Rayleigh resolution limit should not surprise us. In fact, even Lord Rayleigh stressed that his criterion is rather a heuristically defined one and may depend on the system or application, and on uncertainties (noise) (Rayleigh 1879). In general, the resolution depends on various factors such as noise, prior information, technical conventions, and numerical recovery tools. Our resolution limit derivation can be adapted to account for noise by using, for instance, (1.8).

Before concluding this rather theoretical section, we wish to mention that CS techniques were used to demonstrate the reconstruction of optical signals beyond the classical limits in several works, such as Gazit et al. (2009), Rivenson et al. (2010), and Shechtman et al. (2011).

References

Barrett, H. H. and K. J. Myers. 2003. *Foundations of Image Science.* Wiley-VCH, 1584pp., October 2003.

Brady, D. J. 2009. *Optical Imaging and Spectroscopy.* John Wiley & Sons, Hoboken, NJ.

Donoho, D. L. and M. Elad. 2003. Optimally sparse representation in general (non-orthogonal) dictionaries via ℓ1 minimization. *Proceedings of the National Academy of Sciences of the United States of America* 100(5): 2197–2202.

Elad, M. 2010. *Sparse and Redundant Representations: From Theory to Applications in Signal and Image Processing.* Springer, New York.

Eldar, Y. C. and G. Kutyniok. 2012. *Compressed Sensing: Theory and Applications.* Cambridge University Press, New York.

Gazit, S., A. Szameit, Y. C. Eldar, and M. Segev. 2009. Super-resolution and reconstruction of sparse sub-wavelength images. *Optics Express* 17(25): 23920–23946.

Goodman, J. W. 1996. *Introduction to Fourier Optics*. McGraw-Hill, New York.

Rayleigh, L. 1879. XXXI. Investigations in optics, with special reference to the spectroscope. *The London, Edinburgh, and Dublin Philosophical Magazine and Journal of Science* 8(49): 261–274.

Rivenson, Y., A. Stern, and B. Javidi. 2010. Single exposure super-resolution compressive imaging by double phase encoding. *Optics Express* 18(14): 15094–15103.

Shechtman, Y., Y. C. Eldar, A. Szameit, and M. Segev. 2011. Sparsity based sub-wavelength imaging with partially incoherent light via quadratic compressed sensing. *Optics Express* 17: 23920–23946.

Stern, A. and B. Javidi. 2007. Random projections imaging with extended space-bandwidth product. *Journal of Display Technology* 3(3): 315–320.

Compressive Sensing Theory for Optical Systems Described by a Continuous Model

Albert Fannjiang

Contents

Introduction

A monochromatic wave u propagating in a heterogeneous medium is governed by the following Helmholtz equation:

$$\Delta u(\mathbf{r}) + \omega^2(1 + \nu(\mathbf{r}))u(\mathbf{r}) = 0, \quad \mathbf{r} \in \mathbb{R}^d, d = 2, 3 \tag{3.1}$$

where $\nu \in \mathbb{C}$ describes the medium heterogeneities. For simplicity, we choose the physical units such that the wave velocity is unity and the wavenumber equals the frequency ω.

The data used for imaging are the scattered field $u^s = u - u^i$ governed by

$$\Delta u^s + \omega^2 u^s = -\omega^2 \nu u, \tag{3.2}$$

or equivalently the Lippmann–Schwinger integral equation:

$$u^s(\mathbf{r}) = \omega^2 \int_{\mathbb{R}^3} \nu(\mathbf{r}')\,(u^i(\mathbf{r}') + u^s(\mathbf{r}'))G(\mathbf{r},\mathbf{r}')d\mathbf{r}'. \tag{3.3}$$

Here

$$G(\mathbf{r},\mathbf{r}') = \begin{cases} \dfrac{e^{i\omega|\mathbf{r}-\mathbf{r}'|}}{4\pi|\mathbf{r}-\mathbf{r}'|}, & d = 3 \\ \dfrac{i}{4}H_0^{(1)}(\omega|\mathbf{r}-\mathbf{r}'|), & d = 2, \end{cases} \tag{3.4}$$

is the Green function for the background propagator $(\Delta + \omega^2)^{-1}$, where $H_0^{(1)}$ is the zeroth-order Hankel function of the first kind.

We consider two far-field imaging geometries: paraxial and scattering. In the former, both the object plane and the image plane are orthogonal to the optical axis, while in the latter emission and detection of light can take any directions. We take u^s as the measured data in the former and the scattering amplitudes (see Equation 3.7) as the measured data in the latter (Figure 3.1).

- *Paraxial geometry*: For simplicity, let us state the 2D version. Let $\{z = z_0\}$ be the object line and $\{z = 0\}$ the image line. With $\mathbf{r} = (x, z_0)$, $\mathbf{r}' = (x', 0)$, we have

$$u^s(x, z_0) = Ce^{i\omega x^2/(2z_0)} \int_{\mathbb{R}} \nu(x',0)(u^i(x',0) + u^s(x',0))e^{i\omega(x')^2/(2z_0)}e^{-i\omega xx'/z_0}dx', \tag{3.5}$$

where C is a complex number.

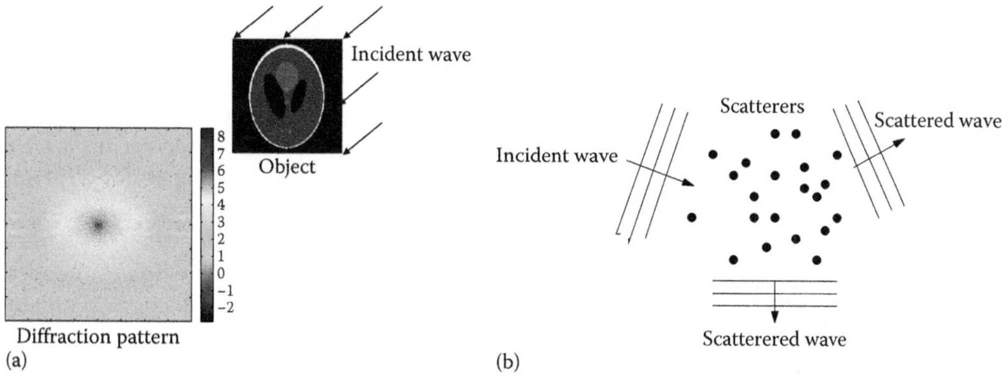

Figure 3.1 Two imaging geometries: (a) diffraction and (b) scattering.

- *Scattering geometry*: The scattered field has the far-field asymptotic (Born and Wolf 1999)

$$u^s(\mathbf{r}) = \frac{e^{i\omega|\mathbf{r}|}}{|\mathbf{r}|^{(d-1)/2}}\left(A(\hat{\mathbf{r}},\hat{\mathbf{d}} + \mathcal{O}\left(\frac{1}{|\mathbf{r}|}\right)\right), \quad \hat{\mathbf{r}} = \frac{\mathbf{r}}{|\mathbf{r}|}, \quad d = 2,3, \tag{3.6}$$

where the scattering amplitude A has the dimension-independent form

$$A(\hat{\mathbf{r}},\hat{\mathbf{d}}) = \frac{\omega^2}{4\pi}\int_{\mathbb{R}^d} v(\mathbf{r}')(u^{\mathrm{i}}(\mathbf{r}') + u^{\mathrm{s}}(\mathbf{r}'))e^{-i\omega\mathbf{r}'\cdot\hat{\mathbf{r}}}d\mathbf{r}'. \tag{3.7}$$

Note that since u in (3.5) and (3.7) is part of the unknown due to multiple scattering, the inverse problem is a nonlinear one. To deal with multiple scattering effects in compressive sensing, it is natural to split the inverse problem into two stages: In the first stage, we recover the *masked* objects

$$V(x) = v(x,0)(u^{\mathrm{i}}(x,0) + u^{\mathrm{s}}(x,0))e^{i\omega x^2/(2z_0)} \quad \text{(paraxial geometry)},$$

$$V(\mathbf{r}) = v(\mathbf{r})(u^{\mathrm{i}}(\mathbf{r}) + u^{\mathrm{s}}(\mathbf{r})) \quad \text{(scattering geometry)},$$

with the Fourier-like integrals in (3.5) and (3.7) as the sensing operators. In the second stage, we recover the true objects from the masked objects.

For most of this chapter, however, we will focus on the first stage or make the Born approximation to linearize the imaging problem and turn to the multiple scattering effect only in the "Inverse Multiple Scattering" section.

Outline

In the "Review of Compressive Sensing" section, we review the basic elements of compressive sensing theory including basis pursuit (BP) and greedy algorithms (orthogonal matching pursuit [OMP], in particular). We place greater emphasis on the incoherence properties than on the restricted isometry property because the former is much easier to estimate than the latter, even though the latter can also be established in several settings as we will see throughout this chapter. One thing to keep in mind about incoherence is that it is far beyond the standard notion of coherence parameter, which is the worst case metric (see Equation 3.17). The incoherence properties are fully expressed in the Gram matrix of the sensing matrix, also known as the coherence pattern. The second thing noteworthy about incoherence is that the standard performance guarantees expressed in terms of the coherence parameter often underestimate the actual performance of algorithms. Its usefulness primarily lies in providing a guideline for designing measurement schemes.

In the "Fresnel Diffraction with Pixel Basis" section, we consider the Fresnel diffraction with the pixel basis. The pixel basis, having a finite, definite size, is emphatically *not* suitable for point-like objects. Indeed, in order to build incoherence in the sensing matrix, it is imperative that the wavelength be shorter than the grid spacing. In other words, the pixel basis is suitable only for objects that are decomposable into "smooth" parts relative to the wavelength. The sparsity priors then come in two kinds: (1) there are few such parts with 1-norm as proxy and (2) there are few changes from part to part with the total variation as proxy (the "Total Variation Minimization" section). In the context of Fourier measurement, we introduce the notion of constrained joint sparsity to connect these two sparse priors and discuss basis pursuit (the "BPDN for Joint Sparsity" section) and orthogonal matching pursuit for joint sparsity (the "OMP for Joint Sparsity" section).

In contrast to the pixelated objects, point objects naturally do not live on grids. Such a problem arises in applications, for example, discrete spectral estimation among others. There is this fundamental tradeoff in using a grid to image point objects with the standard theory of compressive sensing: the finer the grid, the better the point objects are captured but the worse the coherence parameter becomes. In the "Fresnel Diffraction with Point Objects" section, we use the notion of coherence band to analyze the coherence pattern and design new compressive sensing algorithms for imaging well separated, off-grid point objects. In addition to off-grid point objects, the coherence-band techniques are also useful for imaging objects that admit a sparse representation in highly redundant dictionaries. One celebrated example is the single-pixel camera (SPC) discussed briefly in the "Redundant Dictionaries" section.

In the "Fresnel Diffraction with Littlewood–Paley Basis" section, we discuss Fresnel diffraction with sparse representation on the Littlewood–Paley basis, which is a slowly decaying wavelet basis in stark contrast to the pixel basis and the point-like objects. In this basis, the sensing matrix has hierarchical structures completely decoupled over different scales. In the "Near-Field Diffraction with Fourier Basis" section, we discuss near-field diffraction in terms of angular spectrum, which works out nicely with the Fourier basis.

In the "Inverse Scattering" section, we consider inverse scattering with the pixelated as well as point objects. Here, we focus on the design of sampling schemes (the "Sampling Schemes" section) and various coherence bounds for different schemes (the "Coherence Bounds" section).

In the "Inverse Multiple Scattering" section, we discuss multiple scattering of point objects and the appropriate techniques for solving the nonlinear inverse problem. The keys are the combination of the coherence-band and the joint sparsity techniques developed earlier.

In the "Inverse Scattering with Zernike Basis" section, we discuss inverse scattering with extended objects sparsely represented in the Zernike basis. In the "Interferometry with Incoherent Sources" section, we discuss interferometry with incoherent sources in astronomy. As a consequence of the celebrated van Cittert–Zernike theorem, the resulting sensing matrix has a similar structure to that for scattering with multiple inputs and outputs. The difference between them lies in the fact that for interferometry the inputs and outputs are necessarily correlated while for scattering the inputs and outputs can be independent. As a result, the (in)coherence properties of interferometry are more subtle and it is an ongoing problem to search for the optimal sensor arrays in optical interferometry in astronomy.

Review of Compressive Sensing

A distinctive advantage of compressive sensing is accounting for the finite, discrete nature of measurement by appropriately discretizing the object domain.

By a slight abuse of notation, we use $\|\cdot\|_p$ to denote the p-norm ($p \geq 1$) of functions as well as vectors, that is,

$$\|f\|_p = \left(\int |f(\mathbf{r})|^p d\mathbf{r}\right)^{1/p}, \quad f \in L^p, \tag{3.8}$$

$$\|\mathbf{f}\|_p = \left(\sum_{j=1}^{N} |f_j|^p\right)^{1/p}, \quad \mathbf{f} \in \mathbb{C}^N, \tag{3.9}$$

and $\|\mathbf{f}\|_0$ (the sparsity) denotes the number of nonzero components in a vector $\mathbf{f}$.

By discretizing the right-hand side of (3.5) or (3.7) and selecting a discrete set of data on the left-hand side, we rewrite the continuous models in the form of linear inversion

$$\mathbf{g} = \mathbf{\Phi f} + \mathbf{e}, \tag{3.10}$$

where the error vector $\mathbf{e} \in \mathbb{C}^M$ is the sum of the external noise $\mathbf{n}$ and the discretization error $\mathbf{d}$ due to model mismatch. By definition, the discretization error $\mathbf{d}$ is given by

$$\mathbf{d} = \mathbf{g} - \mathbf{n} - \mathbf{\Phi f}. \tag{3.11}$$

Consider the principle of basis pursuit denoising (BPDN) convex program

$$\min \|\mathbf{h}\|_1, \quad \text{s.t. } \|\mathbf{g} - \mathbf{\Phi h}\|_2 \leq \|\mathbf{e}\|_2 = \epsilon. \tag{3.12}$$

When $\epsilon = 0$, (3.12) is called basis pursuit. With the right choice of the parameter λ, BPDN is equivalent to the unconstrained convex program that is called Lasso (Tibshirani 1996)

$$\min_{\mathbf{z}} \frac{1}{2} \| \mathbf{g} - \mathbf{\Phi z} \|_2^2 + \lambda\varepsilon \| \mathbf{z} \|_1, \tag{3.13}$$

Both BPDN (3.12) and Lasso (3.13) are convex programs and have numerically efficient solvers (Chen et al. 2001, Boyd and Vandenberghe 2004, Bruckstein et al. 2009).

A fundamental notion in compressed sensing under which BP yields a unique exact solution is the restrictive isometry property (RIP) due to Candès and

Tao (2005). Precisely, let the restricted isometry constant (RIC) δ_s be the smallest nonnegative number such that the inequality

$$\kappa(1-\delta_s)\,\|\mathbf{h}\|_2^2 \le \|\Phi\mathbf{h}\|_2^2 \le \kappa(1+\delta_s)\,\|\mathbf{h}\|_2^2$$

holds for all $\mathbf{h} \in \mathbb{C}^N$ of sparsity at most s and some constant $\kappa > 0$. RIP means a sufficiently small δ_{2s} (see Equation 3.14).

Now we recall a standard performance guarantee under RIP.

Theorem 3.1 *(Candès 2008) Suppose the RIC of* $\mathbf{\Phi}$ *satisfies the inequality*

$$\delta_{2s} < \sqrt{2} - 1 \tag{3.14}$$

with $\kappa = 1$. *Then the solution* $\mathbf{f}_*$ *of BPDN (3.12) satisfies*

$$\|\mathbf{f}_* - \mathbf{f}\|_2 \le C_1\, s^{-1/2}\, \|\mathbf{f} - \mathbf{f}^{(s)}\|_1 + C_2\epsilon \tag{3.15}$$

for some constants C_1 *and* C_2. *where* $\mathbf{f}^{(s)}$ *consists of the s largest components, in magnitude, of* $\mathbf{f}$.

Remark 3.1 For general $\kappa \neq 1$, we consider the normalized version of (3.10)

$$\frac{1}{\sqrt{\kappa}}\mathbf{g} = \frac{1}{\sqrt{\kappa}}\Phi\mathbf{f} + \frac{1}{\sqrt{\kappa}}\mathbf{e}$$

and obtain from (3.15) that

$$\|\mathbf{f}_* - \mathbf{f}\|_2 \le C_1 s^{-1/2}\, \|\mathbf{f} - \mathbf{f}^{(s)}\|_1 + C_2\frac{\epsilon}{\sqrt{\kappa}}. \tag{3.16}$$

Note however that neither BPDN nor Lasso is an algorithm by itself and there are many different algorithms for solving these convex programs. Some solvers are available online, for example, YALL1 and the open-source code *L1-MAGIC* (http://users.ece.gatech.edu/~justin/l1magic/).

Besides convex programs, greedy algorithms are an alternative approach to sparse recovery. A widely known greedy algorithm is the OMP (Pati et al. 1993, Davis et al. 1997).

Algorithm 3.1 Orthogonal Matching Pursuit (OMP)

Input: $\mathbf{\Phi}$, $\mathbf{g}$.

Initialization: $\mathbf{f}^0 = 0$, $\mathbf{r}^0 = \mathbf{g}$ and $S^0 = \emptyset$

Iteration: For $j = 1, \ldots, s$

(1) $i_{\max} = \arg\max_i |\langle \mathbf{r}^{j-1}, \mathbf{\Phi}_i \rangle|,\ i \notin S^{j-1}$

(2) $S^j = S^{j-1} \cup \{i_{\max}\}$

(3) $\mathbf{f}^j = \arg\min_h \|\mathbf{\Phi h} - \mathbf{g}\|_2$ s.t. $\text{supp}(\mathbf{h}) \subseteq S^j$

(4) $\mathbf{r}^j = \mathbf{g} - \mathbf{\Phi f}^j$

Output: $\mathbf{f}^s$.

OMP has a performance guarantee in terms of the coherence parameter defined by

$$\mu(\mathbf{\Phi}) = \max_{k \neq l} \mu(k,l), \quad \mu(k,l) = \frac{|\Phi_k^\dagger \Phi_l|}{\|\Phi_k\| \|\Phi_l\|} \tag{3.17}$$

where Φ_k is the kth column of $\mathbf{\Phi}$, $\mu(k, l)$ is the pairwise coherence parameter and the totality $[\mu(k, l)]$ is the *coherence pattern* of the sensing matrix $\mathbf{\Phi}$. Here and in the following, $\dagger$ denotes the conjugate transpose.

Theorem 3.2 (*Donoho et al. 2006*) *Suppose that the sparsity s of the signal vector* $\mathbf{f}$ *satisfies*

$$\mu(\mathbf{\Phi})(2s-1) + 2\frac{\|\mathbf{e}\|_2}{f_{\min}} < 1 \tag{3.18}$$

where $f_{\min} = \min_k |f_k|$. *Denote by* $\mathbf{f}_*$, *the output of the OMP reconstruction. Then*

$\mathbf{f}_*$ *has the correct support, that is,* $\text{supp}(\mathbf{f}_*) = \text{supp}(\mathbf{f})$ *where* $\text{supp}(\mathbf{f})$ *is the support of* $\mathbf{f}$.

$\mathbf{f}_*$ *approximates the object vector in the sense that*

$$\|\mathbf{f}_* - \mathbf{f}\|_2 \leq \frac{\|\mathbf{e}\|}{\sqrt{1+\mu-\mu s}}. \tag{3.19}$$

Incoherence or RIP often requires randomness in the sensing matrix, which can come from the randomness in sampling as well as in illumination. Between the two metrics, incoherence is far more flexible and easier to verify for a given sensing matrix. However, performance guarantees in terms of the coherence parameter such as (3.18) of Theorem 3.2 tend to be conservative.

Fresnel Diffraction with Pixel Basis

As a first example, we consider the imaging equation (3.5) for Fresnel diffraction. We shall write (3.5) in the discrete form (3.10) by discretizing the right-hand side of (3.5) and selecting a discrete set of scattered field data for the left-hand side.

We approximate the masked object

$$V(x) = v(x)u(x,0)e^{i\omega x^2/(2z_0)} \tag{3.20}$$

by the discrete sum on the scale ℓ

$$V_\ell(x) = \sum_{k=1}^{N} b\left(\frac{x}{\ell} - k\right) V(\ell k), \quad V(\ell k) = v(\ell k) u(\ell k, 0) e^{i\omega \ell^2 k^2/(2z_0)} \tag{3.21}$$

where

$$b(x) = \begin{cases} 1, & x \in \left[-\frac{1}{2}, \frac{1}{2}\right] \\ 0, & \text{else.} \end{cases} \tag{3.22}$$

is the localized pixel "basis." We assume that V_ℓ is a good approximation of the masked object for sufficiently small ℓ in the sense that $\lim_{\ell \to 0} \|V - V_\ell\|_1 = 0$.

Moreover, we assume that V_ℓ is sparse in the sense that relatively few components $V(k\ell)$ are significant compared to the number of grid points N. Note that sparse objects in the pixel basis are *not* point-like. Point objects typically induce large gridding errors and require techniques beyond standard compressive sensing reviewed in the "Review of Compressive Sensing" section (cf. the "Fresnel Diffraction with Point Objects" section).

To proceed, we make the Born approximation and set $u^i(x, 0) = 1$ (i.e., normal incidence of plane wave).

Let x_j, $j = 1, \ldots, M$ be the sampling points on the image/sensor line and define

$$\xi_j = \frac{\omega \ell x_j}{2\pi z_0}, \quad j = 1, \ldots, M. \tag{3.23}$$

Set the discretized, unknown vector $\mathbf{f} \in \mathbb{C}^N$ as

$$f_k = v(\ell k) e^{i\omega \ell^2 k^2/(2z_0)}, \quad k = 1, \ldots, N$$

and the data vector $\mathbf{g} \in \mathbb{C}^M$ as

$$g_j = \frac{u^s(x_j, z_0)}{C\ell \hat{b}(\xi_j)} e^{-i\omega x_j^2/(2z_0)}, \quad j = 1, \ldots, M$$

where

$$\hat{b}(\xi) = \int b(x) e^{-i2\pi x\xi} dx = \frac{\sin(\pi\xi)}{\pi\xi}. \tag{3.24}$$

As a result, (3.5) can be expressed as (3.10) with the sensing matrix

$$\Phi = [\Phi_1 \cdots \Phi_N] \in \mathbb{C}^{M \times N}, \quad \Phi_k = [e^{-2\pi i \xi_j k}]_{j=1}^{M}, \quad k = 1, \ldots, N. \tag{3.25}$$

A sensing matrix whose columns have the same 2-norm (as in (3.25)) tends to enjoy better performance in compressive sensing reconstruction.

When ξ_j are independent uniform random variables on [−1/2, 1/2], (3.25) is the celebrated random partial Fourier matrix that is among a few examples with a relatively sharp bound on the RIP as given in the following.

Theorem 3.3 (*Rauhut 2008*) *Suppose*

$$\frac{M}{\ln M} \geq c\delta^{-2} k \ln^2 k \ln N \ln \frac{1}{\varepsilon}, \quad \varepsilon \in (0,1) \tag{3.26}$$

for given sparsity k where c is an absolute constant. Then the restricted isometry constant of the matrix (3.25) *satisfies the bound*

$$\delta_k < \delta$$

with probability at least $1 - \varepsilon$.

Remark 3.2 To apply Theorem 3.3 in the context of Theorem 3.1, we can set $k = 2s$ and $\delta = \sqrt{2} - 1$. Equation 3.26 then implies that it would take roughly $\mathcal{O}(s)$, modulo some logarithmic factors, amount of measurement data for BPDN to succeed in the sense of (3.15).

On the other hand, the coherence parameter μ typically scales as $\mathcal{O}(M^{-1/2})$ as we will see in Theorem 3.5; therefore, in view of the condition (3.18) in Theorem 3.2, the amount of needed data is $\mathcal{O}(s^2)$, significantly larger than $\mathcal{O}(s)$ for $1 \ll s \ll N$.

While this observation is usually valid in the case of OMP, it needs not apply to other greedy algorithms such as subspace pursuit (BP) whose performance guarantee requires $\mathcal{O}(s)$, up to logarithmic factor, amount of data (Dai and Milenkovic 2009).

The fact that ξ_j are independent uniform random variables on $[-1/2, 1/2]$ implies that x_j are independent uniform random variables on $[-A/2, A/2]$ with

$$A = \frac{2\pi z_0}{\omega \ell} \tag{3.27}$$

in view of (3.23). Viewing ℓ as the resolution length of the imaging setup, we obtain the resolution criterion

$$\ell = \frac{2\pi z_0}{A\omega} \tag{3.28}$$

which is equivalent to the classical Abbe or Rayleigh criterion.

Now let us estimate the discretization error vector $\mathbf{d}$ in (3.11). Define the transformation $\mathcal{T}$ by

$$(\mathcal{T}V)_j = \frac{1}{\ell \hat{b}(\xi_j)} \int V(x') e^{-2\pi i \xi_j x'/\ell} dx',$$

cf. (3.7). By definition

$$\mathbf{d} = \mathcal{T}V - \mathcal{T}V_\ell$$

we have

$$\|\mathbf{d}\|_\infty \leq \frac{\| V - V_\ell \|_1}{\ell \min_j |\hat{b}(\xi_j)|}, \quad \hat{b}(\xi) = \frac{\sin(\pi\xi)}{\pi\xi}. \tag{3.29}$$

For $\xi \in [-1/2, 1/2]$, $\min |\hat{b}(\xi)| = 2/\pi$ and $\max |\hat{b}(\xi)| = 1$. Hence

$$\|\mathbf{d}\|_2 \leq \|\mathbf{d}\|_\infty \sqrt{M} \leq \frac{\pi\sqrt{M}}{2\ell} \| V - V_\ell \|_1, \tag{3.30}$$

and

$$\frac{\|\mathbf{d}\|_2}{\|\mathbf{g}\|_2} \leq \frac{\pi C\sqrt{M}\|V - V_\ell\|_1}{2\sqrt{\sum_{j=1}^{M} |u^s(x_j)|^2}}$$

which can be made arbitrarily small by setting ℓ sufficiently small while holding M fixed and maintaining the relation (3.28).

Total Variation Minimization

If the masked object V is better approximated by a piecewise (beyond the scale ℓ) constant function V_ℓ, then the sparsity prior can be enforced by the discrete total variation

$$\|\mathbf{h}\|_{\text{tv}} \equiv \sum_j |\Delta h(j)|, \quad \Delta h(j) = h_{j+1} - h_j.$$

Instead of (3.12), we consider a different convex program, which is called total variation minimization (TV-min)

$$\min \|\mathbf{h}\|_{\text{tv}}, \quad \text{s.t. } \|\mathbf{g} - \mathbf{\Phi h}\|_2 \leq \varepsilon. \tag{3.31}$$

cf. (Rudin et al. 1992, Rudin and Osher 1994, Chambolle and Lions 1997, Chambolle 2004, Candès et al. 2006).

For 2D objects $h(i, j)$, $i, j = 1, \ldots, n$, let $\mathbf{h} = (h_p)$ be the vectorized version with index $p = j + (i - 1)n$. The 2D discrete (isotropic) total variation is given by

$$\|\mathbf{h}\|_{\text{tv}} \equiv \sum_{i,j} \sqrt{|\Delta_1 h(i,j)|^2 + |\Delta_2 h(i,j)|^2},$$

$$\Delta_1 h(i, j) = (h(i + 1, j) - h(i, j), \quad \Delta_2 h(i, j) = h(i, j + 1) - h(i, j)).$$

Figures 3.2 and 3.3 are a numerical demonstration of TV-min reconstruction of the 2D object (the phantom). Figure 3.2 shows the original image and its gradient, which is sparse compared to the original dimensionality. Figure 3.3 shows the reconstruction with BPDN (a) and TV-min (b). TV-min performs well as expected because the TV-sparsity is the correct prior for the object. On the other hand, BPDN performs poorly because the L1-sparsity is the wrong prior.

BPDN for Joint Sparsity

The close relationship between (3.31) and (3.12) can be seen from the following equation for the 1D setting:

$$\left(e^{2\pi i \xi_j} - 1\right) g_j = \sum_k e^{-2\pi i \xi_j k} (f_{k+1} - f_k).$$

In other words, the new data vector $\tilde{\mathbf{g}} = ((e^{2\pi i \xi_j} - 1) g_j)$, the new noise vector $\tilde{\mathbf{e}} = ((e^{2\pi i \xi_j} - 1) e_j)$, and the new object vector $\tilde{\mathbf{f}} = (f_{k+1} - f_k)$ are related via the

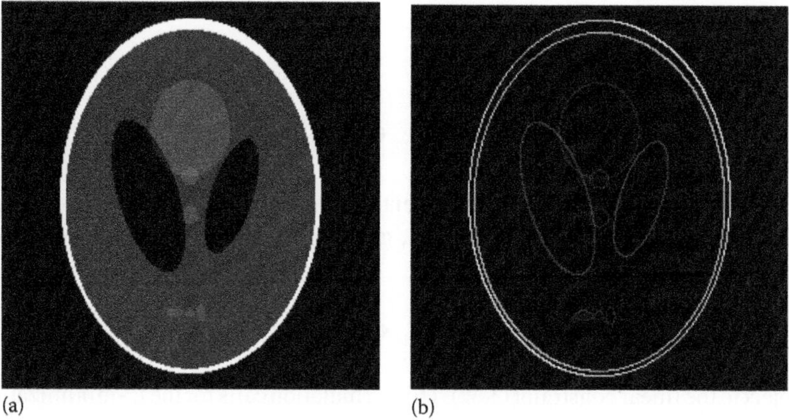

Figure 3.2 The original 256 × 256 Shepp–Logan phantom (a), the Shepp–Logan phantom and the magnitudes of its gradient with sparsity s = 2184 (b). (From Fannjiang, A., *Math. Mech. Complex Syst.*, 1, 81, 2013a. With permission.)

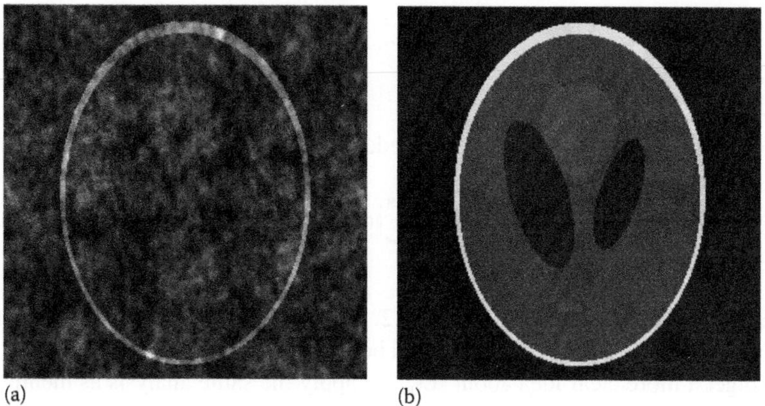

Figure 3.3 BPDN reconstruction without external noise (a) and TV-min reconstruction with 5% noise (b). (From Fannjiang, A., *Math. Mech. Complex Syst.*, 1, 81, 2013a. With permission.)

same sensing matrix as for BPDN. Clearly, $|\tilde{e}_j| \leq 2|e_j|, j = 1,\ldots,M$. Moreover, if e_j are independently and identically distributed, then $\tilde{e}_j$ are also independently and identically distributed with variance

$$\mathbb{E}\,|\,e_j\,|^2 = \mathbb{E}\,|\,e^{2\pi i \xi_j} - 1\,|^2 \times \mathbb{E}\,|\,e_j\,|^2 = 2\mathbb{E}\,|\,e_j\,|^2$$

when ξ_j is the uniform random variable over [−1/2, 1/2]. Hence, for large M, the new noise magnitude is $\|\tilde{\mathbf{e}}\|_2 \approx \sqrt{2}\,\|\mathbf{e}\|_2$. Here and in the following, $\mathbb{E}$ denotes the expected value.

A similar relationship exists in the 2D case. Let $\mathbf{f}_j = \Delta_j \mathbf{f}$ that satisfies the linear constraint

$$\Delta_1 \mathbf{f}_2 = \Delta_2 \mathbf{f}_1. \tag{3.32}$$

Define

$$\mathbf{g}_1 = [(e^{2\pi i \xi_j} - 1)g_j], \quad \mathbf{g}_2 = [(e^{2\pi i \eta_j} - 1)g_j]$$
$$\mathbf{e}_1 = [(e^{2\pi i \xi_j} - 1)e_j], \quad \mathbf{e}_2 = [(e^{2\pi i \eta_j} - 1)e_j]$$

where ξ_j, η_j, $j = 1, \ldots, M$ are independent uniform random variables over [–1/2, 1/2]. Then $\mathbf{F} = [\mathbf{f}_1, \mathbf{f}_2] \in \mathbb{C}^{N\times 2}$, $\mathbf{G} = [\mathbf{g}_1, \mathbf{g}_2] \in \mathbb{C}^{M\times 2}$, and $\mathbf{E} = [\mathbf{e}_1, \mathbf{e}_2]$ are related through

$$\mathbf{G} = [\mathbf{\Phi f}_1, \mathbf{\Phi f}_2] + \mathbf{E}$$

subject to the linear constraint (3.32). This formulation calls for the L^1-minimization (Fannjiang 2013a)

$$\min \|[\mathbf{h}_1, \mathbf{h}_2]\|_{2,1}, \quad \text{s.t. } \|\mathbf{G} - [\mathbf{\Phi h}_1, \mathbf{\Phi h}_2]\|_F \le \|\mathbf{E}\|_F, \tag{3.33}$$

subject to the constraint

$$\Delta_2 \mathbf{h}_1 = \Delta_1 \mathbf{h}_2 \tag{3.34}$$

where

$\|\cdot\|_F$ is the Frobenius norm

$\|\cdot\|_{2,1}$ is the mixed (2, 1)-norm (Benedek and Panzone 1961, Kowalski 2009)

$$\|\mathbf{X}\|_{2,1} = \sum_j \|\text{row}_j(\mathbf{X})\|_2. \tag{3.35}$$

The reason for minimizing the mixed (2, 1)-norm in (3.33) is that $\mathbf{f}_1$ and $\mathbf{f}_2$ share the same sparsity pattern, which should be enforced.

To get a more clear idea about $\|\mathbf{E}\|_F$, we apply the same analysis as mentioned earlier and obtain

$$\|\mathbf{e}_i\|_2^2 \approx \mathbb{E}\|\mathbf{e}_i\|_2^2 = 2\mathbb{E}\|\mathbf{e}\|_2^2, \quad i = 1, 2,$$

for sufficiently large M.

The convex program (3.33) through (3.34) is an example of BPDN with constrained joint sparsity. More generally, suppose that the columns of the unknown multivectors $\mathbf{F} \in \mathbb{C}^{N\times J}$ share the same support and are related to the data multivectors $\mathbf{G} \in \mathbb{C}^{M\times m}$ and the noise multivectors $\mathbf{E} \in \mathbb{C}^{M\times J}$ via

$$\mathbf{G} = [\mathbf{\Phi}_1\mathbf{f}_1, \mathbf{\Phi}_2\mathbf{f}_2, \ldots, \mathbf{\Phi}_J\mathbf{f}_J] + \mathbf{E} \tag{3.36}$$

subject to the linear constraint $\mathcal{L}\mathbf{F} = 0$.

For this setting, the following formulation of BPDN with joint sparsity is natural

$$\min \|\mathbf{H}\|_{2,1}, \quad \text{s.t. } \|\mathbf{G} - [\mathbf{\Phi}_1\mathbf{h}_1, \mathbf{\Phi}_2\mathbf{h}_2, \ldots, \mathbf{\Phi}_J\mathbf{h}_J]\|_F \le \varepsilon, \quad \text{s.t. } \mathcal{L}\mathbf{H} = 0, \tag{3.37}$$

with $\varepsilon = \|\mathbf{E}\|_F$.

OMP for Joint Sparsity

Next, we present an algorithmic extension of OMP for joint sparsity (Cotter et al. 2005, Chen and Huo 2006, Tropp et al. 2006) to the setting with multiple sensing matrices (3.36) (Fannjiang 2013a).

Algorithm 3.2 OMP for joint sparsity

Input: $\{\mathbf{\Phi}_j\}$, $\mathbf{g}$, $\varepsilon > 0$

Initialization: $\mathbf{f}^0 = 0$, $\mathbf{R}^0 = \mathbf{G}$ and $\mathcal{S}^0 = \emptyset$

Iteration: For $k = 1, 2, 3, \ldots$

(1) $i_{\max} = \arg\max_i \sum_{j=1}^{J} |\Phi_{j,i}^{\dagger} R_j^{k-1}|$, where $\Phi_{j,i}^{\dagger}$ is the conjugate transpose of ith column of $\mathbf{\Phi}_j$,

(2) $\mathcal{S}^k = \mathcal{S}^{k-1} \cup \{i_{\max}\}$

(3) $\mathbf{F}^k = \arg\min \|[\mathbf{\Phi}_1\mathbf{h}_1, \ldots, \mathbf{\Phi}_J\mathbf{h}_J] - \mathbf{G}\|_F$ s.t. $\text{supp}(\mathbf{H}) \subseteq S^k$

(4) $\mathbf{R}^k = \mathbf{G} - [\Phi_1\mathbf{f}_1^k, \ldots, \Phi_J\mathbf{f}_J^k]]$

(5) Stop if $\sum_j R_{j2}^k \le \varepsilon$.

Output: $\mathbf{F}^k$.

Note that the linear constraint $\mathcal{L}$ is not enforced in Algorithm 3.2. The idea is to first find the support of the multivectors without taking into account of the linear constraint, and, in the second stage, follow the support recovery with least squares

$$\mathbf{F}_* = \arg\min_{\mathbf{H}} \|\mathbf{G} - [\Phi_1\mathbf{h}_1, \ldots, \Phi_J\mathbf{h}_J]\|_F, \quad \text{s.t. supp}(\mathbf{H}) \subseteq \text{supp}(\mathbf{F}^\infty), \; \mathcal{L}\mathbf{H} = 0 \quad (3.38)$$

where $\mathbf{F}^\infty$ is the output of Algorithm 3.2.

For more discussion and applications of constrained joint sparsity, the reader is referred to Fannjiang (2013a) where the performance guarantees similar to Theorems 3.1 and 3.2 are proved for constrained joint sparsity.

Fresnel Diffraction with Point Objects

A major problem with discretizing the object domain shows up when the objects are point-like. In this case, it is unrealistic to assume that the objects are located exactly on the grid as the forceful matching between the point objects and the grid can create detrimental errors. Without additional prior information, the gridding error due to the mismatch between the point object locations and the grid points can be as large as the data itself, resulting in a low signal-to-noise ratio (SNR).

We call the grid spacing ℓ given in (3.28), the *Resolution length* (RL), which is the natural unit for resolution analysis. In the RL unit, the object domain grid becomes a subset of the integer grid $\mathbb{Z}$.

In the case of point objects, to refine the standard grid and reduce discretization error, we consider a fractional grid

$$\mathbb{Z}/F = \{j/F : j \in \mathbb{Z}\} \quad (3.39)$$

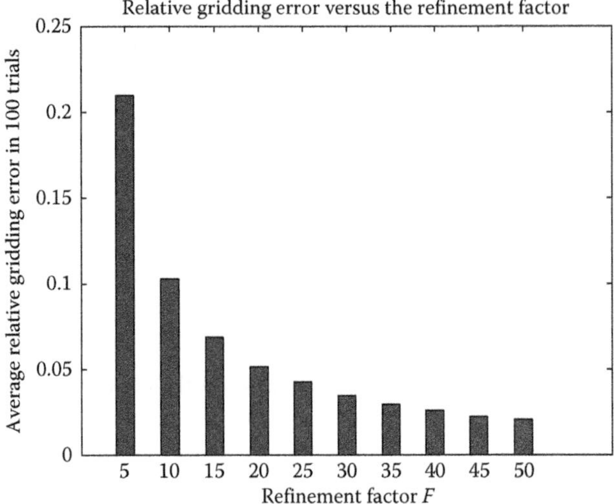

Figure 3.4 The relative gridding error is roughly inversely proportional to the refinement factor. (From Fannjiang, A. and Liao, W., *SIAM J. Imaging Sci.*, 5, 179, 2012a. Copyright © 2012 Society for Industrial and Applied Mathematics. Reprinted with permission.)

where $F \in \mathbb{N}$ is called *the refinement factor*. The random partial Fourier matrix (3.25) now takes the form

$$\Phi = \left[e^{-i2\pi\xi_j k/F} \right], \tag{3.40}$$

where $\xi_j \in [-1/2,1/2]$ are independent uniform random variables. In the following numerical examples, we shall consider both deterministic (see Equation 3.45) as well as random sampling schemes.

As shown in Figure 3.4, the relative gridding error $\|\mathbf{d}\|/\|\Phi\mathbf{f}\|$ is roughly inversely proportional to the refinement factor F.

Figure 3.5 shows the coherence pattern $[\mu(j, k)]$ of a 100×4000 matrix (3.40) with $F = 20$ (Figure 3.5a). The bright diagonal band represents a heightened correlation (pairwise coherence) between a column vector and its neighbors on both sides (about 30). Figure 3.5b shows a half cross section of the coherence band across two RL, averaged over 100 independent trials. In general sparse recovery with large F exceeds the capability of currently known algorithms as the condition number of the 100×30 submatrix corresponding to the coherence band in Figure 3.5 easily exceeds 10^{15}. The high condition number makes stable recovery impossible. While Figure 3.5 is typical of the coherence pattern of 1D sensing matrices, the coherence pattern for two or three dimensions is considerably more complicated depending on how the objects are vectorized.

Band-Excluded, Locally Optimized Orthogonal Matching Pursuit

To overcome the conundrum of a highly coherent sensing matrix due to a refined grid, we have to go beyond the coherence parameter and study the coherence pattern of the sensing matrix.

The coherence pattern of a sensing matrix can be described in terms of the notion of *coherence band* defined in the following. Let $\eta > 0$. Define the η-coherence band of the index k as

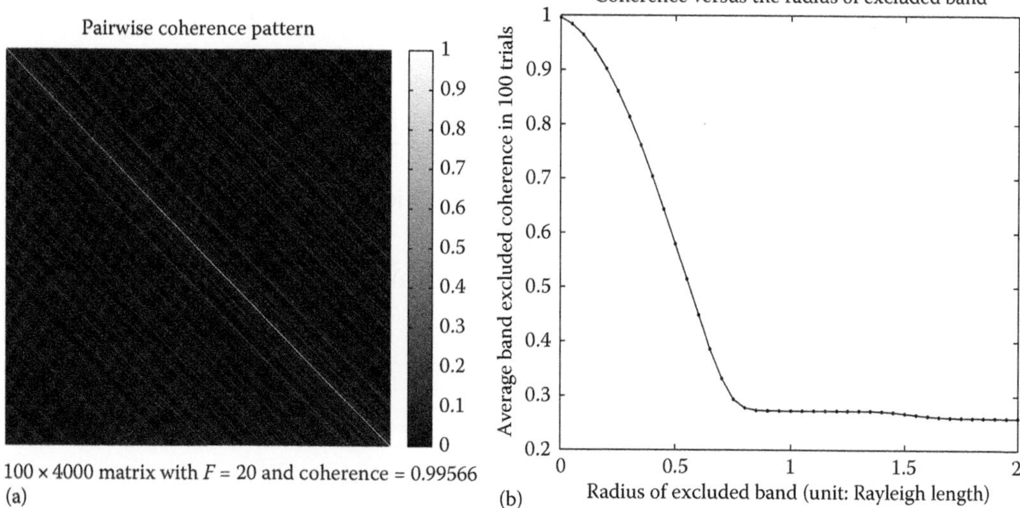

Figure 3.5 Coherence pattern $[\mu(j, k)]$ for the 100×4000 matrix with $F = 20$ (a). The off-diagonal elements tend to diminish as the row number increases. The coherence band near the diagonals, however, persists and has the average profile shown in panel (b) where the vertical axis is the pairwise coherence averaged over 100 independent trials and the horizontal axis is the distance between two object points. (From Fannjiang, A. and Liao, W., *SIAM J. Imaging Sci.*, 5, 179, 2012a. Copyright © 2012 Society for Industrial and Applied Mathematics. Reprinted with permission.)

$$B_\eta(k) = \{i \mid \mu(i, k) > \eta\}, \tag{3.41}$$

and the double coherence band as

$$B_\eta^{(2)}(k) \equiv B_\eta(B_\eta(k)) = \cup_{j \in B_\eta(k)} B_\eta(j) \tag{3.42}$$

The first technique for taking advantage of the prior information of well separated objects is called band exclusion (BE) and can be easily embedded in the greedy algorithm, OMP.

To imbed BE into OMP, we make the following change to the matching step

$$i_{\max} = \arg\min_i |\langle \mathbf{r}^{n-1}, \Phi_i \rangle|, \quad i \notin B_\eta^{(2)}(S^{n-1}), \; n = 1, 2, \ldots.$$

meaning that the double η-band of the estimated support in the previous iteration is avoided in the current search. This is natural if the sparsity pattern of the object is such that $B_\eta(j), j \in \text{supp}(\mathbf{f})$ are pairwise disjoint. We call the modified algorithm as the band-excluded orthogonal matching pursuit (BOMP), as stated in Algorithm 3.3.

Algorithm 3.3 Band-Excluded Orthogonal Matching Pursuit (BOMP)

Input: $\mathbf{\Phi}$, $\mathbf{g}$, $\eta > 0$

Initialization: $\mathbf{f}^0 = 0$, $\mathbf{r}^0 = \mathbf{g}$, and $S^0 = \emptyset$

Iteration: For $j = 1, \ldots, s$

(1) $i_{\max} = \arg\max_i |\langle \mathbf{r}^{j-1}, \Phi_i \rangle|, i \notin B_\eta^{(2)}(S^{j-1})$

(2) $S^j = S^{j-1} \cup \{i_{\max}\}$

(3) $\mathbf{f}^j = \arg\min_h \|\mathbf{\Phi h} - \mathbf{g}\|_2$ s.t. $\mathrm{supp}(\mathbf{h}) \subseteq S^j$

(4) $\mathbf{r}^j = \mathbf{g} - \mathbf{\Phi f}^j$

Output: $\mathbf{f}^s$.

The following theorem gives a (pessimistic) performance guarantee for BOMP.

Theorem 3.4 *(Fannjiang and Liao 2012a) Let* **f** *be s-sparse. Let* $\eta > 0$ *be fixed. Suppose that*

$$B_\eta(i) \cap B_\eta^{(2)}(j) = \emptyset, \quad \forall i, j \in \mathrm{supp}(\mathbf{f}) \tag{3.43}$$

and that

$$\eta(5s-4)\frac{f_{\max}}{f_{\min}} + \frac{5\|\mathbf{e}\|_2}{2f_{\min}} < 1 \tag{3.44}$$

where

$$f_{\max} = \max_k |f_k|, \quad f_{\min} = \min_k |f_k|.$$

Let $\mathbf{f}^s$ *be the BOMP reconstruction. Then* $\mathrm{supp}(\mathbf{f}^s) \subseteq B_\eta\,(\mathrm{supp}(\mathbf{f}))$, *and moreover every nonzero component of* $\mathbf{f}^s$ *is in the* η*-coherence band of a unique nonzero component of* **f**.

Remark 3.3 Condition (3.43) means that BOMP guarantees to resolve 3 RLs. In practice, BOMP can resolve objects separated by close to 1 RL when the dynamic range is nearly 1.

Remark 3.4 A main difference between Theorems 3.2 and 3.4 lies in the role played by the dynamic range $f_{\max}/f_{\min}$ and the separation condition (3.43).

Another difference is approximate recovery of support in Theorem 3.4 versus exact recovery of support in Theorem 3.2(a). In contrast to the F-independent nature of approximate support recovery, exact support recovery would probably be highly sensitive to the refinement factor F. That is, as F increases, the chance of missing some points in the support set also increases. As a result, the error of reconstruction $\|f^s - f\|_2$ tends to increase with F (as evident in Figure 3.7).

A main shortcoming with BOMP is in its failure to perform even when the dynamic range is even moderately greater than unity. To overcome this problem, we introduce the second technique: the *local optimization* (LO) that is a residual-reduction technique applied to the current estimate S^k of the object support (Fannjiang and Liao 2012a).

Algorithm 3.4 Local Optimization (LO)

Inputs: $\mathbf{\Phi}$, $\mathbf{g}$, $\eta > 0$, $S^0 = \{i_1, \ldots, i_k\}$.

Iteration: For $j = 1, 2, \ldots, k$.

(1) $\mathbf{f}^j = \arg\min_{\mathbf{h}} \|\mathbf{\Phi h} - \mathbf{g}\|_2$, $\text{supp}(\mathbf{h}) = (S^{j-1} \setminus \{i_j\}) \cup \{i'_j\}, i'_j \in B_\eta(\{i_j\})$.

(2) $S^j = \text{supp}(\mathbf{f}^j)$.

Output: S^k.

In other words, given a support estimate S^0, LO fine-tunes the support estimate by adjusting each element in S^0 within its coherence band in order to minimize the residual. The object amplitudes for the improved support estimate are obtained by solving the least-squares problem. Because of the local nature of LO, the computation is efficient.

Embedding LO in BOMP gives rise to the band-excluded, locally optimized orthogonal matching pursuit (BLOOMP).

Algorithm 3.5 Band-excluded, Locally Optimized Orthogonal Matching Pursuit (BLOOMP)

Input: $\mathbf{\Phi}$, $\mathbf{g}$, $\eta > 0$

Initialization: $\mathbf{f}^0 = 0$, $\mathbf{r}^0 = \mathbf{g}$ and $S^0 = \emptyset$

Iteration: For $j = 1, \ldots, s$

(1) $i_{\max} = \arg\max_i |\langle \mathbf{r}^{j-1}, \Phi_i \rangle|$, $i \notin B_\eta^{(2)}(S^{j-1})$

(2) $S^j = \text{LO}\,(S^{j-1} \cup \{i_{\max}\})$, where $\text{LO}(S^{j-1} \cup \{i_{\max}\})$ is the output of Algorithm 3.4 with $(S^{j-1} \cup \{i_{\max}\})$ as input.

(3) $\mathbf{f}^j = \arg\min_{\mathbf{h}} \|\mathbf{\Phi h} - \mathbf{g}\|_2$ s.t. $\text{supp}(\mathbf{h}) \in S^j$

(4) $\mathbf{r}^j = \mathbf{g} - \mathbf{\Phi f}^j$

Output: $\mathbf{f}^s$.

The same BLO technique can be used to enhance the other well-known iterative schemes such as SP, CoSaMP (Needell and Tropp 2009), and compressed iterative hard thresholding (IHT) (Blumensath and Davies 2009, 2010), and the resulting algorithms are denoted by BLOSP, BLOCoSaMP, and BLOIHT, respectively, in the following numerical results. We refer the reader to Fannjiang and Liao (2012a) for the details and descriptions of these algorithms. MATLAB code of Algorithm 3.5 is available on-line at https://www.math.ucdavis.edu/~fannjiang/home/codes/BLOOMPcode.

Band-Excluding Thresholding

A related technique that can be used to enhance BPDN/Lasso for off-grid objects is called the band-excluding, locally optimized thresholding (BLOT).

Algorithm 3.6 Band-Excluding, Locally Optimized Thresholding (BLOT)

Input: $\mathbf{f} = (f_1, \ldots, f_N)$, $\mathbf{\Phi}$, $\mathbf{g}$, $\eta > 0$.

Initialization: $S^0 = \emptyset$.

Iteration: For $j = 1, 2, \ldots, s$.

(1) $i_j = \arg\max |f_k|, k \notin B_\eta^{(2)}(S^{j-1})$.

(2) $S^j = S^{j-1} \cup \{i_j\}$.

Output: $\mathbf{f}^s = \arg\min \|\mathbf{\Phi h} - \mathbf{g}\|_2$, $\text{supp}(\mathbf{h}) \subseteq \text{LO}(S^s)$, where LO is the output of Algorithm 3.4.

Numerical Examples

For numerical demonstration in Figures 3.6 and 3.7, we use *deterministic*, equally spaced sampling with

$$\xi_j = -\frac{1}{2} + \frac{j}{M}, \quad j = 1, \ldots, M \tag{3.45}$$

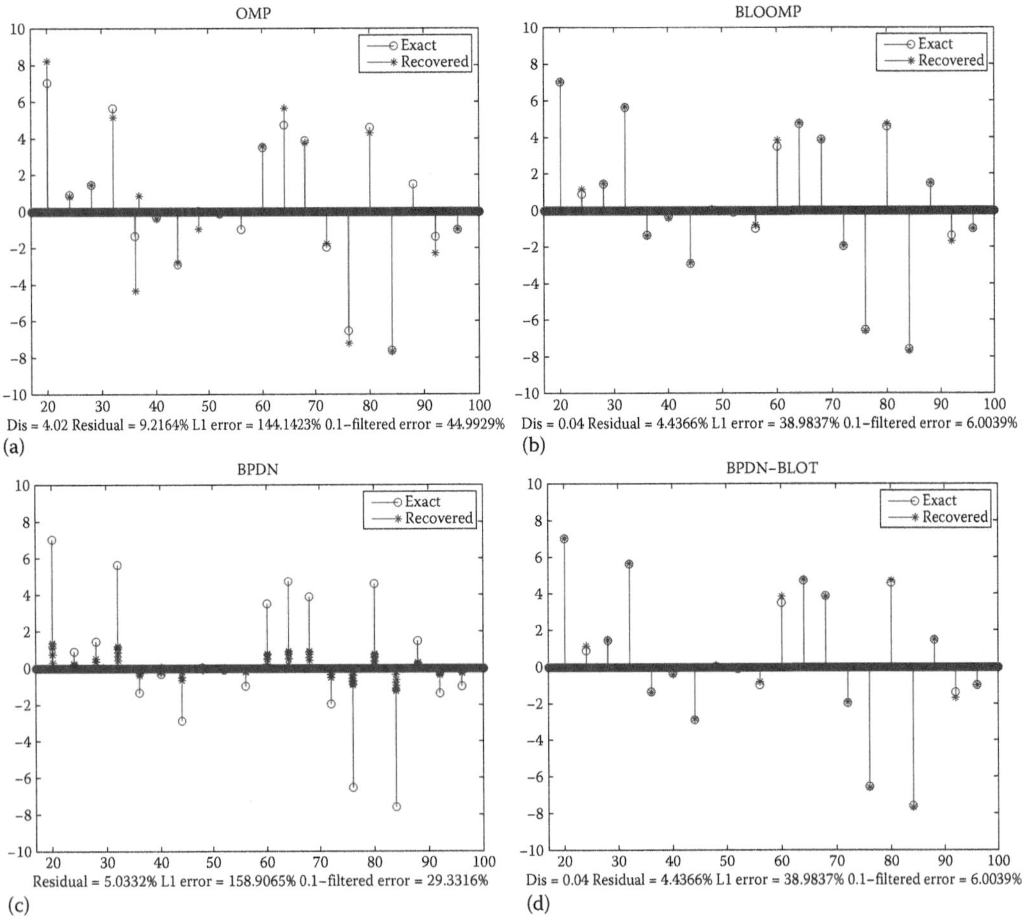

Figure 3.6 Reconstruction by (a) OMP, (b) BLOOMP, (c) BPDN, and (d) BPDN-BLOT of the real part of 20 randomly phased spikes with $F = 50$, SNR = 20. (From Fannjiang, A. and Liao, W., Super-resolution by compressive sensing algorithms, in *IEEE Proceedings of Asilomar Conference on Signals, Systems and Computers*, 2012b. With permission.)

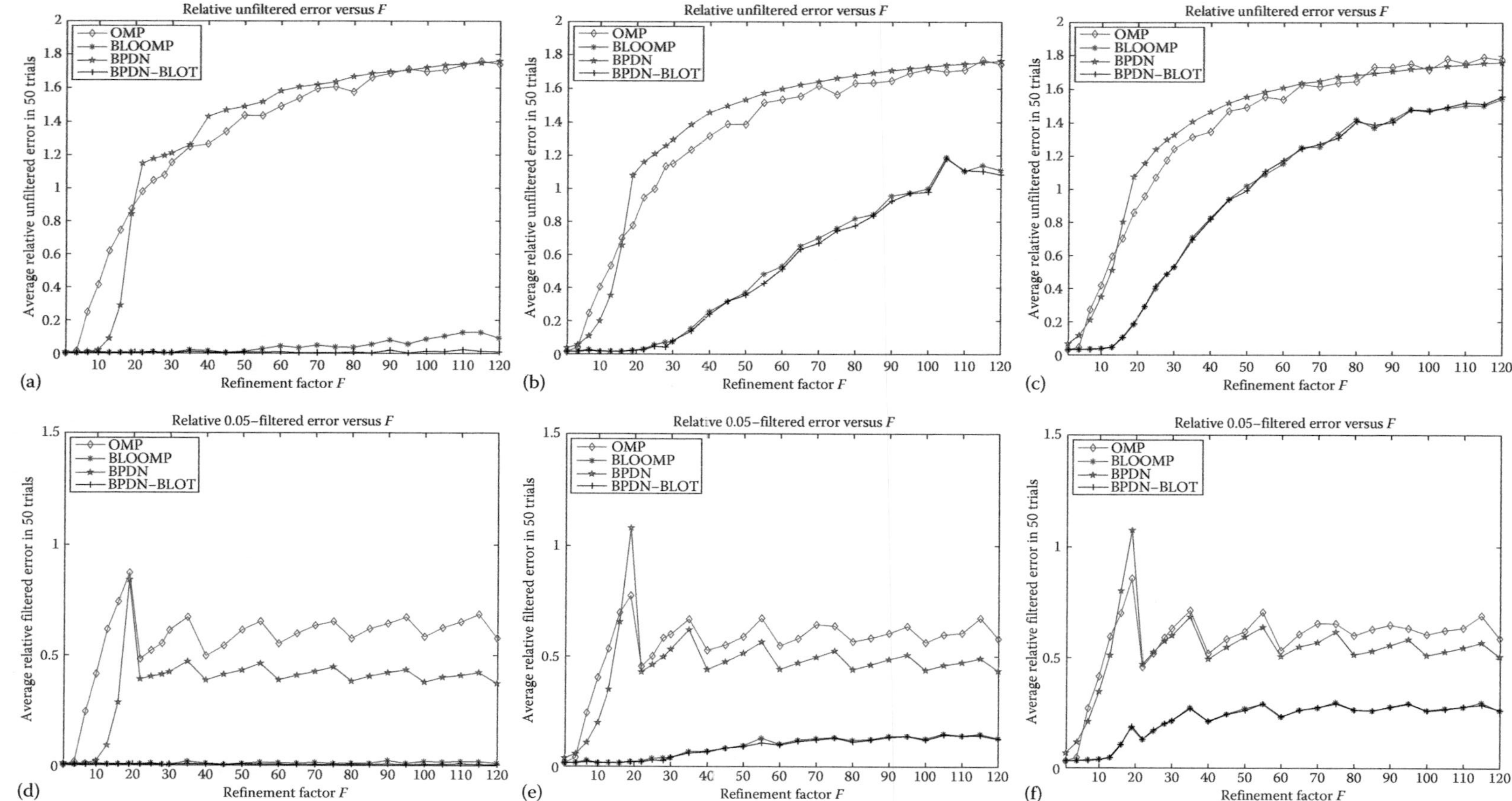

Figure 3.7 Relative errors in reconstruction by OMP, BLOOMP, BP, and BP-BLOT as F varies without (top) or with (bottom) filtering. (a) SNR = 100, $\eta = 0$; (b) SNR = 20, $\eta = 0$; (c) SNR = 10, $\eta = 0$; (d) SNR = 100, $\eta = 0.05\ell$; (e) SNR = 20, $\eta = 0.05\ell$; (f) SNR = 10, $\eta = 0.05\ell$. (From Fannjiang, A. and Liao, W., Super-resolution by compressive sensing algorithms, in *IEEE Proceedings of Asilomar Conference on Signals, Systems and Computers*, 2012b. With permission.)

and $\mathbf{\Phi} \in \mathbb{C}^{M\times FM}$ with $M = 150$, $F = 50$ to recover 20 randomly distributed and randomly phased point objects (spikes) separated by at least 4 RL.

Figure 3.6a and b shows how the BLO technique corrects the error of OMP due to the unresolved grid. In particular, several misses are recaptured and false detections removed. Figure 3.6c and d shows how the BLOT technique improves the BPDN estimate. In particular, BLOT has the effect of "trimming the bushes" and "growing the real trees." Figure 3.7a through c shows the relative error of reconstruction as a function of F by OMP, BPDN, BLOOMP, and BPDN-BLOT with the same setup and three different SNRs. For all SNRs, BLOOMP and BPDN-BLOT produce drastically fewer errors compared to OMP and BPDN.

The growth of relative error with F reflects the sensitivity of the reconstruction error alluded to in Remark 3.4. Note that the reconstruction error in the *discrete* norm cannot distinguish how far off the recovered support is from the true object support. The discrete norm treats any amount of support offset equally. An easy remedy to the injudicious treatment of support offset is to use instead the *filtered error norm* $\left\|\mathbf{f}_\eta^s - \mathbf{f}_\eta\right\|_2$, where $\mathbf{f}_\eta$ and $\mathbf{f}_\eta^s$ are, respectively, $\mathbf{f}$ and $\mathbf{f}^s$ convoluted with an approximate delta-function of width 2η.

Clearly the filtered error norm is more stable to support offset, especially if the offset is less than η. If every spike of $\mathbf{f}^s$ is within η distance from a spike of $\mathbf{f}$ *and* if the amplitude differences are small, then the η-filtered error is small. As shown in Figure 3.7d through f, averaging over $\eta = 5\%$ RL produces acceptable filtered error for any refinement factor relative to the external noise. This suggests that both BPDN-BLOT and BLOOMP recover the object support on average within 5% of 1 RL, a significant improvement over the theoretical guarantee of Theorem 3.4.

Next, we consider the unresolved partial Fourier matrix (3.40) with random sampling points to demonstrate the flexibility of the techniques. Let $\xi_j \in [-1/2, 1/2]$, $j = 1, \ldots, M$ be independent uniform random variables with $M = 100$, $N = 4000$, and $F = 20$. The test objects are 10 randomly phased and distributed objects, separated by at least 3 RL. As in Theorem 3.4, a recovery is counted as a success if every reconstructed object is within 1 RL of the object support.

Figure 3.8 compares the success rates (averaged over 200 trials) of the BLO-enhanced schemes (BLOOMP, BLOSP, BLOCoSaMP, and BLOIHT) and the BLOT-enhanced scheme (Lasso-BLOT). Lasso-BLOT is implemented with the regularization parameter (Chen et al. 2001)

$$\lambda = 0.5\sqrt{\log N} \quad \text{(black curves with diamonds)} \tag{3.46}$$

or

$$\lambda = \sqrt{2\log N} \quad \text{(black curves with stars)} \tag{3.47}$$

The empirically optimal choice (3.46) (labeled as Lasso-BLOT (0.5)) has a much improved performance over the choice (3.47). Clearly, BLOOMP is the best performer in noise stability and dynamic range among all tested algorithms.

Highly Redundant Dictionaries

Our discussion in the "Fresnel Diffraction with Point Objects" section so far is limited to point-like objects. But the methods presented earlier are also applicable to a wide variety of cases where the objects have sparse representations by redundant dictionaries, instead of orthogonal bases.

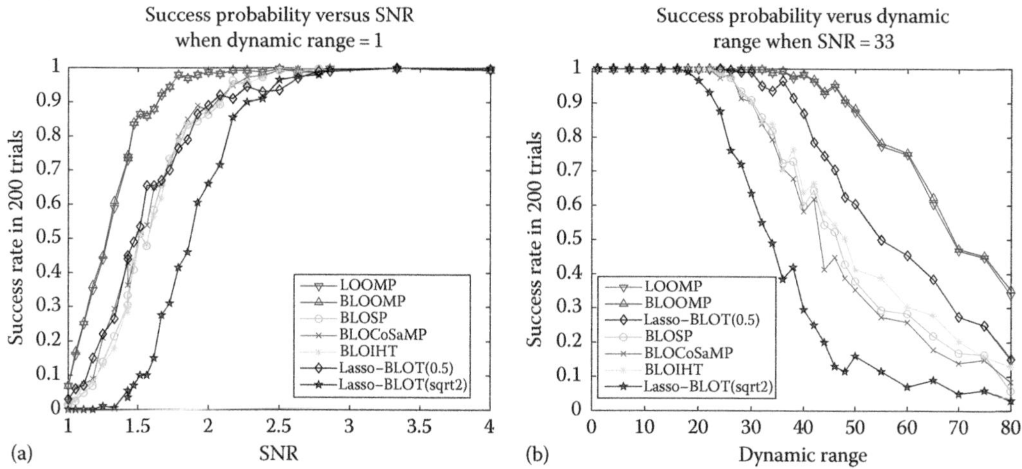

Figure 3.8 Success probability versus (a) SNR for dynamic range = 1 and (b) dynamic range for SNR = 33. Here, LOOMP is a simplified version of BLOOMP and has nearly identical performance curves. (From Fannjiang, A. and Liao, W., *SIAM J. Imaging Sci.*, 5, 179, 2012a. With permission.)

Suppose that the object is sparse in a highly redundant dictionary, which by definition tends to represent an object by fewer number of elements than a nonredundant one does. For example, one can combine different orthogonal bases into a dictionary that can sparsify a wider class of objects than any individual base can. On the other hand, a redundant dictionary tends to produce a larger coherence parameter and be ill suited for compressive sensing. This is the same kind of conundrum about off-grid point-like objects.

One of the most celebrated examples of optical compressive sensing is the SPC depicted in Figure 3.9. In the single-pixel camera (SPC), measurement diversity comes entirely from the digital micromirror device (DMD) instead of the sensor array. The DMD consists of an array of electrostatically actuated micromirrors. Each mirror can be positioned in one of the two states (±12°). Light reflected from mirrors in the +12-state only is then collected and focused by the lens and subsequently detected by a single optical sensor. For each and every measurement, the DMD is randomly and independently reconfigured. The resulting measurement matrix **A** has independently and identically distributed entries.

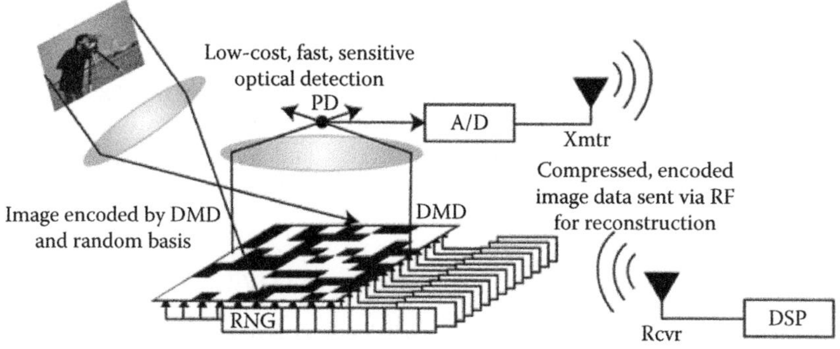

Figure 3.9 Single-pixel camera block diagram. (Courtesy of Rice Single-Pixel Camera Project, http://dsp.rice.edu/cscamera.)

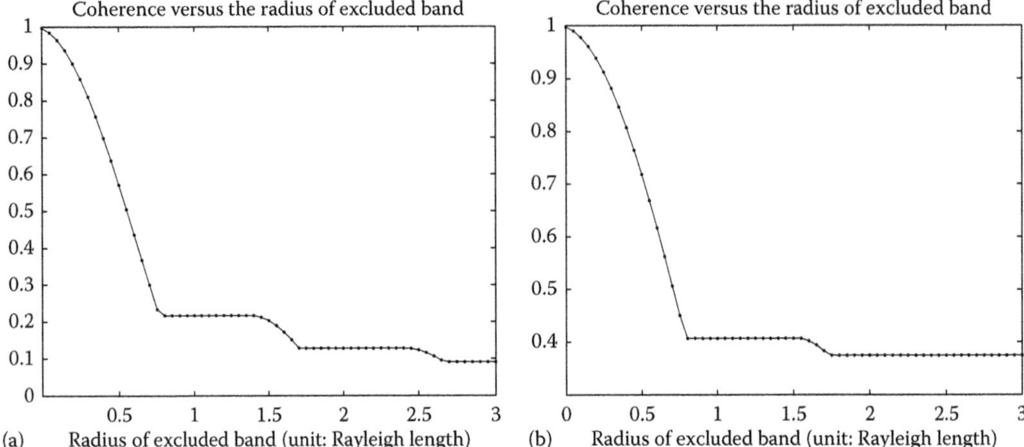

Figure 3.10 The coherence bands of the redundant Fourier frame $\boldsymbol{\Psi}$ (a) and $\boldsymbol{\Phi} = \mathbf{A}\boldsymbol{\Psi}$ (b), the latter being averaged over 100 realizations of **A**. (From Fannjiang, A. and Liao, W., *SIAM J. Imaging Sci.*, 5, 179, 2012a. Copyright © 2012 Society for Industrial and Applied Mathematics. Reprinted with permission.)

Suppose that the object is sparse in terms of a highly redundant dictionary. For simplicity of presentation, consider an 1D object sparse in an overcomplete Fourier frame (i.e., a dictionary that satisfies the frame bounds [Daubechies 1992]) with entries

$$\Psi_{k,j} = \frac{1}{\sqrt{R}} e^{-2\pi i((k-1)(j-1)/RF)}, \quad k = 1,\ldots,R,\ j = 1,\ldots,RF, \tag{3.48}$$

that includes harmonic as well as nonharmonic modes as its columns, where F is the redundant factor and R is a large integer. In other words, the object can be written as $\boldsymbol{\Psi}\mathbf{f}$ with a sufficiently sparse vector $\mathbf{f}$. The final sensing matrix then becomes

$$\boldsymbol{\Phi} = \mathbf{A}\boldsymbol{\Psi}. \tag{3.49}$$

The coherence bands of $\boldsymbol{\Psi}$ and $\boldsymbol{\Phi}$ are shown in Figure 3.10 from which we see that, as in Figure 3.5, the coherence radius is less than 1 RL. The same BLO- and BLOT-based techniques can be applied to (3.49); see Fannjiang and Liao (2012a) for numerical results and performance comparison with other techniques for off-grid objects (Candès et al. 2011, Candès and Fernandez-Granda 2013, 2014, Duarte and Baraniuk 2013, Tang et al. 2013).

Fresnel Diffraction with Littlewood–Paley Basis

Opposite to the localized pixel basis, the Littlewood–Paley basis is slowly decaying nonlocal modes based on the wavelet function

$$\psi(x) = (\pi x)^{-1}(\sin(2\pi x) - \sin(\pi x)), \tag{3.50}$$

which has a compactly supported Fourier transform

$$\hat{\psi}(\xi) = \int \psi(x) e^{-i2\pi\xi x}\, dx = \begin{cases} 1, & \frac{1}{2} \leq |\xi| \leq 1 \\ 0, & \text{otherwise.} \end{cases} \tag{3.51}$$

The following functions

$$\psi_{p,q}(x) = 2^{-p/2}\psi(2^{-p}x - q), \quad p, q \in \mathbb{Z} \tag{3.52}$$

form an orthonormal wavelet basis in $L^2(\mathbb{R})$ (Daubechies 1992). Expanding the masked object V (3.20) in the Littlewood–Paley basis, we write

$$V(x) = \sum_{p,q \in \mathbb{Z}} V_{p,q} \psi_{p,q}(x). \tag{3.53}$$

The main point of the subsequent discussion is to design a sampling scheme such that the resulting sensing matrix has desirable compressive sensing properties (Fannjiang 2009).

Let $\{2^p: p = -p_*, -p_* + 1, \ldots, p_*\}$ be the dyadic scales present in (3.53), $\{q: |q| \leq N_p\}$ the modes present on the scale 2^p, and $2M_p + 1$ the number of measurements corresponding to the scale 2^p. Let

$$k = \sum_{j=-p_*}^{p'-1} (2M_j + 1) + q', \quad |q'| \leq M_{p'}, \quad |p'| \leq p_* \tag{3.54}$$

be the index for the sampling points. Throughout this section, k is determined by p', q' by (3.54). Let x_k be the sampling points and set the normalized coordinates

$$\frac{x_k \omega \ell}{2\pi z_0} = \xi_k, \quad k = 1, \ldots, M \tag{3.55}$$

where, as shown in the following, ℓ is a resolution parameter and $\xi_k \in [-1/2, 1/2]$ are determined in the following; cf. (3.23). This means that the aperture (i.e., the sampling range of x_k) is again given by (3.27).

Let $\mathbf{g} = (g_k)$ be the data vector with

$$g_k = C^{-1} u^s(x_k, z_0) e^{-i\omega x_k^2/(2z_0)}.$$

Direct calculation with (3.5) and (3.55) then gives

$$g_k = \sum_{p,q \in \mathbb{Z}} 2^{p/2} V_{p,q} e^{-i2\pi\xi_k \ell^{-1} 2^p q} \hat{\psi}(\xi_k \ell^{-1} 2^p), \quad k = 1, \ldots, M. \tag{3.56}$$

Let $\mathbf{f} = (f_l)$ be the object vector with

$$f_l = (-1)^q 2^{p/2} V_{p,q}$$

where the indices are related by

$$l = \sum_{j=-p_*}^{p-1} (2N_j + 1) + q.$$

Suppose that

$$\ell \le 2^{-p_*-1} \tag{3.57}$$

That is, 2ℓ is less than or equal to the smallest scale in the wavelet presentation (3.53).

Let $\zeta_{p',q'}$ be independent, uniform random variables on [–1/2, 1/2] and let

$$\xi_k = \frac{\ell}{2^{p'}} \cdot \begin{cases} 1/2 + \zeta_{p',q'}, & \zeta_{p',q'} \in [0,1/2] \\ -1/2 + \zeta_{p',q'}, & \zeta_{p',q'} \in [-1/2,0]. \end{cases} \tag{3.58}$$

where k is determined by (3.54). By the assumption (3.57), we have

$$\xi_k \in [-/2,\, 1/2], \quad \forall p' \ge -p_*.$$

More specifically, by (3.55), we have

$$x_k \in \frac{2\pi z_0}{\omega 2^{p'}}\left(\left[-1, -\frac{1}{2}\right] \cup \left[\frac{1}{2}, 1\right]\right),$$

that is, the sampling regions for different dyadic scales indexed by p' are disjoint with the ones for the smaller scales on the outer skirt of the aperture, taking up a bigger portion of the aperture. The resulting sampling points are geometrically concentrated near (but not exactly at) the center of the aperture.

Let the sensing matrix elements be

$$\Phi_{k,l} = (-1)^q \hat{\psi}(\xi_k 2^p \ell^{-1}) e^{-i2\pi\xi_k 2^p q/\ell}. \tag{3.59}$$

We claim that $\Phi_{k,l} = 0$ for $p \neq p'$. This is evident from (3.58) and the following calculation:

$$\ell^{-1}\xi_k 2^p = 2^{p-p'} \cdot \begin{cases} 1/2 + \zeta_{p',q'}, & \zeta_{p',q'} \in [0,1/2] \\ -1/2 + \zeta_{p',q'}, & \zeta_{p',q'} \in [-1/2,0]. \end{cases} \tag{3.60}$$

For $p \neq p'$, the absolute value of (3.60) is either greater than 1 or less than 1/2 and hence (3.60) is outside the support of $\hat{\psi}$.

On the other hand, for $p = p'$, (3.60) is inside the support of $\hat{\psi}$ and so

$$\Phi_{k,l} = e^{-i2\pi q\zeta_{p,q'}}, \quad |q'| \le M_p, \quad |q| \le N_p, \tag{3.61}$$

which constitute the same random partial Fourier matrix that we have seen before. In other words, under the assumption (3.57) the sensing matrix $\mathbf{\Phi} = [\Phi_{k,l}] \in \mathbb{C}^{M\times N}$, with $N = \sum_{|p|\le p_*}(2N_p+1)$ and $M = \sum_{|p|\le p_*}(2M_p+1)$, is block diagonal with each block (indexed by p) in the form of the random partial Fourier matrix, representing the sensing matrix on the dyadic scale 2^p.

Near-Field Diffraction with Fourier Basis

Consider near-field diffraction by a periodic, extended object (e.g., diffraction grating) where the evanescent modes as well as the propagation modes are taken into account. Since we cannot apply the paraxial approximation, we resort to the Lippmann–Schwinger equation (3.3).

Suppose the masked object function is sparse in the Fourier basis

$$V(x) = \sum_{j=-\infty}^{\infty} \hat{V}_j e^{i2\pi jx/L} \tag{3.62}$$

where L is the period and only s modes have nonzero amplitudes. Suppose that $\hat{V}_j = 0$ for $j \neq 1, \ldots, N$.

The 2D Green function can be expressed by the Sommerfeld integral formula (Born and Wolf 1999)

$$G(\mathbf{r}) = \frac{i}{4\pi} \int e^{i\omega(|z|\beta(\alpha)+x\alpha)} \frac{d\alpha}{\beta(\alpha)}, \quad \mathbf{r} = (z, x) \tag{3.63}$$

where

$$\beta(\alpha) = \begin{cases} \sqrt{1-\alpha^2}, & |\alpha| < 1 \\ i\sqrt{\alpha^2 - 1}, & |\alpha| > 1 \end{cases}. \tag{3.64}$$

The integrand in (3.63) with real-valued β (i.e., $|\alpha| < 1$) corresponds to the homogeneous wave, and that with imaginary-valued β (i.e., $|\alpha| > 1$) corresponds to the evanescent (inhomogeneous) wave, which has an exponential-decay factor $e^{-\omega|z|\sqrt{\alpha^2-1}}$. Likewise, the 3D Green function can be represented by the Weyl integral formula (Born and Wolf 1999).

The signal arriving at the sensor located at $(0, x)$ is given by the Lippmann–Schwinger equation with (3.63)

$$\int G(z_0, x - x')V(x')dx' = \frac{i}{2\omega} \sum_j \frac{\hat{V}_j}{\beta_j} e^{i\omega z_0 \beta_j} e^{i\omega\alpha_j x} \tag{3.65}$$

where

$$\alpha_j = \frac{2\pi j}{L\omega}, \quad \beta_j = \beta(\alpha_j). \tag{3.66}$$

The subwavelength structure is encoded in $\hat{V}_j$ with $\alpha_j > 1$ corresponding to the evanescent modes.

Let $(0, x_k)$, $x_k = \xi_k L$, $k = 1, \ldots, M$ be the coordinates of the sampling points where $\xi_k \in [-1/2, 1/2]$. In other words, L is also the aperture (i.e., the sampling range for x_k). To set the problem in the framework of compressed sensing, we set the vector $\mathbf{f} = (f_j) \in \mathbb{C}^N$ as

$$f_j = \frac{ie^{i\omega z_0 \beta_j}}{2\omega\beta_j} \hat{V}_j. \tag{3.67}$$

To avoid a vanishing denominator in (3.67), we assume that $\alpha_j \neq 1$ and hence $\beta_j, \neq 0, \forall j \in \mathbb{Z}$. This is the case, for instance, when $L\omega/(2\pi)$ is irrational.

This gives rise to the sensing matrix $\mathbf{\Phi}$ with the entries

$$\Phi_{kj} = e^{i\omega\alpha_j x_k} = e^{i2\pi j\xi_k}, \quad k = 1,\dots,M,\ j = 1,\dots,N \tag{3.68}$$

which again is the random partial Fourier matrix.

A source of instability lurks in the expression (3.67) where β_j may be complex valued, corresponding to the evanescent modes. Stability in inverting the relationship (3.67) requires limiting the number of the evanescent modes involved in (3.67). Here, the transition is however not clear-cut. For example, if we demand that

$$\left|e^{i\omega z_0\beta_j}\right| \geq e^{-2\pi} \tag{3.69}$$

as the criterion for stable modes, then the stable modes include $|\alpha_j| \leq 1$ as well as $|\alpha_j| > 1$ such that

$$\omega|\beta_j|z_0 \leq 2\pi \tag{3.70}$$

or equivalently

$$\frac{|j|}{L} \leq \sqrt{\frac{\omega^2}{4\pi^2} + \frac{1}{z_0^2}} \tag{3.71}$$

In other words, the number of *stably* resolvable modes is proportional to the probe frequency and inversely proportional to the distance z_0 between the sensor array and the object. As z_0 drops below the wavelength, the subwavelength Fourier modes of the object can be stably recovered. This is the idea behind the near-field imaging systems such as scanning microscopy.

Inverse Scattering

In the inverse scattering theory, the scattering amplitude is the observable data, and the main objective then is to reconstruct ν from the knowledge of the scattering amplitude.

Pixel Basis

To obtain a sensing matrix with compressive sensing properties, we first make the Born approximation in (3.7) and neglect the scattered field u^s on the right-hand side of (3.7). Our purpose here is to demonstrate how to coordinate the incidence direction and the sampling direction and create a favorable sensing matrix.

Consider the incidence field

$$u^{\mathrm{i}}(\mathbf{r}) = e^{i\omega\mathbf{r}\cdot\hat{\mathbf{d}}} \tag{3.72}$$

where $\hat{\mathbf{d}}$ is the incident direction. Under the Born approximation, we have from (3.7) that

$$A(\hat{\mathbf{r}},\hat{\mathbf{d}}) = A(\mathbf{s}) = \frac{\omega^2}{4\pi}\int_{\mathbb{R}^d} \nu(\mathbf{r}')e^{-i\omega\mathbf{r}'\cdot\mathbf{s}}\,d\mathbf{r}' \tag{3.73}$$

where $\mathbf{s} = \hat{\mathbf{r}} - \hat{\mathbf{d}}$ is the scattering vector.

We proceed to discretize the continuous system (3.73) as before. Consider the discrete approximation of the extended object ν:

$$\nu_\ell(\mathbf{r}) = \sum_{\mathbf{q}\in\mathbb{Z}_N^2} b\left(\frac{\mathbf{r}}{\ell} - \mathbf{q}\right)\nu(\ell\mathbf{q}) \tag{3.74}$$

where

$$b(\mathbf{r}) = \begin{cases} 1, & \mathbf{r} \in \left[-\frac{1}{2}, \frac{1}{2}\right]^2 \\ 0, & \text{else.} \end{cases} \tag{3.75}$$

is the pixel basis.

Define the target vector $\mathbf{f} = (f_j) \in \mathbb{C}_N$ with $f_j, = \nu(\ell\mathbf{p})$, $\mathbf{p} = (p_1, p_2) \in \mathbb{Z}_N^2$, $j = (p_1 - 1)\sqrt{N} + p_2$. Let ω_l and $\hat{\mathbf{d}}_l$ be the probe frequencies and directions, respectively, and let $\hat{\mathbf{r}}_l$ be the sampling directions for $l = 1, \ldots, M$. Let $\mathbf{g}$ be the data vector with

$$g_l = \frac{4\pi A(\hat{\mathbf{r}}_l - \hat{\mathbf{d}}_l)}{\omega^2\hat{b}((\ell\omega_l/2\pi)(\hat{\mathbf{r}}_l - \hat{\mathbf{d}}_l))}.$$

Then the sensing matrix takes the form

$$\Phi_{lj} = e^{i\omega_l\ell\mathbf{q}\cdot(\hat{\mathbf{d}}_l - \hat{\mathbf{r}}_l)}, \quad \mathbf{q} = (q_1, q_2) \in \mathbb{Z}_N^2, \quad j = (q_1 - 1)\sqrt{N} + q_2. \tag{3.76}$$

Sampling Schemes

Our strategy is to construct a sensing matrix analogous to the random partial Fourier matrix. To this end, we write the (l, j)-entry of the sensing matrix in the form

$$e^{i\pi(j_1\xi_l + j_2\zeta_l)}, \quad j = (j_1 - 1)\sqrt{N} + j_2, \quad j_1, j_2 = 1, \ldots, \sqrt{N}, l = 1, \ldots, M$$

where ξ_l, ζ_l are independently and uniformly distributed in $[-1, 1]$. We write (ξ_l, ζ_l) in the polar coordinates ρ_l, ϕ_l as

$$(\xi_l, \zeta_l) = \rho_l(\cos\phi_l, \sin\phi_l), \quad \rho_l = \sqrt{\xi_l^2 + \zeta_l^2} \le \sqrt{2} \tag{3.77}$$

and set

$$\omega_1(\cos\theta_l - \cos\tilde{\theta}_l) = \sqrt{2}\rho_l\Omega\cos\phi_l$$

$$\omega_1(\sin\theta_l - \sin\tilde{\theta}_l) = \sqrt{2}\rho_l\Omega\sin\phi_l$$

where Ω is a parameter to be determined later (3.91). Equivalently, we have

$$-\sqrt{2}\omega_l \sin\frac{\theta_l - \tilde{\theta}_l}{2}\sin\frac{\theta_l + \tilde{\theta}_l}{2} = \Omega\rho_l\cos\phi_l \tag{3.78}$$

$$-\sqrt{2}\omega_l \sin\frac{\theta_l - \tilde{\theta}_l}{2}\cos\frac{\theta_l + \tilde{\theta}_l}{2} = \Omega\rho_l\sin\phi_l. \tag{3.79}$$

This set of equations determines the single-input-(θ_l, ω_l)-single-output-$\tilde{\theta}_l$ mode of sampling.

The following implementation of (3.78) through (3.79) is natural. Let the sampling angle $\tilde{\theta}_l$ be related to the incident angle θ_l via

$$\theta_l + \tilde{\theta}_l = 2\phi_l + \pi, \tag{3.80}$$

and set the frequency ω_l to be

$$\omega_l = \frac{\rho_l \Omega}{\sqrt{2}\sin((\theta_l - \tilde{\theta}_l)/2)}. \tag{3.81}$$

Then the entries (3.76) of the sensing matrix $\mathbf{\Phi}$ have the form

$$e^{i\sqrt{2}\Omega\ell(j_1\xi l + j_2\xi l)}, \quad l = 1,\ldots,n,\ j_1, j_2 = 1,\ldots,\sqrt{N}. \tag{3.82}$$

By the square symmetry of the problem, it is clear that the relation (3.80) can be generalized to

$$\theta_l + \tilde{\theta}_l = 2\phi_l + \eta\pi, \quad \eta \in \mathbb{Z}. \tag{3.83}$$

On the other hand, the symmetry of the square lattice should not play a significant role and hence we expect the result to be insensitive to any *fixed* $\eta \in \mathbb{R}$, independent of l, as long as (3.81) holds. Indeed, this is confirmed by numerical simulations.

Let us focus on two specific measurement schemes.

Backward Sampling

This scheme employs Ω—band limited probes, that is, $\omega_l \in [-\Omega, \Omega]$. This and (3.81) lead to the following constraint:

$$\left|\sin\frac{\theta_l - \tilde{\theta}_l}{2}\right| \geq \frac{\rho_l}{\sqrt{2}}. \tag{3.84}$$

The simplest way to satisfy (3.80) and (3.84) is to set

$$\phi_l = \tilde{\theta}_l = \theta_l + \pi, \tag{3.85}$$

$$\omega_l = \frac{\rho_l \Omega}{\sqrt{2}} \tag{3.86}$$

$l = 1, \ldots, n$. In this case, the scattering amplitude is always sampled in the backscattering direction. This resembles the synthetic aperture imaging, which has been previously analyzed under the paraxial approximation in Fannjiang et al. (2010). In contrast, the forward scattering direction with $\tilde{\theta}_l = \theta_l$ almost surely violates the constraint (3.84).

Forward Sampling

This scheme employs single frequency probes no less than Ω:

$$\omega_l = \gamma\Omega, \quad \gamma \geq 1, l = 1, \ldots, n. \tag{3.87}$$

To satisfy (3.83) and (3.81), we set

$$\theta_l = \phi_l + \frac{\eta\pi}{2} + \arcsin\frac{\rho_l}{\gamma\sqrt{2}} \tag{3.88}$$

$$\tilde{\theta}_l = \phi_l + \frac{\eta\pi}{2} - \arcsin\frac{\rho_l}{\gamma\sqrt{2}} \tag{3.89}$$

with $\eta \in \mathbb{Z}$. The difference between the incident angle and the sampling angle is

$$\theta_l - \tilde{\theta}_l = 2\arcsin\frac{\rho_l}{\gamma\sqrt{2}}, \tag{3.90}$$

which diminishes as $\gamma \to \infty$. In other words, in the high frequency limit, the sampling angle approaches the incident angle. This resembles the setting of x-ray tomography.

In summary, let ξ_l, ζ_l be independently and uniformly distributed in $[-1, 1]$ and let (ρ_l, ϕ_l) be the polar coordinates of (ξ_l, ζ_l), that is,

$$(\xi_l, \zeta_l) = \rho_l (\cos\phi_l, \sin\phi_l).$$

Then with

$$\Omega\ell = \frac{\pi}{\sqrt{2}}, \tag{3.91}$$

both forward and backward samplings give rise to the random partial Fourier sensing matrix.

Coherence Bounds for Single Frequency

As in the "Fresnel Diffraction with Point Objects" section, we let the point scatterers be continuously distributed over a finite domain, not necessarily on a grid. Any computational imaging would involve some underlying, however refined, grid. Hence, let us assume that there is an underlying, possibly highly refined and unresolved, grid of spacing $\ell \ll \omega_l^{-1}$ (the reciprocal of probe frequency).

We shall focus on the monochromatic case with $\omega_l = \omega$, $l = 1, \ldots, M$. We recall that the sensing matrix is continuous of the form (3.76), which now becomes

$$\phi_{lj} = e^{i\omega\ell\mathbf{p}\cdot(\hat{\mathbf{d}}_l - \hat{\mathbf{r}}_l)}, \quad j = (p_1 - 1)\sqrt{N} + p_2, \quad \mathbf{p} \in \mathbb{Z}_N^2. \tag{3.92}$$

In other words, the measurement diversity comes entirely from the variations of the incidence and detection directions. We assume that the n incident directions and the m detection directions are each independently chosen according to some distributions with the total number of data $M = nm$ fixed.

Theorem 3.5 (2D case) *Suppose the incident and sampling angles are randomly, independently, and identically distributed according to the probability density functions* $f^i(\theta) \in C^1$ *and* $f^s(\theta) \in C^1$, *respectively. Suppose*

$$N \leq \frac{\varepsilon}{8} e^{K^2/2}, \quad \varepsilon, K > 0. \tag{3.93}$$

Set $L = \ell|\mathbf{p}-\mathbf{p}|$ *for any* $\mathbf{p}, \mathbf{q} \in \mathbb{Z}_N^2$. *Then the sensing matrix satisfies the pairwise coherence bound*

$$\mu_{\mathbf{p},\mathbf{q}} < \left(\bar{\mu}^i + \frac{\sqrt{2K}}{\sqrt{n}}\right)\left(\bar{\mu}^s + \frac{\sqrt{2K}}{\sqrt{m}}\right). \tag{3.94}$$

with probability greater than $(1-\varepsilon)^2$, *where*

$$\bar{\mu}^i \leq c(1+\omega L)^{-1/2} \sup_\theta \left\{ |f^i(\theta)|, \left|\frac{d}{d\theta} f^i(\theta)\right| \right\}, \tag{3.95}$$

$$\bar{\mu}^s \leq c(1+\omega L)^{-1/2} \sup_\theta \left\{ |f^s(\theta)|, \left|\frac{d}{d\theta} f^s(\theta)\right| \right\}, \tag{3.96}$$

with a constant c.

In 3D, the coherence bound can be improved with a faster decay rate in terms of $\omega L \gg 1$ as stated in the following.

Theorem 3.6 (3D case) *Assume* (3.93). *Suppose the incidence and sampling directions, parameterized by the polar angle* $\theta \in [0, \pi]$ *and the azimuthal angle* $\phi \in [0, 2\pi]$, *are randomly, independently, and identically distributed. Let* $f^i(\theta) \in C^1$ *and* $f^s(\theta) \in C^1$ *be the marginal density functions of the incident and sampling polar angles, respectively.*

Let $L = \ell|\mathbf{p}-\mathbf{p}|$. *Then the sensing matrix satisfies the pairwise coherence bound*

$$\mu_{\mathbf{p},\mathbf{q}} < \left(\bar{\mu}^i + \frac{\sqrt{2K}}{\sqrt{n}}\right)\left(\bar{\mu}^s + \frac{\sqrt{2K}}{\sqrt{m}}\right). \tag{3.97}$$

with probability greater than $(1-\varepsilon)^2$, *where*

$$\bar{\mu}^i \leq c(1+\omega L)^{-1} \sup_\theta \left\{ |f^i(\theta)|, \left|\frac{d}{d\theta} f^i(\theta)\right| \right\} \tag{3.98}$$

$$\bar{\mu}^s \leq c(1+\omega L)^{-1} \sup_\theta \left\{ |f^s(\theta)|, \left|\frac{d}{d\theta} f^s(\theta)\right| \right\}. \tag{3.99}$$

Remark 3.5 The original statements of the theorems (Fannjiang 2010b, Theorems 1 and 6) have been adapted to the present context of off-grid objects. The original proofs, however, carry over here verbatim upon minor change of notation.

Remark 3.6 When the sampling directions are randomized and the incidence directions are deterministic, then the coherence bounds (3.94) and (3.97) hold with the first factor on the right-hand side removed.

According to Remark 3.6, we have the pairwise coherence bound:

$$\text{(2D)} \quad \mu_{p,q} \le c(1+\omega L)^{-1/2} \sup_{\theta}\left\{\left|f^s(\theta)\right|, \left|\frac{d}{d\theta} f^s(\theta)\right|\right\} + \frac{\sqrt{2}K}{\sqrt{M}} \tag{3.100}$$

$$\text{(3D)} \quad \mu_{p,q} \le c(1+\omega L)^{-1} \sup_{\theta}\left\{\left|f^s(\theta)\right|, \left|\frac{d}{d\theta} f^s(\theta)\right|\right\} + \frac{\sqrt{2}K}{\sqrt{M}} \tag{3.101}$$

which is an estimate of the coherence pattern of the sensing matrix. Hence, if L is unresolvable (i.e., $\omega L \le 1$), the corresponding pairwise coherence parameter is high, and if L is well resolved (i.e., $\omega L \gg 1$), the corresponding pairwise coherence parameter is low. A typical coherence band has a coherence radius $\mathcal{O}(\omega^{-1})$ according to (3.100) and (3.101).

Therefore, if the point objects are well separated in the sense that any pair of objects are larger than ω^{-1}, then the same BLO- and BLOT-based techniques discussed in the "Fresnel Diffraction with Point Objects" section can be used to recover the masked object support and amplitudes. For a simple illustration, Figure 3.11 shows two instances of reconstruction by BOMP. The recovered objects (blue asterisks) are close to the true objects (red circles) well within the coherence bands (yellow patches).

Inverse Multiple Scattering

In this section, we present an approach to compressive imaging of multiply scattering point scatterers. First consider the multiple scattering effect with just a single illumination, that is, $n = 1$ and $M = m$.

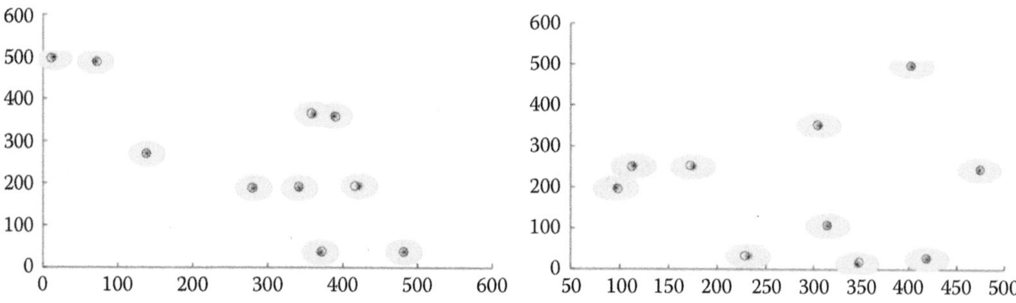

Figure 3.11 Two instances of BOMP reconstruction: red circles are the exact locations, blue asterisks are recovered locations, and the yellow patches are the coherence bands around the objects.

Note that the original object support is the same as the masked object support. With the support accurately recovered, let us consider how to unmask the objects and recover the true objects.

Define the incidence and full field vectors at the locations of the objects:

$$\mathbf{u}^i = (u^i(\mathbf{r}_1),\ldots,u^i(\mathbf{r}_s))^T \in \mathbb{C}^s$$

$$\mathbf{u} = (u(\mathbf{r}_1),\ldots,u(\mathbf{r}_s))^T \in \mathbb{C}^s.$$

Let $\mathbf{\Gamma}$ be the $s \times s$ matrix

$$\mathbf{\Gamma} = [(1 - \delta_{jl})G(\mathbf{r}_j, \mathbf{r}_l)]$$

and $\mathcal{V}$ the diagonal matrix

$$\mathcal{V} = \mathrm{diag}(v_1, \ldots, v_s).$$

The full field is determined by the Foldy–Lax equation (Mishchenko et al. 2006)

$$\mathbf{u} = \mathbf{u}^i + \omega^2\mathbf{\Gamma}\mathcal{V}\mathbf{u} \tag{3.102}$$

from which we obtain the full field

$$\mathbf{u} = (\mathbf{I} - \omega^2\mathbf{\Gamma}\mathcal{V})^{-1}\mathbf{u}^i \tag{3.103}$$

and the masked objects

$$\begin{aligned}\mathbf{f} &= \mathcal{V}\mathbf{u} = \mathcal{V}(\mathbf{I} - \omega^2\Gamma\mathcal{V})^{-1}\mathbf{u}^i \\ &= (\mathbf{I} - \omega^2\mathcal{V}\Gamma)^{-1}\mathcal{V}\mathbf{u}^i\end{aligned} \tag{3.104}$$

provided that ω^{-2} is not an eigenvalue of $\mathbf{\Gamma}\mathcal{V}$.

Hence, by solving (3.104) we have

$$(\mathbf{I} - \omega^{-2}\mathcal{V}\mathbf{\Gamma})\mathbf{f} = \mathcal{V}\mathbf{u}^i. \tag{3.105}$$

The true objects ν can then be recovered by solving (3.105) as

$$\nu = \frac{\mathbf{f}}{\omega^2\mathbf{\Gamma}\mathbf{f}+\mathbf{u}^i} \tag{3.106}$$

where the division is carried out entrywise (Hadamard product).

Joint Sparsity

With the total number of data $M = nm$ fixed, the coherence bounds (3.94) and (3.97) are optimized with $n \sim m \sim \sqrt{M}$. To take advantage of this result, we should deploy multiple incidence fields for which the formula (3.106) is no longer valid.

Multiple illuminations give rise to multiple data vectors $\mathbf{g}_j$, and multiple masked object vectors $\mathbf{f}_j$, $j = 1, \ldots, n$, each of which is masked by an unknown field $\mathbf{u}_j$. However, all masked object vectors give rise to the same sensing matrix

$$\Phi_{lj} = e^{-i\omega\ell\mathbf{p}\cdot\hat{\mathbf{r}}l}, \quad j = (p_1 - 1)\sqrt{N} + p_2, \mathbf{p} \in \mathbb{Z}_N^2.$$

Since every masked object vector shares the same support as the true object vector, this is a suitable setting for the application of joint sparsity techniques discussed in the "BPDN for joint sparsity" and "OMP for joint sparsity" sections.

Compiling the masked object vectors as $\mathbf{F} = [\mathbf{f}_1, \ldots, \mathbf{f}_n] \in \mathbb{C}^{m\times n}$ and the data vectors as $\mathbf{G} = [\mathbf{g}_1, \ldots, \mathbf{g}_n] \in \mathbb{C}^{m\times n}$, we obtain the imaging equations

$$\mathbf{G} = \mathbf{\Phi}\mathbf{F} + \mathbf{E} \tag{3.107}$$

where $\mathbf{E}$ accounts for noise. When the true objects are widely separated, we have two ways to proceed as follows.

BPDN-BLOT for Joint Sparsity

In the first approach, we use BPDN for joint sparsity (3.37) with $\mathbf{\Phi}_j = \mathbf{\Phi}, \forall j, \mathcal{L} = 0$ to solve the imaging equation (3.107). Let $\mathbf{F}_* = (\mathbf{f}_{1*}, \ldots, \mathbf{f}_{n*})$ be the solution. We then apply the BLOT technique (Algorithm 3.5) to improve $\mathbf{F}_*$ (i.e., trim the bushes and grow the trees). In order to enforce the joint sparsity structure, we modify Algorithm 3.5 as follows.

First, we modify the LO algorithm to account for joint sparsity.

Algorithm 3.7 LO for Joint Sparsity

Inputs: $\mathbf{\Phi}_1, \ldots, \mathbf{\Phi}_n, \mathbf{G}, \eta > 0, S^0 = \{i_1, \ldots, i_s\}$.

Iteration: For $k = 1, 2, \ldots, s$.

(1) $\mathbf{F}^k = \arg\min \|[\mathbf{\Phi}_1\mathbf{h}_1, \ldots, \mathbf{\Phi}_n\mathbf{h}_n] - \mathbf{G}\|_F$ s.t. $\cup_j$, $\mathrm{supp}(\mathbf{h}_j) \subseteq (S^{k-1} \setminus \{i_k\}) \cup \{i'_k\}$, $i'_k \in B_\eta(\{i_k\})$.

(2) $S^k = \mathrm{supp}(\mathbf{F}^k)$.

Output: S^s.

Next, we modify the BLOT algorithm to account for joint sparsity.

Algorithm 3.8 BLOT for Joint Sparsity

Input: $\mathbf{f}_1, \ldots, \mathbf{f}_n, \mathbf{\Phi}_1, \ldots, \mathbf{\Phi}_n, \mathbf{G}, \eta > 0$.

Initialization: $S^0 = \emptyset$.

Iteration: For $k = 1, 2, \ldots, s$.

(1) $i_k = \arg\max_j \|\mathbf{f}_j\|_2, k \notin B^{(2)}_\eta(S^{k-1})$.

(2) $S^k = S^{k-1} \cup \{i_k\}$.

Output: $\mathbf{F}_* = \arg\min \|[\mathbf{\Phi}_1\mathbf{h}_1, \ldots, \mathbf{\Phi}_n\mathbf{h}_n] - \mathbf{G}\|_F$, $\cup_j \mathrm{supp}(\mathbf{h}_j) \subseteq \mathrm{JLO}(S^s)$, where $\mathrm{JLO}(S^s)$ is the output of Algorithm 3.7 with the sth iterate S^s of BLOT as input.

BLOOMP for Joint Sparsity

In the second approach, we propose the following joint sparsity version of BLOOMP.

Algorithm 3.9 BLOOMP for Joint Sparsity

Input: $\mathbf{\Phi}_1, \ldots, \mathbf{\Phi}_1, \mathbf{G}, \eta > 0$

Initialization: $\mathbf{F}^0 = 0$, $\mathbf{R}^0 = \mathbf{G}$ and $S^0 = \emptyset$

Iteration: For $k = 1, \ldots, s$

(1) $i_{\max} = \arg\max_i \sum_{j=1}^{J} |\Phi_{j,i}^{\dagger}\mathbf{r}_j^{k-1}|, i \notin B_\eta^{(2)}(S^{k-1})$, where $\Phi_{j,i}^{\dagger}$ = conjugate transpose of col$_i$ ($\mathbf{\Phi}_j$).

(2) $S^k = \mathrm{JLO}(S^{k-1} \cup \{i_{\max}\})$, where JLO is the output of Algorithm 3.7.

(3) $[\mathbf{f}_1^k,\ldots,\mathbf{f}_n^k] = \arg\min_{\mathbf{H}} \|[\mathbf{\Phi}_1\mathbf{h}_1, \ldots, \mathbf{\Phi}_n\mathbf{h}_n] - \mathbf{G}\|_{\mathrm{F}}$ s.t. $\cup_j \mathrm{supp}(\mathbf{h}_j) \subseteq S^k$

(4) $[\mathbf{r}_1^k,\ldots,\mathbf{r}_n^k] = \mathbf{G} - [\Phi_1\mathbf{f}_1^k,\ldots,\Phi_n\mathbf{f}_n^k]$

Output: $\mathbf{F}_* = [\mathbf{f}_1^s,\ldots,\mathbf{f}_n^s]$.

After the first stage of either approach, we obtain an estimate of the object support as well as the amplitudes of masked objects. In the second stage, we estimate the true object amplitudes. If we use the formula (3.106) for each incident wave $\mathbf{u}_j^{\mathrm{i}}$, we end up with n amplitude estimates

$$\frac{\mathbf{f}_{j*}}{\omega^2\Gamma\mathbf{f}_{j*} + \mathbf{u}_j^{\mathrm{i}}}, \quad j = 1,\ldots,n$$

that are typically inconsistent. Least squares is the natural way to solve this overdetermined system and obtain the object estimate.

$$v_* = \arg\min_{\mathbf{v}} \sum_{j=1}^{n} \|(\omega^2\Gamma\mathbf{f}_{j*} + \mathbf{u}_j^i)\mathbf{v} - \mathbf{f}_{j*}\|_2^2.$$

Inverse Scattering with Zernike Basis

In this section, we discuss a basis for representing extended objects in the scattering geometry and its application to compressive inverse scattering. We shall make the Born approximation.

A well-known orthogonal basis for representing an extended object with a compact support (e.g., the unit disk) is the product of Zernike polynomials R_n^m and trigonometric functions

$$V_n^m(x,y) = V_n^m(\rho\cos\theta, \rho\sin\theta) = R_n^m(\rho)e^{im\theta}, \quad x^2 + y^2 \le 1 \tag{3.108}$$

where $m \in \mathbb{Z}$, $n \in \mathbb{N}$, $n \ge |m|$, and $n - |m|$ are even. We refer to V_n^m as the Zernike functions of order (m, n) (Born and Wolf 1999). These Zernike functions are very useful in optics because the lowest few terms of a Zernike expansion have a simple optical interpretation (Dai and Milenkovic 2008). In addition, a Zernike expansion usually has a superior rate of convergence (hence sparser) compared with other expansions such as a Bessel-Fourier or Chebyshev-Fourier expansion (Boyd and Yu 2011; Boyd and Petschek 2014).

We show now that the Zernike basis also results in a better coherence parameter (hence better resolution) than the pixel basis. The Zernike polynomials are given explicitly by the formula

$$R_n^m(\rho) = \frac{1}{\left(\frac{n-|m|}{2}\right)\rho^{|m|}}\left[\frac{d}{d(\rho^2)}\right]^{\frac{n-|m|}{2}}\left[(\rho^2)^{\frac{n+|m|}{2}}(\rho^2-1)^{\frac{n-|m|}{2}}\right] \tag{3.109}$$

which are nth degree polynomials in ρ and normalized such that $R_n^m(1)=1$ for all permissible values of m, n. The Zernike polynomials satisfy the following properties:

$$\int_0^1 R_n^m(\rho)R_{n'}^m(\rho)\rho\, d\rho = \frac{\delta_{nn'}}{2(n+1)} \tag{3.110}$$

$$\int_0^1 R_n^m(\rho)J_m(u\rho)\rho\, d\rho = (-1)^{\frac{n-m}{2}}\frac{J_{n+1}(u)}{u} \tag{3.111}$$

where J_{n+1} is the $(n + 1)$-order Bessel function of the first kind. As a consequence of (3.110), the Zernike functions satisfy the orthogonality property

$$\int_{x^2+y^2\le 1} \overline{V_n^m(x,y)V_{n'}^{m'}(x,y)}dx\,dy = \frac{\pi}{n+1}\delta_{mm'}\delta_{nn'}. \tag{3.112}$$

Writing $\mathbf{s} = s(\cos\phi, \sin\phi)$, let us compute the matrix element for the scattering amplitude (3.73) as follows:

$$\begin{aligned}\int_{x^2+y^2\le 1} \overline{V_n^m(x,y)}e^{-i\omega\mathbf{s}\cdot(x,y)}dx\,dy &= \int_0^1\int_0^{2\pi} e^{i\omega s\rho\cos(\phi+\theta)R_n^m(\rho)e^{-im\theta}}\,d\theta\rho\,d\rho \\ &= \int_0^1\int_0^{2\pi} e^{i\omega s\rho\cos\theta}e^{\ im\theta}\,d\theta R_n^m(\rho)\rho\,d\rho e^{im\phi} \\ &= 2\pi i^n e^{im\phi}\int_0^1 J_m(\omega s\rho)R_n^m(\rho)\rho\,d\rho\end{aligned} \tag{3.113}$$

by the definition of Bessel function

$$J_m(z) = \frac{1}{\pi i^m}\int_0^{\pi} e^{iz\cos\theta}\cos(m\theta)d\theta.$$

Using the property (3.111), we then obtain from (3.113) that

$$\int_{x^2+y^2\le 1} \overline{V_n^m(x,y)} e^{-i\omega \mathbf{s}\cdot(x,y)} dx\, dy = 2\pi i^m (-1)^{\frac{n-m}{2}} e^{im\phi} \frac{J_{n+1}(\omega s)}{\omega s} \tag{3.114}$$

which are the sensing matrix elements with all permissible m, n. Note that the columns of the sensing matrix are indexed by the permissible $m \in \mathbb{Z}$, $n \in \mathbb{N}$ with the constraint that $n \ge |m|$ and $n - |m|$ are even.

Let the scattering vector $\mathbf{s} = \hat{\mathbf{r}} - \hat{\mathbf{d}}$ be parameterized as

$$\mathbf{s}_{jk} = s_j(\cos\phi_k, \sin\phi_k), \quad j,k = 1,\ldots,\sqrt{M}$$

such that $\{\phi_k\}$ are independently and identically distributed uniform random variables on $[0, 2\pi]$ according to the uniform distribution and that $\{s_j\}$ are independently distributed on $[0, 2\omega]$ according to the linear density function $f(r) = r/2$. As a result, $z_j = \omega s_j$ are independently and identically distributed on $[0, 2\omega]$ according to a linear density function.

Calculation of the coherence parameter between the columns corresponding to $(m, n) \ne (m', n')$ gives the following expression:

$$\left(\frac{1}{\sqrt{M}} \sum_{j=1}^{\sqrt{M}} \frac{J_{n+1}(\omega s_j)}{\omega s_j} \frac{J_{n'+1}(\omega s_j)}{\omega s_j}\right)\left(\frac{1}{\sqrt{M}} \sum_{k=1}^{\sqrt{M}} e^{i(m-m')\phi_k}\right).$$

We recall that for $p, q \in \mathbb{N}$ (Abramowitz and Stegun 1972, formula 11.4.6)

$$\int_0^\infty J_p(z) J_q(z) \frac{dz}{z} = \begin{cases} 0, & p \ne q \\ \frac{1}{2p}, & p = q \end{cases}. \tag{3.115}$$

For $M \gg 1$, we have by the law of large numbers

$$\begin{aligned} \frac{1}{\sqrt{M}} \sum_{j=1}^{\sqrt{M}} \frac{J_{n+1}(\omega s_j)}{\omega s_j} \frac{J_{n'+1}(\omega s_j)}{\omega s_j} &\sim \mathbb{E}\left[\frac{J_{n+1}(\omega r)}{\omega r} \frac{J_{n'+1}(\omega r)}{\omega r}\right] \\ &= \frac{1}{2\omega^2} \int_0^{2\omega} J_{n+1}(z) J_{n'+1}(z) \frac{dz}{z} \end{aligned} \tag{3.116}$$

and

$$\begin{aligned} \frac{1}{\sqrt{M}} \sum_{k=1}^{\sqrt{M}} e^{i(m-m')\phi_k} \sim \mathbb{E} e^{i(m-m')\phi} &= \int_0^{2\pi} e^{i(m-m')\phi} g(\phi) d\phi \\ &= \delta_{mm'}. \end{aligned} \tag{3.117}$$

When $m \ne m'$, the two columns are orthogonal and the pairwise coherence parameter is zero. When $n \ne n'$, the right-hand side of (3.116) becomes $\mathcal{O}(\omega^{-3})$ in view of (3.115) and the fact that the Bessel functions $J_n(z)$ decay like $z^{-1/2}$ for $z \gg 1$. From

(3.115) and (3.116) with $n = n'$, we see that the 2-norm of the columns is $\mathcal{O}(\omega^{-2})$. After dividing (3.116) with $n \neq n'$ by the 2-norm of the columns, the coherence parameter scales at worst like ω^{-1} (for $m = m'$, $n \neq n'$).

Notice that this decay date of the coherence parameter is faster than the $\omega^{-1/2}$ behavior in (3.94) through (3.95). Hence, imaging with the Zernike basis possesses better resolution capability than with the pixel basis, all else being equal.

Interferometry with Incoherent Sources

In this last section, we discuss the compressive sensing application to optical interferometry in astronomy, which has a similar mathematical structure to that of the inverse scattering (3.92) under the Born approximation.

In astronomy, interferometry often deals with signals emitted from incoherent sources. In this section, we present the compressive sensing approach to such a problem. With the help of the van Cittert–Zernike theorem, the sensing matrix has a structure not unlike what we discussed earlier.

Suppose the field of view is small enough to be identified with a planar patch of the celestial sphere $\mathcal{P} \subset \mathbb{R}^2$, called the object plane. Let $I(\mathbf{s})$ be the radiation intensity from the point $\mathbf{s}$ on the object plane $\mathcal{P}$. Let n antennas be located in a square of size L on the sensor plane parallel to $\mathcal{P}$ with locations $L\mathbf{r}_j$, $j = 1, \ldots, n$, where $\mathbf{r}_j \in [0, 1]^2$. Then by the van Cittert–Zernike theorem (Born and Wolf 1999), the measured visibility $v(\mathbf{r}_j - \mathbf{r}_k)$ is given by the Fourier integral

$$v(\mathbf{r}_j - \mathbf{r}_k) = \int_{\mathcal{P}} I(\mathbf{s}) e^{i\omega \mathbf{s}\cdot(\mathbf{r}_j - \mathbf{r}_k)L} d\mathbf{s}. \tag{3.118}$$

Consider the discrete approximation of the extended object I with the pixel basis on the grid $\ell\mathbb{Z}_N^2$:

$$I_\ell(\mathbf{r}) = \sum_{\mathbf{q}\in\mathbb{Z}_N^2} b\left(\frac{\mathbf{r}}{\ell} - \mathbf{q}\right) I(\ell\mathbf{q}) \tag{3.119}$$

where b is given in (3.75) and

$$\mathbb{Z}_N^2 = \{\mathbf{p} = (p_1, p_2) : p_1, p_2 = 1, \ldots, \sqrt{N}\}. \tag{3.120}$$

Substituting (3.119) into (3.118), we obtain the discrete sum

$$v(\mathbf{r}_j - \mathbf{r}_k) = \ell^2 \hat{b}\left(\frac{\omega\ell L}{2\pi}(\mathbf{r}_k - \mathbf{r}_j)\right) \sum_{l=1}^{N} I_l e^{i\omega \mathbf{p}\cdot(\mathbf{r}_j - \mathbf{r}_k)\ell L}, \tag{3.121}$$

where l, $\mathbf{p}$ are related by $l = (p_1 - 1)\sqrt{N} + p_2$ and

$$\hat{b}(\xi, \eta) = \frac{\sin(\pi\xi)}{\pi\xi} \frac{\sin(\pi\eta)}{\pi\eta}.$$

For every pair (j, k) of sensors, we measure and collect the interferometric datum $v(\mathbf{r}_j - \mathbf{r}_k)$ and we want to determine I from the collection of $n(n - 1)$ real-valued data.

Let us rewrite Equation (3.121) in the form (3.10). In contrast to (3.28), we set

$$\ell = \frac{\pi}{\omega L} \tag{3.122}$$

to account for the "two-way" structure in the imaging equation (3.121). Note that ℓ is the resolution length on the celestial sphere and hence dimensionless.

Let $\mathbf{f} = (f_i) \in \mathbb{R}^N$ be the unknown object vector, that is, $f_i = \ell^2 I_i$. Let $\mathbf{g} = (g_l) \in \mathbb{R}^M$, $M = n(n-1)/2$

$$g_l = \frac{1}{\hat{b}((\mathbf{r}_k - \mathbf{r}_j)/2)} \times \begin{cases} \Re[v(\mathbf{r}_j - \mathbf{r}_k)], & l = (2n-j)(j-1)/2 + k, \quad j < k = 1,\ldots,n \\ \Im[v(\mathbf{r}_j - \mathbf{r}_k)], & l = n(n-1)/2 + (2n-j)(j-1)/2 + k, \quad j < k = 1,\ldots,n \end{cases}$$

be the data vector where $\Re$ and $\Im$ stand for, respectively, the real and imaginary parts. The sensing matrix $\mathbf{\Phi} \in \mathbb{R}^{M\times N}$ now takes the form

$$\Phi_{il} = \begin{cases} \cos[2\pi\mathbf{p}_l \cdot (\mathbf{r}_j - \mathbf{r}_k)], & i = (2n-j)(j-1)/2 + k, \quad j < k \\ \sin[2\pi\mathbf{p}_l \cdot (\mathbf{r}_j - \mathbf{r}_k)], & i = n(n-1)/2 + (2n-j)(j-1)/2 + k, \quad j < k \end{cases} \tag{3.123}$$

which is no longer the simple random partial Fourier matrix for 2D as the baselines $\mathbf{r}_j - \mathbf{r}_k$ are related to one another. Nevertheless, (3.123) has a similar structure to that of the inverse scattering (3.92) when the transmitters and receivers are colocated. Note that as $(\mathbf{r}_k - \mathbf{r}_j)/2 \in [-1/2, 1/2]^2$, the denominator $\hat{b}((\mathbf{r}_k - \mathbf{r}_j)/2)$ in the definition of g_l does not vanish.

Next, we give an upper bound for the coherence parameter. For the pairwise coherence for columns i, i' corresponding to $\mathbf{p}, \mathbf{p} \in \mathbb{Z}_N^2$, we have the following calculation:

$$\mu(i,i') = \frac{2}{n(n-1)} \left| \sum_{j<k} \cos[2\pi\mathbf{p} \cdot (\mathbf{r}_j - \mathbf{r}_k)]\cos[2\pi\mathbf{p}' \cdot (r_j - r_k)] + \sin[2\pi\mathbf{p} \cdot (\mathbf{r}_j - \mathbf{r}_k)]\sin[2\pi\mathbf{p}' \cdot (\mathbf{r}_j - \mathbf{r}_k)] \right|$$

$$= \frac{2}{n(n-1)} \left| \sum_{j<k} \cos[2\pi(\mathbf{p} - \mathbf{p}') \cdot (\mathbf{r}_j - \mathbf{r}_k)] \right|$$

$$= \frac{1}{n(n-1)} \left| \sum_{j\neq k} \cos[2\pi(\mathbf{p} - \mathbf{p}') \cdot (\mathbf{r}_j - \mathbf{r}_k)] \right|$$

First, we claim

$$\mu(i,i') = \frac{1}{n(n-1)} \left| \left| \sum_{j=1}^{n} e^{i2\pi(\mathbf{p}-\mathbf{p}')\cdot\mathbf{r}_j} \right|^2 - n \right|.$$

This follows from the calculation

$$\left|\sum_{j=1}^{n} e^{i2\pi(\mathbf{p}-\mathbf{p}')\cdot\mathbf{r}_j}\right|^2 - n = \sum_{j\neq k} e^{i2\pi(\mathbf{p}-\mathbf{p}')\cdot(\mathbf{r}_j-\mathbf{r}_k)}$$

$$= \sum_{j\neq k} \cos\left[2\pi(\mathbf{p}-\mathbf{p}')\cdot(\mathbf{r}_j-\mathbf{r}_k)\right] + i\sin\left[2\pi(\mathbf{p}-\mathbf{p}')\cdot(\mathbf{r}_j-\mathbf{r}_k)\right]$$

$$= \sum_{j\neq k} \cos\left[2\pi(\mathbf{p}-\mathbf{p}')\cdot(\mathbf{r}_j-\mathbf{r}_k)\right]$$

Some modification of the arguments for Theorems 3.5 and 3.6 leads to the following coherence bound.

Theorem 3.7 *Assume that the total number of grid point N satisfies the bound*

$$N \leq \frac{\varepsilon}{2} e^{K^2/2}. \tag{3.124}$$

with some constants δ *and K. Suppose that the sensor locations* $\mathbf{r}_j$, $j = 1,\ldots,n$ *are independent uniform random variables on* $[0, 1]^2$. *Then the coherence parameter* μ *satisfies the bound*

$$\mu(\Phi) \leq \frac{|2K^2 - 1|}{n-1}. \tag{3.125}$$

with probability greater than $1 - 2\varepsilon$.

In other words, with high probability, the coherence parameter for the uniform distribution decays as n^{-1}. A central problem in interferometry is the design of an optimal array, see Fannjiang (2013b) for a discussion from the perspective of compressed sensing.

Acknowledgments

The research is supported in part by NSF grant DMS-1413373 and Simons Foundation grant 275037.

References

Abramowitz, M. and I. Stegun, *Handbook of Mathematical Functions* (New York: Dover, 1972).

Benedek, P. and R. Panzone, The space ℓ^p with mixed norm, *Duke Math. J.* 28 (1961): 301–324.

Blumensath, T. and M.E. Davies, Iterative hard thresholding for compressed sensing, *Appl. Comput. Harmon. Anal.* 27 (2009): 265–274.

Blumensath, T. and M.E. Davies, Normalized iterative hard thresholding: Guaranteed stability and performance, *IEEE J. Sel. Top. Signal Process.* 4 (2010): 298–309.

Born, M. and E. Wolf, *Principles of Optics*, 7th edn. (Cambridge, U.K.: Cambridge University Press, 1999).

Boyd, J.P. and R. Petschek, The relationships between Chebyshev, Legendre and Jacobi polynomials: The generic superiority of Chebyshev polynomials and three important exceptions, *J. Sci. Comput.* 59 (2014): 1–27.

Boyd, J.P. and F. Yu, Comparing six spectral methods for interpolation and the Poisson equation in a disk: Radial basis functions, Logan-Shepp ridge polynomials, Fourier-Bessel, Fourier-Chebyshev, Zernike polynomials, and double Chebyshev series, *J. Comput. Phys.* 230 (2011): 1408–1438.

Boyd, S. and L. Vandenberghe, *Convex Optimization*. (Cambridge, U.K.: Cambridge University Press, 2004).

Bruckstein, A.M., D.L. Donoho, and M. Elad, From sparse solutions of systems of equations to sparse modeling of signals, *SIAM Rev.* 51 (2009): 34–81.

Candès, E.J., The restricted isometry property and its implications for compressed sensing, *C. R. Acad. Sci., Paris, Ser. I.* 346 (2008): 589–592.

Candès, E.J., Y.C. Eldar, D. Needell, and P. Randall, Compressed sensing with coherent and redundant dictionaries, *Appl. Comput. Harmon. Anal.* 31 (2011): 59–73.

Candès, E.J. and C. Fernandez-Granda, Super-resolution from noisy data, *J. Fourier Anal. Appl.* 19(6) (2013): 1229–1254.

Candès, E.J. and C. Fernandez-Granda, Towards a mathematical theory of super-resolution, *Commun. Pure Appl. Math.* 67(6) (2014): 906–956.

Candès, E.J., J. Romberg, and T. Tao, Robust uncertainty principles: Exact signal reconstruction from highly incomplete frequency information, *IEEE Trans. Inf. Theory* 52 (2006): 489–509.

Candès, E.J. and T. Tao, Decoding by linear programming, *IEEE Trans. Inf. Theory* 51 (2005): 4203–4215.

Chambolle, A., An algorithm for total variation minimization and applications, *J. Math. Imaging Vis.* 20 (2004): 89–97.

Chambolle, A. and P.-L. Lions, Image recovery via total variation minimization and related problems, *Numer. Math.* 76 (1997): 167–188.

Chen, J. and X. Huo, Theoretical results on sparse representations of multiple-measurement vectors, *IEEE Trans. Signal Process.* 54 (2006): 4634–4643.

Chen, S.S., D.L. Donoho, and M.A. Saunders, Atomic decomposition by basis pursuit, *SIAM Rev.* 43 (2001): 129–159.

Cotter, S.F., B.D. Rao, K. Engan, and K. Kreutz-Delgado, Sparse solutions to linear inverse problems with multiple measurement vectors, *IEEE Trans. Signal Process.* 53 (2005): 2477–2488.

Dai, G.-M. and V.N. Mahajan, Orthonormal polynomials in wavefront analysis: error analysis, *Appl. Opt.* 47 (2008): 3433–3445.

Dai, W. and O. Milenkovic, Subspace pursuit for compressive sensing: Closing the gap between performance and complexity, *IEEE Trans. Inf. Theory* 55 (2009): 2230–2249.

Davis, G.M., S. Mallat, and M. Avellaneda, Adaptive greedy approximations, *J. Construct. Approx.* 13 (1997): 57–98.

Daubechies, I., *Ten Lectures on Wavelets*. (Philadelphia, PA: SIAM, 1992).

Donoho, D.L., M. Elad, and V.N. Temlyakov, Stable recovery of sparse overcomplete representations in the presence of noise, *IEEE Trans. Inf. Theory* 52 (2006): 6–18.

Duarte, M.F. and R.G. Baraniuk, Spectral compressive sensing, *Appl. Comput. Harmon. Anal.* 35 (2013): 111–129.

Fannjiang, A., Compressive imaging of subwavelength structures, *SIAM J. Imaging Sci.* 2 (2009): 1277–1291.

Fannjiang, A., Compressive inverse scattering I. High-frequency SIMO/MISO and MIMO measurements, *Inverse Probl.* 26 (2010a): 035008.

Fannjiang, A., Compressive inverse scattering II. SISO measurements with Born scatterers, *Inverse Probl.* 26 (2010b): 035009.

Fannjiang, A., TV-min and greedy pursuit for constrained joint sparsity and application to inverse scattering, *Math. Mech. Complex Syst.* 1 (2013a): 81–104.

Fannjiang, A. and W. Liao, Coherence-pattern guided compressive sensing with unresolved grids, *SIAM J. Imaging Sci.* 5 (2012a): 179–202.

Fannjiang, A. and W. Liao, Super-resolution by compressive sensing algorithms, in *IEEE Proceedings of Asilomar Conference on Signals, Systems and Computers*, 2012b.

Fannjiang, A., T. Strohmer, and P. Yan, Compressed remote sensing of sparse objects, *SIAM J. Imaging Sci.* 3 (2010): 596–618.

Fannjiang, C., Optimal arrays for compressed sensing in snapshot-mode interferometry, *Astron. Astrophys.* 559 (2013b): A73–A84.

Kowalski, M., Sparse regression using mixed norms, *Appl. Comput. Harmon. Anal.* 27 (2009): 303–324.

Mishchenko, M.I., L.D. Travis, and A.A. Lacis, *Multiple Scattering of Light by Particles: Radiative Transfer and Coherent Backscattering.* (Cambridge, U.K.: Cambridge University Press, 2006).

Needell, D. and J.A. Tropp, CoSaMP: Iterative signal recovery from incomplete and inaccurate samples, *Appl. Comput. Harmon. Anal.* 26 (2009): 301–329.

Pati, Y.C., R. Rezaiifar, and P.S. Krishnaprasad, Orthogonal matching pursuit: Recursive function approximation with applications to wavelet decomposition, in *Proceedings of the 27th Asilomar Conference in Signals, Systems and Computers*, 1993.

Rauhut, H., Stability results for random sampling of sparse trigonometric polynomials, *IEEE Trans. Inf. Theory* 54 (2008): 5661–5670.

Rudin, L. and S. Osher, Total variation based image restoration with free local constraints, in *Proceedings of IEEE ICIP*, Vol. 1, 1994, pp. 31–35.

Rudin, L.I., S. Osher, and E. Fatemi, Nonlinear total variation based noise removal algorithms, *Phys. D* 60 (1992): 259–268.

Tang, G., B. Bhaskar, P. Shah, and B. Recht, Compressed sensing off the grid, *IEEE Trans. Inf. Theory* 59 (2013): 7465–7490.

Tibshirani, R., Regression shrinkage and selection via the lasso, *J. R. Stat. Soc. Ser. B* 58 (1996): 267–288.

Tropp, J.A., Greed is good: Algorithmic results for sparse approximation, *IEEE Trans. Inf. Theory* 50 (2004): 2231–2242.

Tropp, J.A., A.C. Gilbert, and M.J. Strauss, Algorithms for simultaneous sparse approximation. Part I: Greedy pursuit, *Signal Process* (*Special Issue on Sparse Approximations in Signal and Image Processing*) 86 (2006): 572–588.

Special Aspects of the Application of Compressive Sensing in Optical Imaging and Sensing

Challenges, Some Solutions, and Open Questions

Adrian Stern

Contents

Introduction

The theory of compressive sensing (CS) has opened up new opportunities in the field of optical sensing and imaging. However, its implementation in this field is often not straightforward. In this chapter, we discuss the special issues encountered in the integration of CS in practical optical systems. We also discuss the implementation challenges that might arise in the design of optical CS systems and present some solutions to overcome them.

The principles of optical CS system design may differ drastically from the principles used for conventional optical sensing and imaging. For instance, traditional incoherent imaging seeks to perform isomorphic mapping ("Interpretation of the Compressive Sensing Model" section in Chapter 2), that is, to create images that are exact replica of the object (up to a scale factor). Ideally, each object point is mapped to a single sensor pixel so that, besides being a simple geometrical transformation (e.g., mirroring), the captured image is a precise copy of the object. In contrast, CS acquisition guidelines require some way of mixing the information so that multiple image points are projected onto a single sensor pixel. Classical optical engineering analytical tools and components are not optimal for such conditions.

When applying the CS framework within optical imaging and sensing, one needs to consider the special characteristics of the optical data collection systems. Our main purpose in this chapter is to highlight the special aspects of the application of CS in optical imaging and sensing and to point out the main challenges faced in optical CS system design. In the "Special Aspects of the Application of CS in Optical Sensing" section, we overview the special aspects and implementation difficulties arising in the application of CS in optical imaging and sensing. Next, in the "Some Feasible Implementation of the CS Sampling Matrix" section, we present examples of solutions that overcome some limitations in the realization of the CS matrix. Finally, in the "Ten Challenges and Open Questions" section, we summarize some specific realization issues and list 10 open questions that confront optical CS designers. The questions range from the theoretical to the computational. Answers to these questions will significantly enrich the optical designer's toolkit.

Before starting, we would like to add a word of caution: the forthcoming discussion may be somewhat biased by the author's own experience and perspective, and the ideas discussed should be taken as such.

Special Aspects of the Application of CS in Optical Sensing

Compressing Sensing versus Optical Physical Models

Let us consider the basic CS sensing model introduced in the "A Historical Perspective" section in Chapter 1:

$$\mathbf{g} = \Phi\mathbf{f}. \tag{4.1}$$

This is a discrete-to-discrete (D-D) linear model; that is, it relates discrete finite inputs to discrete finite measurements through a linear transform. Such a model can be applied only as an approximation to optical sensing systems because most optical systems and measurements are not linear and they do not relate discrete optical inputs to discrete outputs.

Linearity

Let us consider first the linearity of optical sensing systems. Most optical systems are nonlinear to some extent due to the nonlinear behavior of their optical, electro-optical, or acousto-optical components. Even more fundamental nonlinearity is induced by the nature of the optical inputs and outputs and the relation between them; the only optical observables are second-order statistical features of the electromagnetic fields (irradiance, mutual intensity, and more general mutual coherence), which are quadratically related to the input optical fields. Fortunately, linear relations can be practically applied for the two common extremes of coherence, that is, for (completely) spatial incoherent sensing and for coherent sensing. Incoherent imaging refers to the case where the light from any two distinct points of the input are uncorrelated. Practically, this applies when spatial coherence cross section at the input is much smaller than the input's features. In the spatial incoherent case, a linear relation can be expressed relating the input *intensity*, f_{ic}, to the output *intensity* g_{ic}:

$$g_{ic}(\mathbf{u}) = \int h_{ic}(\mathbf{u},\mathbf{x}) f_{ic}(\mathbf{x}) dx, \tag{4.2}$$

where h_{ic} is the *incoherent impulse response*, also known as the "point spread function."

The other coherence extreme refers to the case where the spatial coherence cross section is larger than the input, and in such a case, a linear relation exists between the input and output *fields*, $f_c(\mathbf{x})$ and $g_c(\mathbf{u})$, respectively, given by

$$g_c(\mathbf{u}) = \int h_c(\mathbf{u},\mathbf{x}) f_c(\mathbf{x}) dx, \tag{4.3}$$

where h_{ic} is the *coherent impulse response*. Although the output field of a coherent system, $g_c(\mathbf{u})$, cannot be measured directly, it can be measured indirectly by means of interferometry (as, e.g., is done in digital holography, discussed in Chapter 8).

In the case of partially coherent illumination, it is still possible to express linear models relating the input and output *second-order statistics* of the optical fields:

$$G(\mathbf{x}_1,\mathbf{x}_2) = \iint h_c^*(\mathbf{x}_1;\mathbf{x}_1') h_c(\mathbf{x}_2;\mathbf{x}_2') F(\mathbf{x}_1',\mathbf{x}_2') d\mathbf{x}_1' d\mathbf{x}_2', \tag{4.4}$$

where $F(\mathbf{x}_1', \mathbf{x}_2'), G(\mathbf{x}_1, \mathbf{x}_2)$ are the mutual intensities (Goodman 1996; Brady 2009) at the input and output planes positions $(\mathbf{x}_1', \mathbf{x}_2'), (\mathbf{x}_1, \mathbf{x}_2)$ of the quasi-monochromatic input and output fields, respectively. For the model in (4.4), a more general CS model than that in (4.1) is applicable, namely, a CS model of low-rank matrices (Recht et al. 2010, 471–501; Shechtman et al. 2011, 23920–23946). This kind of problem is considered in Chapter 13.

Discrete-Discrete Model

Most physical optical systems follow a continuous-to-continuous (C-C) model, such as in Equations 4.2 and 4.3. In practice, the output is also often discretized by pixelated electro-optical sensors, and thus, a general optical sensing system is described by a continuous-to-discrete (C-D) sensing process. Hence, we now make the important observation that real optical systems are represented by a C-D or C-C model rather by the D-D model underlying the CS theorem. In order to utilize the D-D approach, we need to *artificially* sample (discretize) the input. In general, this is done by *regular uniform* sampling, that is, by taking the discrete values of the input $f(x)$ at a constant sampling rate. This is the most common way of discretization, often done automatically. However, this is not necessarily the most efficient discretization method (Brady 2009). One of the artifacts of such a discretization method is that it gives rise to the so-called gridding error discussed in Chapter 3.

Values and Dimensions

The entries of the matrix and vectors in the D-D model (4.1) depend on the type of the optical signal measured. For instance, in incoherent imaging and spectroscopy, we measure optical intensities and, therefore, $\mathbf{f} \in \Re^N$, $\mathbf{g} \in \Re^M$ and $\mathbf{\Phi} \in \Re^{M \times N}$, with all entries being nonnegative and real, $g_i, f_j, \phi_{i,j} \geq 0$, $i = 1, \ldots, M$, $j = 1, \ldots, N$. On the other hand, in coherent imaging we measure the complex field amplitude and, therefore, $\mathbf{f} \in C^N$, $\mathbf{g} \in C^M$, and $\mathbf{\Phi} \in C^{M \times N}$.

In the following, we shall consider the particular features of the components of the D-D approach (1.1) in the context of optical imaging and sensing.

Input Signal

In optical sensing, the input signal **f** represents the features of the "object," such as the spatial, spatiotemporal, spectral, or polarimetric distributions of the electromagnetic field or of the radiant power. We shall list the special features of **f** and their consequences in the following sections.

Sparsity

In most optical sensing and imaging scenarios, the object is indeed sparse or highly compressible after decomposing it in another domain (e.g., discrete cosine transform, wavelet, or some other, tailored, dictionary), as required for CS. For instance, 2D images in the visible may be compressible by a factor of 10–50 as we experience in common digital compression applications (e.g., JPEG, JPEG2000) and 3D images and hyperspectral images may be compressible by even higher factors.

Physical Representation Dimensions

The object is typically represented as a multidimensional image or a multidimensional array. Planar objects are 2D, while volumetric objects and video sequences are typically arranged in the form of 3D distributions. Hyperspectral data are arranged in 3D datacubes (two spatial dimensions and one spectral). More generally, optical information can be represented by "plenoptic functions" or "ray phase space" (Stern and Javidi 2005, 141–150), which may have as many as seven dimensions, accounting for the spatial location of the light ray (3D), its direction (2D), its temporal dependence (1D), and its spectral composition (1D). Therefore, in order to adjust to the matrix–vector formalism of (4.1), the signal needs to be converted into the form of a vector by lexicographic ordering. Once reordered in vector form, the matrix–vector model (4.1) can be applied together with the rich theoretical and computational kit available for it, such as reconstruction guarantees and bounds, noise robustness analysis, and prescription of appropriate algorithms. Unfortunately, with lexicographical reordering inherent a priori information available within the structure of the signal is lost. For instance, local 2D correlations in planar images reduce to only 1D correlation in a lexicographic vector. In general, multidimensional signal models are richer than 1D ones. Without bookkeeping the structural information and accounting for it in the reconstruction step, the overall performance is expected to be poorer. Also, for efficient implementation of CS one should attempt to define sparsifying operators, $\boldsymbol{\Psi}$, that account for the internal structure and local correlations that exist in their original representation. In other words, the concepts of sparsity and sparsifying need to be generalized to account for the input signal structure. Some approaches in this direction are considering structured sparse supports and, more generally, union of signal subspaces (Duarte and Eldar 2011, 4053–4085).

Size

The signal $\mathbf{f}$ and measurements $\mathbf{g}$ are typically large. For example, in incoherent imaging in the visible spectrum, N can easily be of the order of 10^7 and in multidimensional imaging (such as in 3D images, hyperspectral images, and videos) it can be orders of magnitude larger. Current imaging performance paradigms relate to gigapixel (10^{12}) size frames, ultraspectral imagers (10^3 spectral bands), and frame rates of a hundred billion frames per seconds (Gao et al. 2014, 74–77), which implies signal representations in the order of 10^{11}–10^{14} pixels per second. Obviously, this yields computational implications in terms of storage requirements and reconstruction speed.

Nonnegativity

In many optical applications (e.g., incoherent imaging, spectrometry), we are interested in measuring the optical intensity of the input. In all such applications, the signal $\mathbf{f}$ is nonnegative. The common CS framework assumes bipolar input signals. For efficient CS, this fact should be considered in the reconstruction process by introducing appropriate constraints in the reconstruction problem or by working with centralized signals (with the average subtracted).

System Matrix

Size of the Matrix

In the D-D setup (4.1), the size of the system matrix is $M \times N$, where N and M may be in the order of 10^5–10^9 in conventional imaging tasks. Therefore, the size of the system matrix is huge, having giga (10^{12}) to peta (10^{15}) entries. Moreover, if one wishes to better approximate the continuous sensing processes by using off-grid techniques (see Chapter 3), one needs to pay by increasing by a factor of ~10^4 the system matrix dimensions for each signal coordinate. The matrix dimensionality leads to the following significant challenges:

- *Computational*—$\mathbf{\Phi}$ may require hundreds of gigabytes of storage; the execution of reconstruction algorithms with such large matrices is very difficult and time-consuming.
- *Optical realization*—Attempting to realize the common CS measuring model, which often involves random measuring matrices, may be extremely challenging. First, the distribution of the realizable matrix entries is obviously limited to one with a finite range; therefore, it cannot be a common CS matrix with normal identically independent distribution (i.i.d.) entries. Moreover, realization of random $\mathbf{\Phi}$ requires building an imaging system with a space bandwidth product (SBP) larger than $N \times M$. Therefore, the optical system needs to have $N \times M$ almost independent modes or degrees of freedom. It is not trivial to design a system with such a large SBP. For example, spatial light modulators that are commonly used in compressive imaging have an SBP of $O(N)$. Therefore, in order to realize a SBP that is about $\times M$ times larger ~M sequential measurements are required (see Chapter 5). This has obvious implications in terms of performance (acquisition time, limited frame rate) and complexity of the system.
- *Optical calibration*—Sensing systems with a large SBP often require exhaustive and time-consuming calibration processes. The calibration process refers to the task of specifying $\mathbf{\Phi}$ by measuring its response to a high resolution. Typically, in order to calibrate $\mathbf{\Phi}$, one needs to measure N impulse responses (each determining one column of $\mathbf{\Phi}$), each having M samples.

Nonnegativity

In many optical applications, such as incoherent imaging, it is impossible to realize a system matrix $\mathbf{\Phi}$ with negative entries. This means that $\mathbf{\Phi}$ spans only the positive orthant. As a result, the coherence parameter of $\mathbf{\Phi}$, defined in the "CS Solutions" section in Chapter 1, is lower, indicating lower compressibility. This problem may be addressed by applying preconditioning in the reconstruction process (Bruckstein et al. 2008, 4813–4820) or by doubling the number of measurements to generate measurements equivalent to those of a bipolar system matrix (Goodman 1996).

Realization of Sensors with Stable Response

The performance of CS may be highly sensitive to the accuracy of the knowledge of the measurement matrix. If the system sensing matrix needs to be characterized by a calibration process, then the physical realization should be as static as possible; that is, it should be sufficiently insensitive to mechanical vibration, thermal changes, etc.

Measured Signal

Size

Although the dimension of the measured image $\mathbf{g}$ is smaller than that of the signal $\mathbf{f}$ ($M < N$), in typical optical CS systems it is still large. Therefore, similar computation issues to those already discussed in the context of the input signal **f** are also relevant for the measurement signal **g**.

Realness and Nonnegativity

Electro-optical sensors usually measure irradiance (power per unit area), which is real and nonnegative. Negative and complex values can be measured only indirectly, typically by acquiring multiple measurements. For example, in compressive holography (Chapter 8) the complex field amplitude is measured using temporal or spatial multiplexing.

Dynamic Range and Quantization

The dynamic range of optical sensors is always limited. For example, conventional, uncooled optoelectronic sensors in the visible spectrum have a dynamic range of 8–12 bits. At longer wavelengths, the dynamic range may be even smaller. This may impose significant limitations, particularly in incoherent imaging, where $\mathbf{\Phi}$ is nonnegative. In such a case, we are actually measuring small fluctuations (determined by $\mathbf{\Phi f}$) around some constant offset (determined by $\|\mathbf{f}\|_1$). Therefore, conventional optoelectronic sensors that are designed to be linear and to perform linear quantization are not well suited for an output signal **g** that has a relatively narrow distribution around some offset value. If the range of the optoelectronic sensors is set to be too small, the elements of **g** may fall outside it. On the other hand, if the sensor range is set to be large, then precision is lost because only a few quantization levels will cover the signal **g**. In Stern et al. (2013b, 1069–1077), it is shown that straightforward application of the CS sensing framework (4.1) in an imaging system requires sensors with $O(\log_2 2\sqrt{N})$ more bits than in a conventional imaging system that generates similar image quality. In practice, this may be a prohibitive requirement in many applications. Although there exist ways to reduce significantly this quantization depth overhead (Stern et al. 2013b, 1069–1077), they require nontrivial optical implementations.

Some Feasible Implementations of the CS Sampling Matrix

The first step in optical CS system design is the optical implementation of the CS sensing operator. Technically, the optical engineer has to design an optical scheme that realizes the CS sensing matrix $\mathbf{\Phi}$. Although the desired sensing matrices are well specified in CS theory, in practice the optical engineer may not have the privilege to choose $\mathbf{\Phi}$ as he/she pleases. In practice, the sensing matrix is often dictated by the physical properties of the sensing process and of the optical devices.

CS matrices can be categorized into two main groups: matrices generated from random basis ensembles (RBE) and random modulation matrices (e.g., see Chapter 1 and Eldar and Kutyniok 2012). Randomization of the measurement scheme is a preferred approach from the analytical point of view because it provides theoretical guarantees that are otherwise hard to obtain. Random basis ensembles are obtained by taking an $N{\times}N$ orthogonal matrix and sampling M of its rows at random. Some examples of RBE that are relevant for optics are the partial Fourier ensemble, the partial Fresnel ensemble, and the partial Hadamard ensemble. Fortunately, the optical engineer's toolkit contains optical components and schemes that can be modeled or at least approximated by an RBE. Two such examples are free space propagation, used in holography and discussed in

Chapter 8, and spectral modulators, used with thin tunable layer devices, which are discussed in the "Compressive Sensing System with Spectral Encoding in the Spectral Domain" section in Chapter 9).

The second category of CS sensing mechanisms uses random modulation matrices. In this group, the matrix is generated by drawing its entries from an i.i.d. distribution, such as i.i.d. Gaussian, Rademacher (±1), Bernoulli (0/1), or, more generally, any other sub-Gaussian distribution.* This kind of sensing models is less natural for optical implementation. The implementation challenges mentioned in the "The System Matrix" section are predominant for this type of CS sensors.

The challenges associated with common CS sensing operators described in the "How Appropriate Is a Sensing Matrix for CS" section in Chapter 2 can be partially alleviated by intelligently making compromises with regard to the guidelines for optimal universal CS. For instance, instead of using random projections one may use some kind of *structured* pseudorandom projection scheme (Duarte and Eldar 2011, 4053–4085). In this section, we bring two such optical realization examples in the "Optical Radon Projections for Compressive Imaging" and "Optical Radon Projections for Motion Tracking" sections. The CS implementation challenges may also be bypassed if a specific task system is to be designed. For example, if the task is to track motion in the scene, the technique described in the "Optical Radon Projections for Motion Tracking" section can be efficiently applied.

Separable Sensing Matrix

One way to alleviate the complexity associated with implementing CS systems with random projections is by designing sensing operators $\mathbf{\Phi}$ that are *separable* in the natural coordinates of the optical signal (Rivenson and Stern 2009a, 449–452). Optical systems and devices often operate separately in the spatial dimensions of the optical field, the spectral components, the state of polarization, and time. For example, to capture a typical 2D image one may use a sampling operator that is separable in the x–y directions, as illustrated in Figure 4.1. Mathematically, such a sensing operator can be expressed by means of the Kronecker product of the sensing operators in each Cartesian direction, $\mathbf{\Phi}=\mathbf{\Phi}_x \otimes \mathbf{\Phi}_y$ (Figure 4.1). For CS, the sensing operators in each direction, $\mathbf{\Phi}_x$, $\mathbf{\Phi}_y$, can be designed to perform some random projections. Obviously, the entries of these matrices cannot be realized to have a Gaussian distribution due to the finite dynamic range of realizable systems, but they can be chosen to have an approximate sub-Gaussian distribution. Common choices are the Bernoulli distribution and Hadamard codes.

The SBP of an x–y separable $\mathbf{\Phi}$ is $O(\sqrt{N \cdot M})$; thus, the matrix storage requirements and the optical sensing complexity are reduced from $O(M \cdot N)$ to $O(\sqrt{N \cdot M})$. The calibration process (see the "The System Matrix" section) of such a system would be reduced by the same ratio. Employing separable $\mathbf{\Phi}$ can also be useful in the reconstruction step, as it permits using much faster block-iterative algorithms.

However, there is no benefit without an accompanying cost! The aforementioned advantages of using separable sensing operators come with the cost of lower compressibility performance. For instance, a theoretical analysis in Rivenson and Stern (2009a, 449–452) showed that for 2D images, approximately $\log(\sqrt{N})$ times more samples are needed to achieve similar performance as achieved with

* A random variable X is called sub-Gaussian if there exits $c > 0$ such that $E[e^{Xt}] \le e^{(ct)^2/2}$ for all real t. Examples include the Gaussian, Bernoulli and Rademacher random variables, as well any bounded random variable.

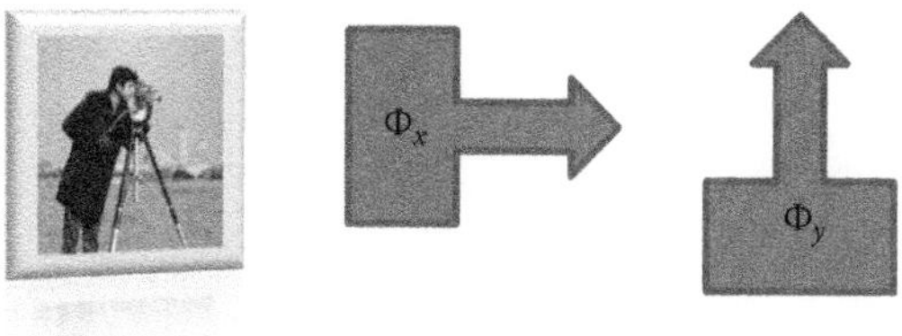

$$\boldsymbol{\Phi} = \boldsymbol{\Phi}_x \otimes \boldsymbol{\Phi}_y = \begin{bmatrix} (\phi_x)_{11}\boldsymbol{\Phi}_y & \cdot & \cdot & \cdot & (\phi_x)_{1N}\boldsymbol{\Phi}_y \\ \cdot & & \cdot & & \cdot \\ \cdot & & & \cdot & \cdot \\ \cdot & & & \cdot & \cdot \\ (\phi_x)_{M1}\boldsymbol{\Phi}_y & \cdot & \cdot & \cdot & (\phi_x)_{MN}\boldsymbol{\Phi}_y \end{bmatrix}$$

Figure 4.1 Separable image processing described by the Kronecker product $\boldsymbol{\Phi}=\boldsymbol{\Phi}_x \otimes \boldsymbol{\Phi}_y$, where $(\phi_x)_{i,j}$ denote the *i,j*, element of matrix $\boldsymbol{\Phi}_x$.

a nonseparable random system matrix. However, an empirical study in Rivenson and Stern (2009c, 1–8) showed significantly more relaxed requirements. An analysis of oversampling requirements for signals that are separable in more than two dimensions may be found in Rivenson and Stern (2009b).

Compressive imaging with a separable sensing operator has been demonstrated for 2D images in Rivenson and Stern (2009a, 449–452) and for hyperspectral imaging in August et al. (2013, D46).

Optical Radon Projections for Compressive Imaging

In the seminal CS paper (Candes and Romberg 2006, 227–254), the authors first demonstrate the concept through an example of image recovery from radon projections. An optical implementation of a compressive imager that uses radon projections was demonstrated in Stern (2007, 3077–3079). The optical setup relays on an anamorphic optical element (e.g., a cylindrical lens) that implements optically radon projections of the object plane on a line array of sensors (Figure 4.2a). The system performs a rotational scan to capture multiple radon projections at various angles during the scanning process. By applying reconstruction algorithms based on l_1 minimization, the image can be reconstructed from many fewer projections than are needed conventionally, for example, with filtered back-projection algorithms.

The compressive imaging approach in Stern (2007, 3077–3079) based on the optical Radon projections exhibits a very good trade-off between acquisition time and system complexity. Referring to the main compressive imaging architectures (see Chapter 5), it allows a much faster scan than with sequential scanning systems (e.g., with the "single-pixel camera") (Figure 3.9) while, on the other hand, its implementation complexity is much lower than with a typical parallel setup, such as that in the "single-shot compressive imaging camera" in Stern and Javidi (2007, 315–320). The imaging approach presented in Stern (2007, 3077–3079) was further improved in Evladov et al. (2012, 4260–4271), where it was shown that angular sampling with golden angle steps allows progressive compressive image acquisition. Gradual improvement of the reconstructed image is obtained by adding new projections to the existing ones without resampling and recalculation. Each new

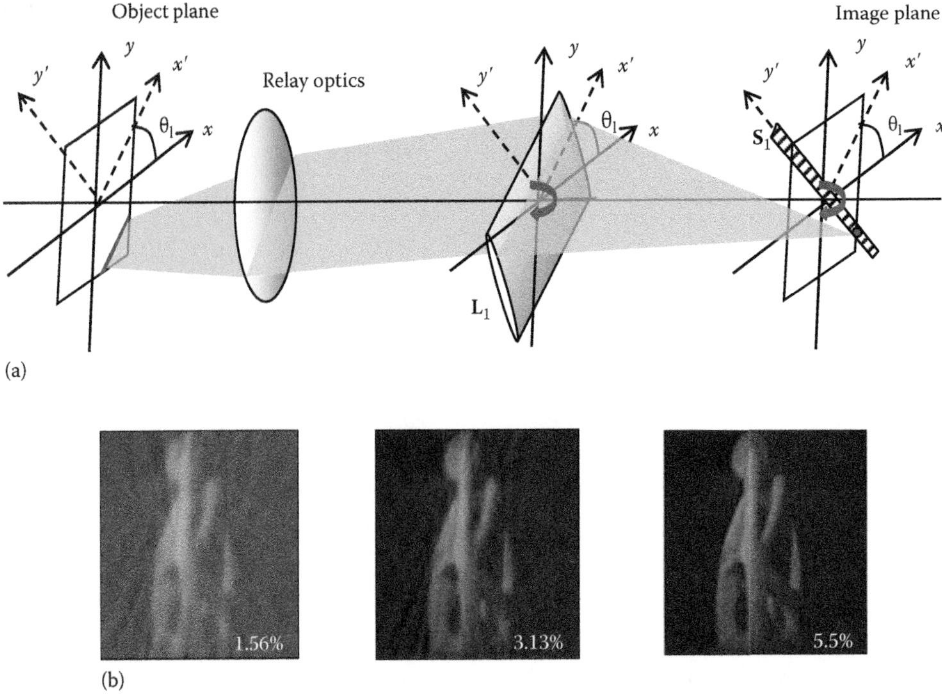

Figure 4.2 (a) Schematics of the optical Radon imager. By means of the cylindrical lens $\mathbf{L}_1$, Radon projections of the image plane are collected on a line array vector $\mathbf{S}_1$. (b) Reconstructions obtained with 1.56% (left), 3.13% (middle), and 5.5% (right) of nominal samples (Nyquist). The image size is 1280×1280 pixels.

measurement increases the quality of the previous reconstruction, as demonstrated in Figure 4.2b. This progressive compressive imaging approach is particularly useful when no prior knowledge of the required number of samples for good reconstruction is available. This means that the progressive radon acquisition scheme is inherently adjustable to the type of the object imaged. The approach is also shown to be immune to sudden stopping of the scanning process, which otherwise would be intolerable with the uniform angular sampling scheme. An additional advantage of the approach is that it facilitates compressive imaging of large size images by employing "ordered sets" of reconstruction algorithms on subsets of the data, thus remedying otherwise severe computation issues (Evladov et al. 2012, 4260–4271). Note, for example, that the images in Figure 4.2 are of megapixel size.

Optical Radon Projections for Motion Tracking

In the case where the task of the acquisition system is the detection of change or motion tracking, the signal is extremely sparse. Consider, for example, the task of tracking a point object during 10 seconds with a temporal resolution of 20 milliseconds in a field of view of 1 megapixel. With conventional imagers, 500 megapixels are acquired for this task, while the trajectory of the moving point can be described by only 500 pairs of Cartesian coordinates; thus, $k/N = 2 \cdot 10^{-6}$. Cartesian coordinates of moving objects can be obtained by measuring the temporal differences of two perpendicular radon projections. As mentioned in the previous section, Radon projections can be obtained optically with anamorphic

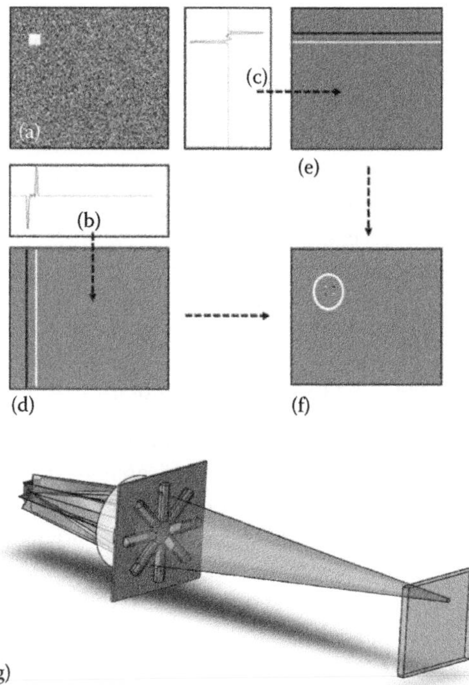

Figure 4.3 Motion detection from projections. (a) Original frame out of two consecutive frames; (b and c) difference between projections of two consecutive frames; (d and e) back projection of the frame difference; (f) intersection of the *x*,*y* back projections. The detected object is marked with a white circle. (g) Optical system for change detection and motion tracking.

optical elements such as a cylindrical lens (Figure 4.2a). Figure 4.3 depicts the concept behind change detection from two radon projections. Consecutive temporal projections are subtracted from one another, indicating the projected location of the changes (Figure 4.3b and c). Then, the projections may be back projected to give the location of the changes on a Cartesian grid. Since the signal is extremely sparse, ℓ_1 minimization algorithms are particularly efficient.

In practice, two projections are insufficient for detecting multiple moving objects in arbitrary directions. At least three projections are necessary to track objects moving in an arbitrary direction. The system in Figure 4.3g, proposed in Kashter et al. (2012, 2491–2496), uses a superposition of four projections. Simulative experiments in Kashter et al. (2012, 2491–2496) show that this system is able to track up to 10 moving object points. Experiments showed that objects can be tracked within a field of view of 500×500 pixels with approximately 250 times fewer samples than required by a conventional camera for the same task.

Ten Challenges and Open Questions

In the "Special Aspects of the Applications of CS in Optical Sensing" section, we overviewed specific aspects of applications of conventional CS theory for optical sensing. We also mentioned that some of these specific aspects raise significant technical and theoretical challenges for the practical compressive optical system designer. In this section, we summarize some theoretical and computational open questions that typically confront an optical CS designer. Some of the questions in the list are quite general and well known (Elad 2012, 922–928; Strohmer 2012, 887–893; Stern et al. 2013a, 69–72) but are, nevertheless, listed because of their particular importance in the optical sensing field. In addition, many of the questions listed have already been addressed and progress in solving them has been made in recent years. Nevertheless, there remains a need for further progress to facilitate efficient optical CS engineering.

The optical CS is challenged by the following open questions and issues:

1. *How to handle the dimensionality problem?* How to design universal CS operators that are *computationally* and *optically* feasible for *large-scale* problems (see the "The Input Signal" section)?

2. The "classical" CS framework considers signals in the form of 1D vectors; therefore, the inherent nature of optical signals is not directly reflected, as discussed in the "Physical Representation Dimensions" section. This implies that the concepts of sparsity and sparsifying need to be generalized to account for the input signal structure. Appropriate sparsifying and sensing operators and respective algorithms should be designed. Consider, for example, the task of imaging a 3D object. Most likely, applying a sparsifyer operator on lexicographically reordered 3D points is not the most efficient way to simplify the signal representation. In addition, the sensing process should not perform uniformly in the lateral and longitudinal directions. Finally, it would be better not to apply the algorithms on the huge dimension of the vectorized 3D data. The questions raised are: *How to exploit the inherent structure of optical signals beyond sparsity? What would be the appropriate sparsifyers and sensing operators? What are the information theoretic guarantees of CS for this kind of signal? Lastly, how will the algorithms better exploit the additional prior data besides the sparsity?* Although some answers to these questions have been provided in the past few years, we believe that there is room for further investigation.

3. There is a need for a practical and efficient figure of merit for sensing operators used for optical imaging and sensing and for reconstruction guarantees. The CS performance of a CS matrix is probably best evaluated through the restricted isometry property (RIP) (Chapters 1 through 3) or its variants (Eldar and Kutyniok 2012). Unfortunately, evaluation of RIP bounds is not practical for most realizable optical sensing operators. A much more practical way is to calculate the "coherence parameter," μ (see Chapters 1 through 3). Unfortunately, theoretical results based on the coherence parameter present only the worst-case estimates. It is well known that reconstruction limits based on the mutual coherence are overly pessimistic (Elad 2010). This leads to wide gaps between theoretical predictions and actual performance. For example, the coherence parameter for the applications described in the "Optical Radon Projections for Compressive Imaging" section is approximately 0.7 (Farber et al. 2013, 87500L-1–87500L-9), which definitely cannot guarantee compression of the order of ×10, which we achieved in practice. The quest for system characterization beyond the RIP (and its variants) and the coherence parameter is a general open question in CS theory (Elad 2010). Alongside

the general open question, the optical engineer would greatly benefit from figures of merit for optically realizable sensing matrices, such as those having nonnegative entries or posing some structure.

4. Mainstream CS is formulated for a D-D forward system model, as explained in the "The Discrete–Discrete Model" section. However, most optical sensing operators are described by a C-C model or a C-D operator, that is, by a model relating a continuous physical input to a continuous or discreet output. Hence, for CS implementation the continuous optical system model is converted to a D-D operator Φ, typically by regular uniform sampling. For example, the linear D-D spectral sensing matrix of the system in the "Compressive Liquid Crystal Spectrometers for Imaging" section in Chapter 9 is obtained by sampling at equidistant wavelengths λ. This leads to the following questions: *What would be the best way, in terms of CS, to convert the physical C-C model to a D-D model*? In other words, *what is the best way to discretize a continuous sensing operator*? Or, an even more general question: *Is there a better way to apply CS with a C-C or a C-D sensing model*? A positive answer is provided through the Xampling theorem (Mishali et al. 2011, 4719–4734), which demonstrates the application of CS for C-D models. In general, it would be of great benefit to have a rigorous CS framework formalized with a C-C sensing model that is applicable to a large class of signals and yet preserves the simplicity and power of the standard D-D CS approach.

5. CS was demonstrated to be useful for surpassing traditional optical resolution limits. For example, CS was used to increase the SBP of optical systems (Stern and Javidi 2007, 315–320), to achieve spatial (Fannjiang 2009, 1277–1291; Gazit et al. 2009, 23920–23946; Shechtman et al. 2010, 1148–1150, 2011, 23920–23946) and temporal (Llull et al. 2013, 10526–10545; Gao et al. 2014, 74–77) superresolution imaging. Superresolution is a long-standing problem in the optical sensing field. CS has introduced a new and effective approach to the superresolution problem. Roughly speaking, its power can be attributed to the fact that it behaves as an interpolation method rather than an extrapolation one (Gehm and Brady 2015, C14–C22). Whereas conventional superresolution attempts to extrapolate signal values in unsampled regions (e.g., Fourier space beyond the band limit), most CS systems attempt to interpolate signal values residing between measurements in sparsely sampled regions. However, all these methods involve a conversion of the optical problem to a D-D. This introduces gridding errors (Chapter 3) and model errors. It would be helpful to have an analytical and algorithmic theory that avoids the artifacts that appear due to transferring to the discrete domain.

6. There is a need for computationally efficient algorithms for the large N involved in optical signals. CS reconstruction algorithms are essentially nonlinear, demanding a heavy computation overhead and a large storage memory, especially in the case of optical signals of high dimension. We note that most CS recovery algorithms are defined for the vector sensing problem in (4.1). Specially tailored algorithms should address the fact that the optical data represented is inherently structural (typically 2D or 3D). Taking into account the signal structure, faster and more memory-efficient algorithms can be used. A huge amount of algorithms have

been developed, some more applicable for our purposes and other less. One may ask: given the properties of the optical signals described in the "Special Aspects of the Applications of CS in Optical Sensing" section, *what are the guidelines for choosing the appropriate algorithm from the various families of algorithms used in CS?* While there is plethora of algorithms to choose from, and while some overview papers have been recently published, there is a need for a performance-oriented taxonomy of algorithms. The optical engineer is less interested in the underlying working principle of the algorithm (e.g., greedy, convex relaxation, thresholding); rather, he needs simple guidelines that account for the type of sensing operator (e.g., random modulator, random basis ensemble, structural form), for the sensing conditions (e.g., N and k), and that clearly specify the trade-offs between performance in terms of signal reconstruction fidelity and running time.

7. Current CS is *brittle* (Gehm and Brady 2015, C14–C22); CS is built upon the idea that natural signals are highly redundant and seeks to acquire measurements that contain as little redundancy as possible. However, from an engineering point of view, often redundancy is a desired property in order to ensure reliability. The reliability of CS measurements in terms of robustness to additive noise is well known. However, the reliability loss due variations in the design process has been less investigated. This is of great importance due the fact that optical system specifications are often sensitive to environmental changes (e.g., temperature, mechanical vibration, humidity). It would be beneficial to have guidelines for CS design schemes that include priors on likely system variations. In some sense, this is analogous to the way that coding designs in communication applications are designed to ensure recoverability from the most common error types.

8. *Calibration burden.* As mentioned in the "The System Matrix" section, optical system calibration may be an exhausting process. An interesting area for future exploration is how to design a minimal set of test signals (calibration inputs) subject to constraints on the set of available test signals, such that we acquire the best possible understanding of the system measurement matrix in the given circumstances.

9. *Nonlinear CS.* The vast majority of CS theory focuses on the linear sensing model in (4.1). Although many optical systems can be approximated to be linear, as discussed in the "Linearity" section, there are still many others that are *nonlinear.* One such important example in optics is that of phase retrieval from intensity measurement, which is basically a quadratic problem. Chapter 13 addresses this problem. Other nonlinear optical systems include severely quantized systems (e.g., 1-bit systems), photon counting systems, and systems that rely on nonlinear optical elements (e.g., nonlinear crystals), to name a few.

10. *Superresolution via nonlinear CS.* Classical resolution limits are defined for linear systems. It is well known that such limits can be bypassed with properly designed nonlinear systems. We believe that by properly designing nonlinear systems for sparse signals can outperform even the optical CS superresolution regimes already developed, such as those mentioned earlier.

Conclusions

The harnessing of the CS mathematical model for optical applications is often not trivial. We have overviewed the characteristics of optical imaging that preclude the straightforward application of CS theory to imaging and optical sensing. In many cases, practical and physical limitations force the optical CS designer to deviate from basic CS guidelines. For instance, he/she has to compromise the randomness of the sensing operator required for universal CS by introducing some amount of structure. We have presented two examples to demonstrate this point. In some cases, the implementation limitations are much less severe if a specific task is defined, as we have shown with our compressive motion detection and tracking system, while in others, the particular optical sensing mechanism fits the CS guidelines well.

In view of the special aspects of CS applications in optical imaging, we have summarized some of the main theoretical and applicative questions that make widespread implementation of optical CS difficult. We believe that providing solutions to these questions will open the door to embrace new optical CS applications.

References

August, Y., C. Vachman, Y. Rivenson, and A. Stern. 2013. Compressive hyperspectral imaging by random separable projections in both the spatial and the spectral domains. *Applied Optics* 52: D46.

Brady, D. J. 2009. *Optical Imaging and Spectroscopy*. Wiley-Interscience.

Bruckstein, A. M., M. Elad, and M. Zibulevsky. 2008. On the uniqueness of nonnegative sparse solutions to underdetermined systems of equations. *IEEE Transactions on Information Theory* 54(11): 4813–4820.

Candes, E. J. and J. Romberg. 2006. Quantitative robust uncertainty principles and optimally sparse decompositions. *Foundations of Computational Mathematics* 6(2): 227–254.

Duarte, M. F. and Y. C. Eldar. 2011. Structured compressed sensing: From theory to applications. *IEEE Transactions on Signal Processing* 59(9): 4053–4085.

Elad, M. 2010. *Sparse and Redundant Representations: From Theory to Applications in Signal and Image Processing*. Springer.

Elad, M. 2012. Sparse and redundant representation modeling—What next? *IEEE Signal Processing Letters* 19(12): 922–928.

Eldar, Y. C. and G. Kutyniok. 2012. *Compressed Sensing: Theory and Applications*. Cambridge University Press.

Evladov, S., O. Levi, and A. Stern. 2012. Progressive compressive imaging from radon projections. *Optics Express* 20(4): 4260–4271.

Fannjiang, A. C. 2009. Compressive imaging of subwavelength structures. *SIAM Journal on Imaging Sciences* 2: 1277–1291.

Farber, V., E. Eduard, Y. Rivenson, and A. Stern. 2013. A study of the coherence parameter of the progressive compressive imager based on radon transform. *Proc. SPIE*. 9117–10, p. 87500L.

Gao, L., J. Liang, C. Li, and L. V. Wang. 2014. Single-shot compressed ultrafast photography at one hundred billion frames per second. *Nature* 516(7529): 74–77.

Gazit, S., A. Szameit, Y. C. Eldar, and M. Segev. 2009. Super-resolution and reconstruction of sparse sub-wavelength images. *Optics Express* 17(25): 23920–23946.

Gehm, M. E. and D. J. Brady. 2015. Compressive sensing in the EO/IR. *Applied Optics* 54(8): C14–C22.

Goodman, J. W. 1996. *Introduction to Fourier Optics*. McGraw-Hill.

Kashter, Y., O. Levi, and A. Stern. 2012. Optical compressive change and motion detection. *Applied Optics* 51(13): 2491–2496.

Llull, P., X. Liao, X. Yuan, J. Yang, D. Kittle, L. Carin, G. Sapiro, and D. J. Brady. 2013. Coded aperture compressive temporal imaging. *Optics Express* 21(9): 10526–10545.

Mishali, M., Y. C. Eldar, and A. J. Elron. 2011. Xampling: Signal acquisition and processing in union of subspaces. *IEEE Transactions on Signal Processing* 59(10): 4719–4734.

Recht, B., M. Fazel, and P. A. Parrilo. 2010. Guaranteed minimum-rank solutions of linear matrix equations via nuclear norm minimization. *SIAM Review* 52(3): 471–501.

Rivenson, Y. and A. Stern. 2009a. Compressed imaging with a separable sensing operator. *IEEE Signal Processing Letters* 16(6): 449–452.

Rivenson, Y. and A. Stern. 2009b. An efficient method for multi-dimensional compressive imaging. In *Computational Optical Sensing and Imaging, OSA Technical Digest (CD)* (Optical Society of America, 2009), paper CTuA4.

Rivenson, Y. and A. Stern. 2009c. Practical compressive sensing of large images. In *2009 16th International Conference on Digital Signal Processing*, pp. 1–8. IEEE.

Shechtman, Y., Y. C. Eldar, A. Szameit, and M. Segev. 2011. Sparsity based sub-wavelength imaging with partially incoherent light via quadratic compressed sensing. *Optics Express* 17: 23920–23946.

Shechtman, Y., S. Gazit, A. Szameit, Y. C. Eldar, and M. Segev. 2010. Super-resolution and reconstruction of sparse images carried by incoherent light. *Optics Letters* 35(8): 1148–1150.

Stern, A. 2007. Compressed imaging system with linear sensors. *Optics Letters* 32(21): 3077–3079.

Stern, A., Y. Auguts, and Y. Rivenson. 2013a. Challenges in optical compressive imaging and some solutions. In *10th international conference on Sampling Theory and Applications SampTa*, Vol. 24, pp. 69–72.

Stern, A. and B. Javidi. 2005. Ray phase space approach for 3-D imaging and 3-D optical data representation. *Journal of Display Technology* 1(1): 141–150.

Stern, A. and B. Javidi. 2007. Random projections imaging with extended space-bandwidth product. *Journal of Display Technology* 3(3): 315–320.

Stern, A., Y. Zeltzer, and Y. Rivenson. 2013b. Quantization error and dynamic range considerations for compressive imaging systems design. *Journal of the Optical Society of America A, Optics, Image Science, and Vision* 30(6): 1069–1077.

Strohmer, T. 2012. Measure what should be measured: Progress and challenges in compressive sensing. *IEEE Signal Processing Letters* 19(12): 887–893.

II

Compressive Imaging Systems

5 Optical Architectures for Compressive Imaging

Mark A. Neifeld and Jun Ke

Contents

Introduction

In this chapter, we compare three optical architectures for compressive imaging: sequential, parallel, and photon sharing. Each of these architectures is analyzed using two different types of projection: (1) principal component (PC) projections and (2) pseudorandom (PR) projections. Both linear and nonlinear reconstruction methods are presented. The performance of each architecture–projection combination is quantified in terms of reconstructed image quality as a function of measurement noise strength. Using a linear reconstruction operator, we show that in all cases of (1), there is a measurement noise level above which compressive imaging is superior to conventional imaging. Normalized by the average object pixel brightness, these threshold noise standard deviations are 6.4, 4.9, and 2.1 for the sequential, parallel, and photon-sharing (PS) architectures, respectively. We also show that conventional imaging outperforms compressive imaging using PR projections when linear reconstruction is employed. In all cases, the PS architecture is found to be more photon-efficient than the other two optical implementations and thus offers the highest performance among all compressive methods that we discuss. For example, with PC projections and a linear reconstruction operator, the PS architecture provides at least 17.6% less reconstruction error than either of the other two architectures for a noise strength of 1.6 times the average object pixel brightness. We also demonstrate that nonlinear reconstruction methods can offer additional performance improvements to all architectures for small values of noise.

This chapter is based on work originally reported in 2007 [1]. Since that time, a great deal of attention continues to be directed toward the innovative mathematical framework and resulting novel measurement paradigm of compressive sensing [1–3]. As described elsewhere in this book, the term "compressive" here refers to any measurement process in which the number of measured quantities is significantly smaller than the native dimensionality of the signals of interest. The sparse nature of many/most signals of interest enables high-fidelity reconstructions to be obtained from these compressive measurements. The primary focus of this book, compressive *imaging* therefore refers to any imager in which the number of physical measurements is much smaller than the number of desired/reconstructed image pixels [4]. The imaging hardware community has sometimes referred to these compressive techniques as feature-specific imaging [4–6]. Because typical imagery is known to be sparse (i.e., easily compressible), compressive imaging has become a convenient platform for the application of compressive sensing paradigm. Compared with conventional imaging, compressive imaging offers (1) reduced cost owing to a reduction in the number of photodetectors and the concomitant reduction in camera size, weight, power, etc. and (2) improved detector-noise-limited measurement fidelity because the same total number of photons can be measured using fewer photodetectors.

The measurements that are made by a compressive imager are simply linear projections of the object space. Therefore, common to all compressive imaging techniques is some type of optical hardware that can project the high-dimensional object space onto a low-dimensional measurement space. Although a large number of candidate projections have been studied in this regard (e.g., wavelets, PCs, Hadamard, discrete cosine, PR), there are relatively few candidate optical architectures for compressive imaging. In this chapter, we will quantify the performance of three different optical implementations of compressive imaging. It should be stressed that the purpose of this chapter is not to compare the many and varied algorithmic approaches to compressive sensing but rather to provide a quantitative comparison among the performances of several optical architectures. This chapter is organized as follows. The "Algorithm Descriptions" section provides a description of the two classes of projection (PC and PR) that will be employed in our study as well as the associated optimal linear reconstruction operators. The "Architecture Descriptions" section describes the operation of the three compressive optical architectures: sequential, parallel, and photon sharing. The "Results Using Linear Reconstructions" section presents a quantitative comparison among these architecture–projection combinations using linear reconstruction. The "Results Using Nonlinear Reconstruction" section extends these results to the case of nonlinear reconstruction algorithms. The "Shot-Noise-Limited Performance" section presents some new results obtained in the shot-noise limit and the "Conclusions" section presents the conclusions of our study.

Algorithm Descriptions

Compressive imaging is a special case of computational imaging [7,8]. Computational imagers do not necessarily generate a visually pleasing representation as the output of their measurement process. Instead, they produce a set of numbers (e.g., linear projections) that can be used along with a postprocessing algorithm in order to achieve some overall system objective. The system objective may or may not require the traditional "pretty picture" representation of the object information. In this way, computational imagers may be considered to be more general/flexible than conventional imagers insofar as the measurement process may be made task specific. Within the compressive imaging framework, therefore, the imaging system *task* must define the type and number of projections to be measured as well as the necessary postprocessing. Elsewhere we have discussed the use of compressive imaging for human face recognition [9]. In this chapter, we will be concerned with tasks that depend upon reconstructed image quality (i.e., a pretty picture) as measured by the root mean squared error (RMSE) criterion. Next, we describe two different projection bases and the associated linear postprocessing algorithms that can be used to obtain high-quality image reconstructions. These two algorithmic approaches differ in terms of the type/amount of prior knowledge that is assumed about the object space.

The first algorithmic component of our study (A1) will utilize PC features and the linear minimum mean squared error (LMMSE) reconstruction operator. We begin by defining the measurement process as a linear projection from the N-dimensional object space onto an M-dimensional (M < N) measurement space corrupted by additive white Gaussian noise (AWGN). The measurement vector (**m**) can therefore be written as

$$\mathbf{m} = \mathbf{F}\mathbf{x} + \mathbf{n} \tag{5.1}$$

where **F** is the M × N projection matrix, **x** is the N-dimensional vector that is lexicographically ordered from a $\sqrt{N} \times \sqrt{N}$ pixel object, and **n** is an M-dimensional vector of independent identically distributed (iid) zero-mean AWGN random variables each with variance σ^2. Reconstructed images are obtained via the LMMSE reconstruction operator **W** according to $\hat{\mathbf{x}} = \mathbf{Wm}$, where **W** is the N × M matrix given by

$$\mathbf{W} = \mathbf{R}_x\mathbf{F}^T(\mathbf{F}\mathbf{R}_x\mathbf{F}^T + \mathbf{D}_n)^{-1}, \tag{5.2}$$

$\mathbf{R}_x$ is the object autocorrelation matrix, $\mathbf{D}_n$ is the noise covariance matrix and singular value decomposition is used to invert the matrix $\mathbf{F}\mathbf{R}_x\mathbf{F}^T + \mathbf{D}_n$. The reconstruction RMSE associated with the projection matrix **F** is given by $\text{RMSE} = \sqrt{E\{|\mathbf{x} - \hat{\mathbf{x}}|^2\}}$, where E{} denotes statistical expectation over both the noise statistics and the object class **X1** from whence the objects are drawn. It is well known that for $\sigma = 0$, this RMSE is minimized by setting the rows of **F** equal to the M largest PCs of the object class **X1** [10]. The resulting PC basis vectors are simply the M largest eigenvectors (i.e., those with the M largest eigenvalues) of the autocorrelation matrix associated with **X1**. Despite the suboptimality of PC features in the presence of noise (i.e., for $\sigma \neq 0$), all results reported here for algorithmic component A1 utilize PC features together with the LMMSE reconstruction operator given in Equation 5.2.

We have employed a large commercially available set of human face images to define the class **X1** [4,10,11]. Figure 5.1a shows some example face images from this class. Each image of dimension 80 × 80 contains N = 6400 pixels. A total of 110,241 such images were used to define the object autocorrelation matrix and the resulting PC basis vectors. Figure 5.1b shows the first 5 PCs derived from this image class. These PC bases define the rows of the projection matrix **F**, which in turn defines the optical masks (described further in the "Architecture Descriptions" section) that will be necessary to implement compressive imaging.

The second algorithmic component of our study (A2) will utilize PR features together with the optimal LMMSE reconstruction operator. Once again, we assume the measurement model given by Equation 5.1; however, in this case, the PR basis vectors that define the rows of the projection matrix **F** are simply orthonormalized samples of an N-dimensional multivariate Gaussian random variable. It is important to note that these PR projections do not employ explicit knowledge of the object class to be imaged. The reconstruction algorithm, however, does require object class knowledge. Within algorithmic component A2, we assume

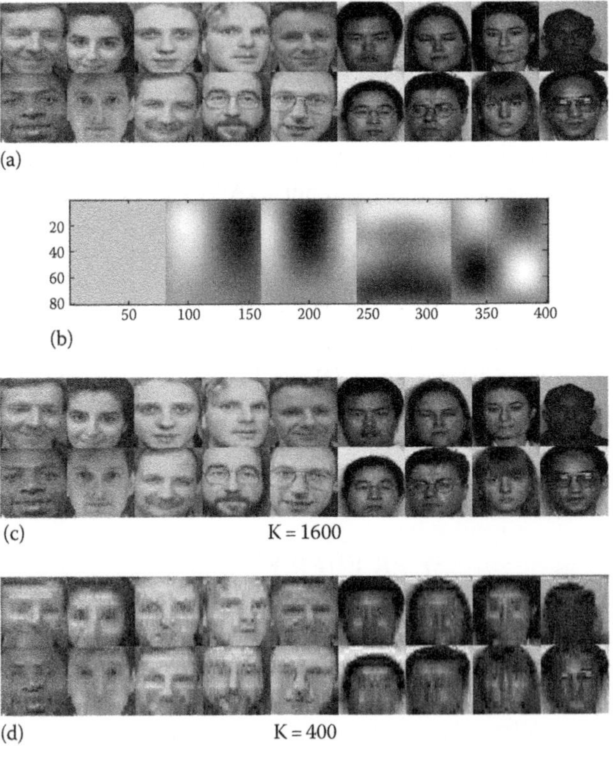

Figure 5.1 Examples used for this study—(a) examples from class **X1**, (b) first five principal component (PC) basis vectors for class **X1**, (c) examples from class **X2** with K = 1600, and (d) examples from class **X2** with K = 400.

that objects are drawn from the class of wavelet-sparse images **X2**. Specifically, we define the class **X2** to consist of all object vectors **x** for which the Haar wavelet transform $\mathbf{v} = \mathbf{Hx}$ has relatively few (e.g., $K \ll N$) nonzero elements, where **H** is the $N \times N$ Haar wavelet transform matrix [12].

In this work, we have generated the class **X2** of wavelet-sparse objects by processing the class **X1** of human face images. An element $\mathbf{x}_2 \in \mathbf{X2}$ is obtained from an element $\mathbf{x}_1 \in \mathbf{X1}$ by (1) computing the wavelet transform $\mathbf{v}_1 = \mathbf{Hx}_1$, (2) setting the smallest N–K elements of $\mathbf{v}_1$ to zero to obtain the sparse wavelet vector $\mathbf{v}_2$, and (3) inverting the wavelet transform to obtain $\mathbf{x}_2 = \mathbf{H}^{-1}\mathbf{v}_2$. Examples of wavelet-sparse objects obtained via this procedure are shown in Figure 5.1c for the case of K = 1600 and in Figure 5.1d for the case of K = 400. Both of these cases will be included in our study. We note that the prior knowledge of object sparsity is only weakly included in the LMMSE operator via the object autocorrelation matrix $\mathbf{R}_x$, whereas the nonlinear reconstruction methods described in the "Results Using Nonlinear Reconstruction" section will make more explicit/stronger use of this prior knowledge.

Architecture Descriptions

We have studied three different architectures for compressive imaging: sequential, parallel, and photon sharing. These architectures may be distinguished by (1) the efficiency with which photons are used to compute the required linear projections and (2) the bandwidth of the noise that corrupts the resulting measurements. The PS architecture will be shown to offer superior reconstruction fidelity by virtue of providing the highest photon efficiency along with the lowest noise bandwidth. It is important to note that each of these architectures may be implemented in a potentially large number of different ways. In this study, we will be interested only in the fundamental performance limits of the architectures themselves and not in the performance degradations that may result from specific device technologies. We therefore assume ideal devices and compute upper bounds on achievable performance.

Sequential

A schematic depiction of the sequential architecture is shown in Figure 5.2a. This architecture has become known as the "Rice single-pixel camera," and it employs a single optical aperture, a single photodetector, and an adaptive optical mask to define its projections. In this architecture, a single value m_i of the measurement vector **m** is measured during each time step. During the ith time step, the mask transmittance will be defined by $\mathbf{f}_i$, the ith row of **F**. Because the photodetector spatially integrates the incident irradiance passed by this mask, the measured photocurrent will be given by the noisy inner product $m_i = \mathbf{f}_i \cdot \mathbf{x} + n$, where n is a scalar AWGN random variable with variance σ^2. Note that if any elements of **F** are negative, a dual-rail measurement system is employed [4,12]. Given the aperture diameter D, total measurement time T, and assuming a uniform division of photons and measurement time among the required M time steps, the number of photons that are available to participate in the ith measurement is given by D^2T/M in appropriately normalized units. Note that the AWGN in each measurement is also affected by the measurement time per feature. If the total data collection time is kept fixed (i.e., independent of the number of features M), then as M is increased, the sequential architecture must allocate less measurement time per feature. This fact impacts both the number of photons that are available

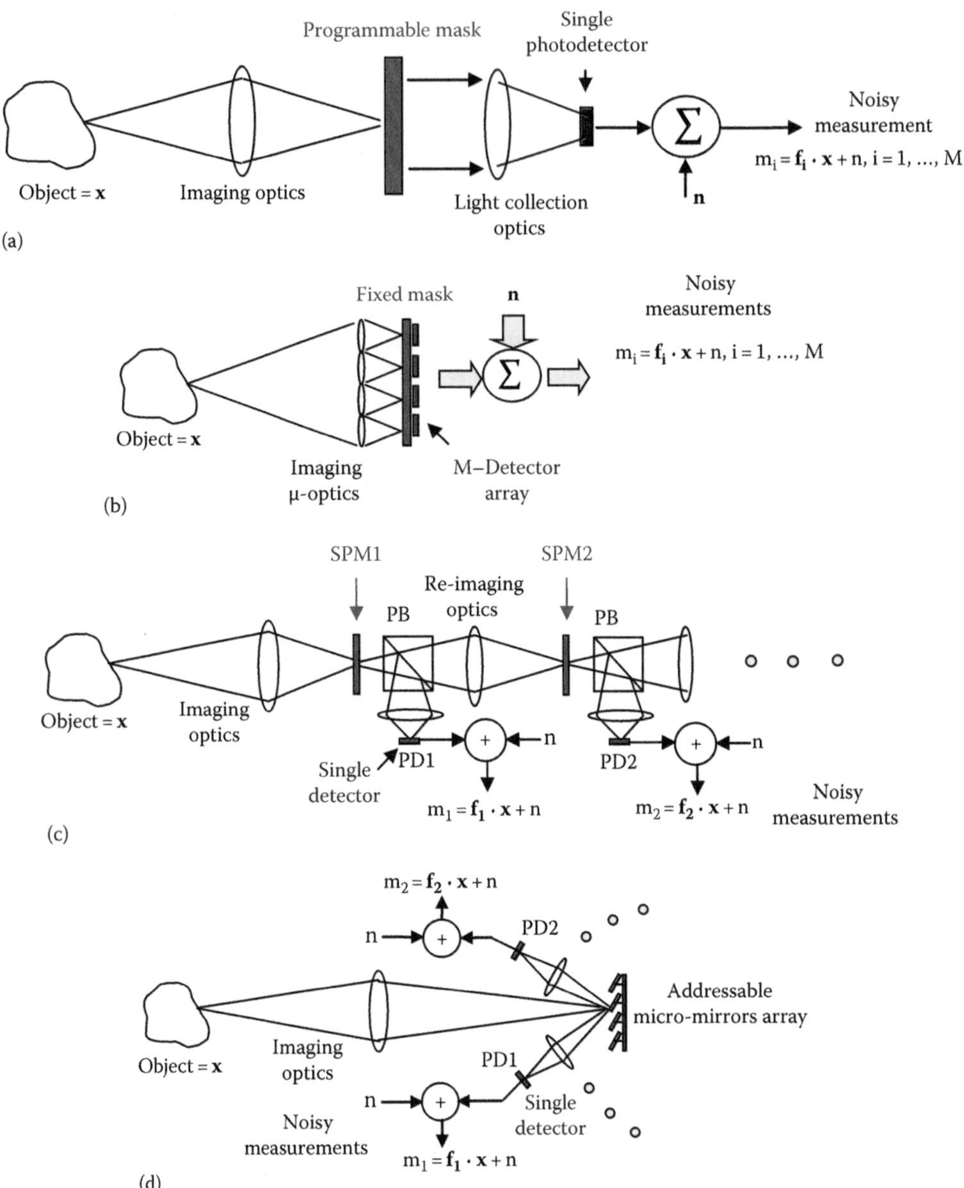

Figure 5.2 Schematic optical system diagram for various compressive imaging architectures: (a) sequential, (b) parallel, (c) space-domain photon sharing, and (d) time-domain photon sharing.

for any single feature measurement as well as the required measurement bandwidth. We define bandwidth as the inverse of the "measurement time per feature" because this is the approximate bandwidth that the associated measurement apparatus (i.e., detectors and supporting electronics) must have in order to realize the required feature measurement rate. In the sequential architecture therefore, the bandwidth increases as M/T. As a result, the measurement noise in the sequential architecture increases linearly with the number of feature measurements M. We can therefore write the resulting measurement noise variance per feature as $\sigma^2 = \sigma_0^2 M/T$, where σ_0 is a constant representing the noise standard deviation per unit bandwidth of the detector and supporting electronics and is also proportional to noise equivalent power (NEP) of the detector. We will refer to this scheme as the uniform-sequential (U.S.) architecture.

The U.S. architecture can be modified in order to improve reconstruction fidelity. Consider a case in which we have *a priori* knowledge that some projections are more important than others with respect to reconstruction RMSE. We would like to provide these important projections with higher measurement fidelity than those that are less important. This prior knowledge can therefore be used to define a nonuniform set of measurement intervals. The number of photons that are available to participate in the ith measurement is now given by $\eta_i D^2 T$, where the allocations $\{\eta_i, i = 1, \ldots, M\}$ are selected to minimize reconstruction RMSE subject to the constraint $\sum \eta_i = 1$. Note that the photodetector and supporting electronics must now support the bandwidth $\max[1/(\eta_i T)]$ and that the measurement noise must be scaled accordingly as $\sigma^2 = \sigma_0^2 \max[1/(\eta_i T)]$. We will refer to this scheme as the nonuniform-sequential (NS) architecture. In all results reported here, the optimum energy allocations have been determined by the use of the Stochastic Tunneling optimization algorithm reported in Wenzel and Hamacher [14].

Parallel

An example of the parallel architecture is shown in Figure 5.2b. This architecture employs an array of M optical apertures and a corresponding array of M photodetectors, each utilizing a fixed mask to measure a different projection. In this architecture, all projections are measured during a single time step. The mask associated with the i'th aperture will have a transmittance function defined by $\mathbf{f}_i$. Once again, we take the total system diameter to be D and the total measurement time to be T: this facilitates a fair comparison of photon utilization among all architectures. If all apertures within the parallel architecture are assumed to have the same diameter $D/\sqrt{M}$, then the number of photons that participate in the computation of the i'th projection will be D^2T/M. This is the same result that was obtained for the U.S. architecture. In the case of the parallel architecture, however, the photodetectors and supporting electronics are required only to operate over a bandwidth of $1/T$ so that $\sigma^2 = \sigma_0^2/T$. Note that this uniform-parallel (UP) architecture can offer reduced noise as compared with the U.S. architecture. Also note that we have restricted our analysis to the domain of geometric optics and therefore do not include the effect of increased diffraction blur for large M.

Once again, nonuniform photon division may be employed to improve the performance of the parallel architecture (e.g., via nonuniform lenslet apertures), albeit with significantly greater implementation complexity. For the resulting nonuniform-parallel (NP) architecture, the number of photons that are available to participate in the ith measurement is once again given by $\eta_i D^2 T$, where the allocations $\{\eta_i, i = 1, \ldots, M\}$ are selected to minimize reconstruction RMSE subject to the constraint $\sum \eta_i = 1$. Note that the nonuniform allocation of photons has no impact on the bandwidth or noise strength associated with measurements in the NP architecture.

Photon Sharing

The PS architecture shown in Figure 5.2c takes the form of a polarization-based optical pipeline processor. Its operation is described in detail in Neifeld and Shankar [5]. A summary of that description is provided here for completeness.

There are M stages of the PS pipeline processor. The i'th stage of the pipeline is responsible for computing the projection of **x** onto the ith row of **F**. The operation of the first stage is as follows. An image of **x** is formed on the first spatial polarization modulator (SPM1). The j'th pixel of SPM1 is designed to rotate the polarization of the incident light by an angle $\theta_{1j} = \cos^{-1}\sqrt{F_{1j}/C}$, where F_{1j} is the j'th element of the first row of **F** and C is the maximum absolute column sum of **F**. The resulting polarization rotation is decomposed into two orthogonal components by the use of a polarizing beam splitter (PBS). One of the orthogonal components is deflected by the PBS and is integrated onto the photodetector PD1 as shown. This process produces a measurement proportional to $m_1 = \mathbf{x}\cdot\mathbf{f}_1 + n_1 = \sum_j x_j \cos^2(\theta_{1j}) + n_1$ as desired. The other polarization component exiting the PBS is reimaged onto SPM2 and the second stage operates in a manner identical to the first; however, the rotation angle θ_2 must be modified in order to account for the photons that were diverted in stage 1. Specifically, we set $\theta_{2j} = \cos^{-1}\sqrt{F_{2j}/(C - F_{1j})}$ so that the measurement in stage 2 will be proportional to $m_2 = \mathbf{x}\mathbf{f}_2 + n_2 = \sum_j x_j[1-\cos^2(\theta_{1j})]\cos^2(\theta_{2j}) + n_2$. Additional stages operate in an analogous fashion. Note that this simplified description assumes that all elements of **F** are positive. In order to include projections for which the elements of **F** can be negative (e.g., PC projections), we assume the use of a dual-arm architecture. Note that this approach will increase the required number of photodetectors (and therefore noise) by a factor of two. The reader is referred to Takhar et al. and Barrett and Myers [4,12] for a more complete description of the dual-arm architecture.

In contrast with the sequential and parallel architectures, the PS architecture discards no useful photons and uses no absorptive masks. Because we can expect absorptive masks to have an average transmittance of 0.5, the PS architecture is roughly 2× more photon efficient than the parallel architecture. We also note that the PS architecture measures all projections in a single time step resulting in a measurement noise strength $\sigma^2 = \sigma_0^2/T$, that is identical to that of the UP and NP architectures. Unlike the parallel architectures, however, nonuniform photon allocation is easily achieved within the PS architecture by simply scaling the rows of **F** by the optimal allocations $\{\eta_i, i = 1, \ldots, M\}$ before converting them to SPM angles $\{\theta_i, i = 1, \ldots, M\}$. Whenever possible, therefore, all of the PS results presented in the next section utilize the optimal photon allocation for achieving minimum reconstruction RMSE.

The PS architecture shown in Figure 5.2c and described earlier employs "photon sharing in the space domain." It is also possible to design a PS architecture that employs "photon sharing in the time domain." Such an architecture is shown in Figure 5.2d. Here, we depict an array of N individually addressable micromirrors, each of which can be pointed in one of M directions. An image of **x** is formed on this micromirror array. The j'th micromirror is programmed to dwell on the ith direction for a time interval proportional to the (i, j)th element of the projection matrix, $\mathbf{F}_{ij}$. In this way, the ith detector accumulates a total number of photons proportional to ($\mathbf{f}_i \cdot \mathbf{x}$) as desired. Once again, no absorptive masks are used (providing high photon efficiency) and the required measurement bandwidth is 1/T (resulting in low noise). Although implementation issues and/or application details may result in a preference for either the space-domain or the time-domain architecture, the fundamental performance limits of both these PS architectures will be identical.

Results Using Linear Reconstructions

In this section, we compare the reconstruction RMSE per pixel of the various compressive imaging architectures that were described in the previous section. A baseline noncompressive (i.e., conventional) imager will also be included in these comparisons. The conventional camera (CC) is assumed to employ an aperture diameter of D and an integration time of T (i.e., it collects the same number of photons as did the compressive imagers) in order to form an N-pixel image $\mathbf{m} = \mathbf{x} + \mathbf{n}$, where once again the noise $\mathbf{n}$ is assumed to derive from AWGN associated with the detector bandwidth 1/T so that $\sigma^2 = \sigma_0^2/T$. Specifically, if $\mathbf{x}$ is an 80 × 80 pixel optical image, the CC uses 80 × 80 = 6400 photon detectors to detect the measurement $\mathbf{m}$ corrupted by an 80 × 80 pixel AWGN image $\mathbf{n}$. Because the compressive imagers have been provided with substantial postprocessing capability, a fair comparison requires that the baseline CC also benefit from postprocessing. We have applied the LMMSE denoising operator $\mathbf{W_c}$ to the measurement $\mathbf{m}$ in order to obtain the estimate $\hat{\mathbf{x}} = \mathbf{W_c m}$, where $\mathbf{W_c} = \mathbf{R_x}(\mathbf{R_x} + \mathbf{D_n})^{-1}$ [13].

There are two sources of reconstruction error within any compressive imaging system. The first results from a failure to measure a sufficient number of projections. This is sometimes referred to as truncation error. Note that there is no truncation error for a CC. The CC does not measure, compute, or otherwise use features in any way and so the RMSE of the CC is independent of M. The second source of reconstruction error arises from noise in the measurement process. This measurement noise produces errors in the measured projections and subsequent errors in reconstruction. In the absence of noise, all three compressive imaging architectures perform equally well. This is because in the noise-free case, no fundamental phenomenon exists to corrupt the optical projections. This means that within a scaling factor, all three architectures make exactly the same measurements, thus performing equally well. Also note that all three compressive imaging architectures can produce reconstruction RMSE = 0 when noise is zero and M = N PC projections are used. The conventional imager also produces RMSE = 0 in the noise-free case. In a more realistic scenario for which $\sigma \neq 0$, the limiting performance RMSE = 0 can no longer be obtained.

All compressive imaging architectures demonstrate a trade-off between truncation error and measurement noise. For example, in the UP architecture, the optical aperture size associated with each detector determines the number of photons available for computing a single projection. Because this aperture scales as $\sim D^2/M$, the signal-to-noise ratio (SNR) at each detector is inversely proportional to M. This causes the component of reconstruction error associated with measurement noise to increase with M. Truncation error, on the other hand, is always reduced by increasing M. As a result, we find that truncation error dominates the RMSE for small M, whereas measurement noise is the main component of RMSE for large M. The same trend holds for the other two compressive imaging architectures as we shall see in the following text.

PC Projections with Linear Reconstruction

We begin with the results of compressive imaging using PC projections. For all results reported here, we take T = 1, D = 1, and consider all object pixel brightness to fall in the range of 0–255. The training set for PC projections contains face image samples in class **X1** as discussed in the "Algorithm Descriptions" section. Testing samples are drawn from the same class. The noise standard deviation $\sigma_0 = 1$ corresponds to $NEP_0 = 0.025\ \text{pw}/\sqrt{\text{Hz}}$. In order to relate this normalization to real-world photodetectors, we note that $\sigma_0 = 133.5$ corresponds to NEP = 3.3 pw/$\sqrt{\text{Hz}}$, which is the average NEP at 900 nm for the New Focus model 2031 large-area photoreceiver on the high gain setting, whereas $\sigma_0 = 222.5$ corresponds to NEP = 5.5 pw/$\sqrt{\text{Hz}}$, equal to the NEP for the Thorlabs model PDA100A amplified silicon detector working at 970 nm with a gain of 40 dB.

Figure 5.3 presents the reconstruction RMSE data as a function of M for three different noise levels. Figure 5.3a is the data obtained for the case of $\sigma_0 = 1335$, whereas Figure 5.3b and c corresponds to noise levels of $\sigma_0 = 750$ and 422, respectively. Each

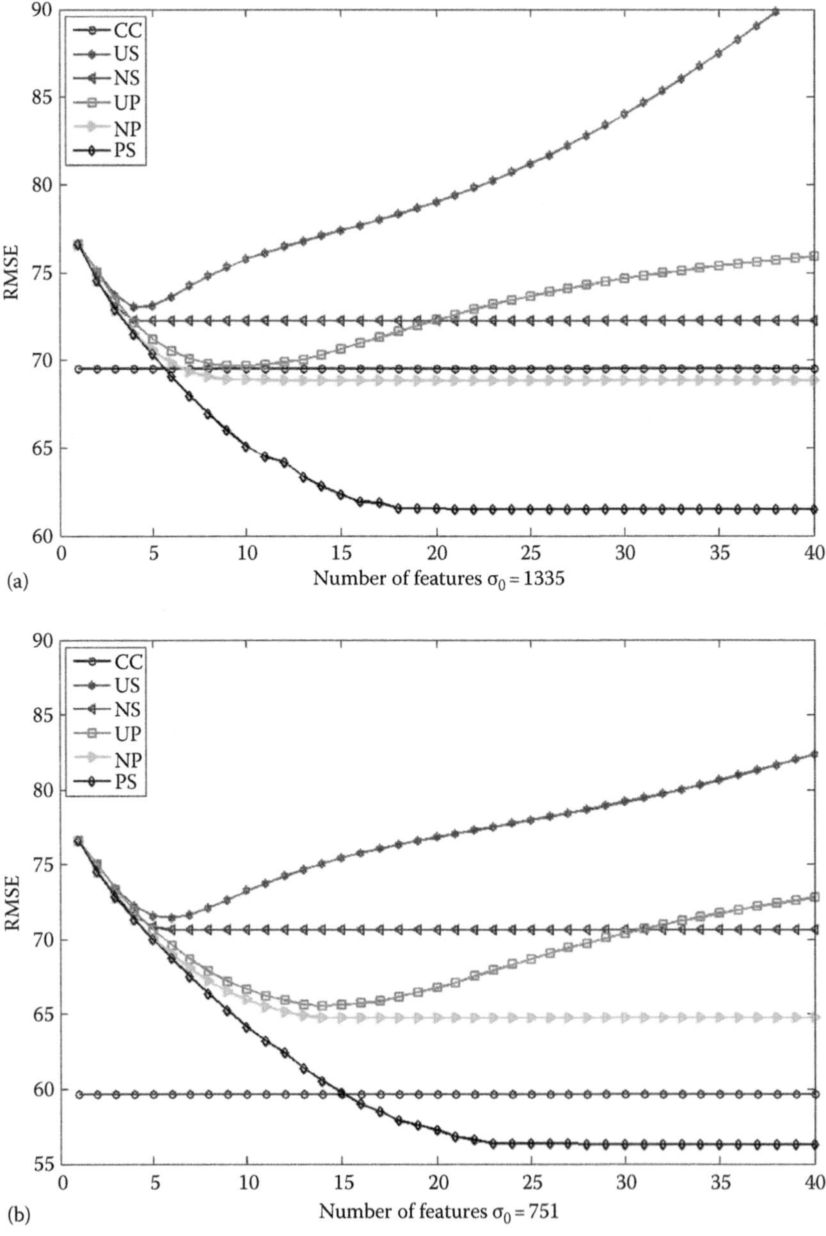

Figure 5.3 Reconstruction root mean squared error (RMSE) versus number of features for compressive imaging based on PC projections and several values of noise—(a) $\sigma_0 = 1335$; (b) $\sigma_0 = 751$. *(Continued)*

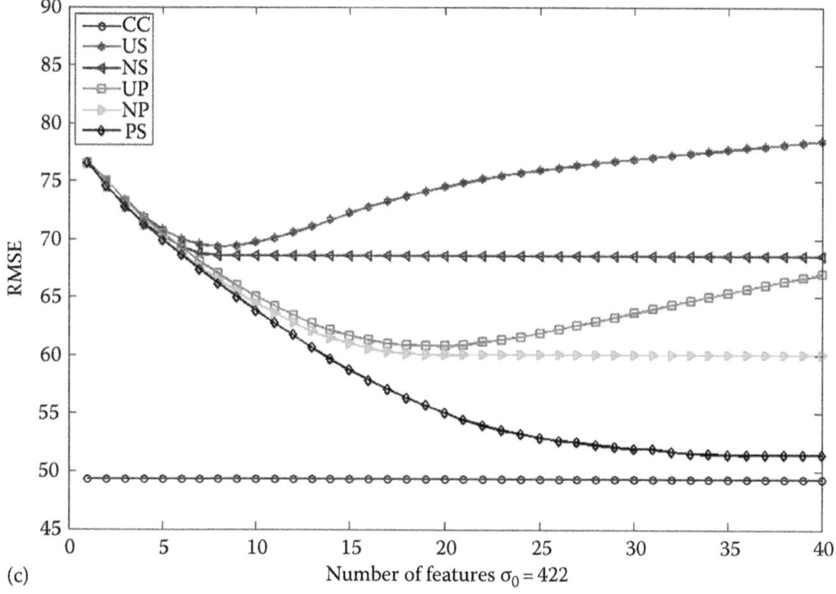

Figure 5.3 ***(Continued)*** Reconstruction root mean squared error (RMSE) versus number of features for compressive imaging based on PC projections and several values of noise—(c) $\sigma_0 = 422$.

of these graphs presents the performance of all six imagers that were described in the previous section: US, NS, UP, NP, PS, and CC. Several observations can be made based on these data. (1) We see that uniform photon allocation gives rise to an optimal number of measurements M_{opt} as a result of the trade-off between truncation error and measurement noise. For example, the UP architecture with $\sigma_0 = 422$ obtains its best performance for $M_{opt} = 20$. For $M < M_{opt}$ truncation error dominates the RMSE, whereas for $M > M_{opt}$ measurement noise dominates. Note that M_{opt} increases monotonically as σ_0 is reduced. (2) Nonuniform photon allocation improves performance and eliminates the existence of an optimum number of measurements. This is because the optimization process for $\{\eta_i, i = 1, \ldots, M\}$ will automatically allocate zero energy to any measurement that would result in an increase in RMSE. This can be easily seen from the plot of η_i versus i in Figure 5.4. These data were generated for the NP architecture and the case M = N. Note that as noise increases, the number of nonzero values for η_i decreases as expected. (3) In low-noise environments ($\sigma_0 < 550$), the CC always provides the best image fidelity. (4) In high-noise environments, the compressive architectures can be superior to CC. (5) PS provides the best RMSE performance among compressive imagers.

The RMSE data shown in Figure 5.3a through c can be summarized by extracting the optimum performance for each curve and plotting this minimum RMSE as a function of noise strength σ_0^2. Figure 5.5 presents the result of this process. Note that the RMSE curves for nonuniform photon allocation (i.e., the NS and NP curves) do not extend below $\sigma_0 = 133.5$. This is because the optimization process does not converge well when the number of nonzero values of η_i becomes large. Because the RMSE performance of the nonuniform solutions is very close to the *minimum* RMSE performance of the uniform solutions, we are confident in interpreting the curves in Figure 5.5 as reasonably tight upper bounds. Note that each compressive imager may be characterized by the value of σ_0^2 above which it

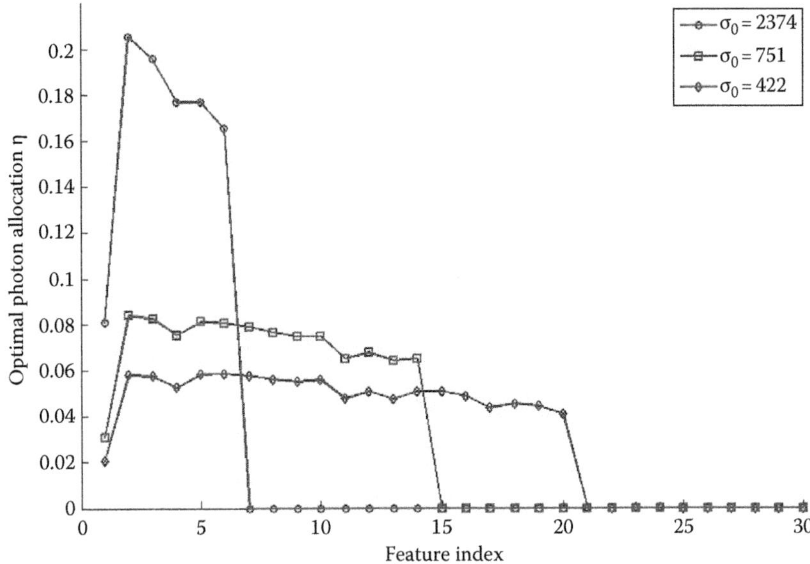

Figure 5.4 Optimal energy allocation η versus feature index for nonuniform-parallel (NP) projections and three values of noise.

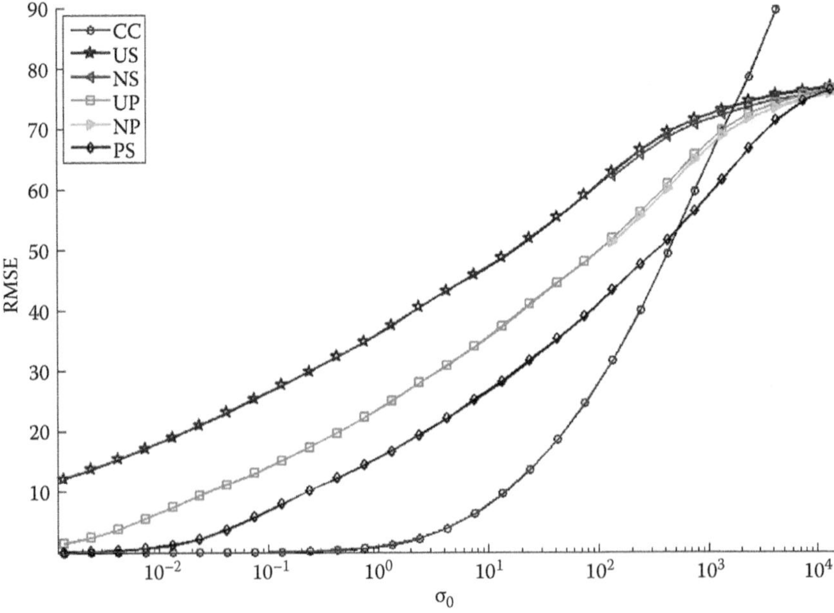

Figure 5.5 Minimum reconstruction RMSE versus σ_0 for all compressive imaging architectures using PC projections.

offers performance superior to the CC. These threshold values of noise strength are presented in Table 5.1. It is interesting to note that for very high-noise environments, compressive imaging can provide image fidelity that is superior to what can be obtained from a CC. In low-noise environment, the PS architecture always provides the best RMSE performance among the three architectures, while the sequential architecture provides the worst. Figure 5.6 presents example reconstruction for the US, UP, and PS architectures at $M_{opt} = 207, 941$, and 2601, respectively, when $\sigma_0 = 0.21$. The reconstructed object quality for these examples are RMSE = 13.96, 10.86, and 7.22 for the US, UP, and PS architectures, respectively.

PR Projections with Linear Reconstruction

In this section, we present architectural comparisons for compressive imaging using PR projections with linear reconstruction. The training samples used to define the autocorrelation matrix $\mathbf{R_x}$ are drawn from the object class **X2**. The testing samples are also drawn from this class. In the work reported here, we have

Table 5.1 Threshold Values of Noise Strength above Which Compressive Imaging Is Superior to Conventional Imaging

Imaging Method	Threshold Value of Noise (σ_0)
PS	530
NP	1250
UP	1290
NS	1640
US	1670

Figure 5.6 Optimal reconstructions for $\sigma_0 = 0.21$ obtained from the (a) sequential architecture ($M_{opt} = 207$), (b) parallel architecture ($M_{opt} = 941$), and (c) pipeline architecture ($M_{opt} = 2601$).

generated PR basis vectors by using normalized samples of an N-dimensional multivariate Gaussian random variable. In order to facilitate comparisons with the results of compressive imaging using PC features, we have first sorted the PR basis vectors. This sorting procedure strives to obtain good photon efficiency from the PR bases and is accomplished by ordering the PR vectors in descending order according to the sum of their projection values in the vector space defined by the principle components of object class **X2**.

Figure 5.7 presents the reconstruction RMSE data as a function of M for three different noise levels and sparsity K = 1600. Figure 5.7a is the data obtained for

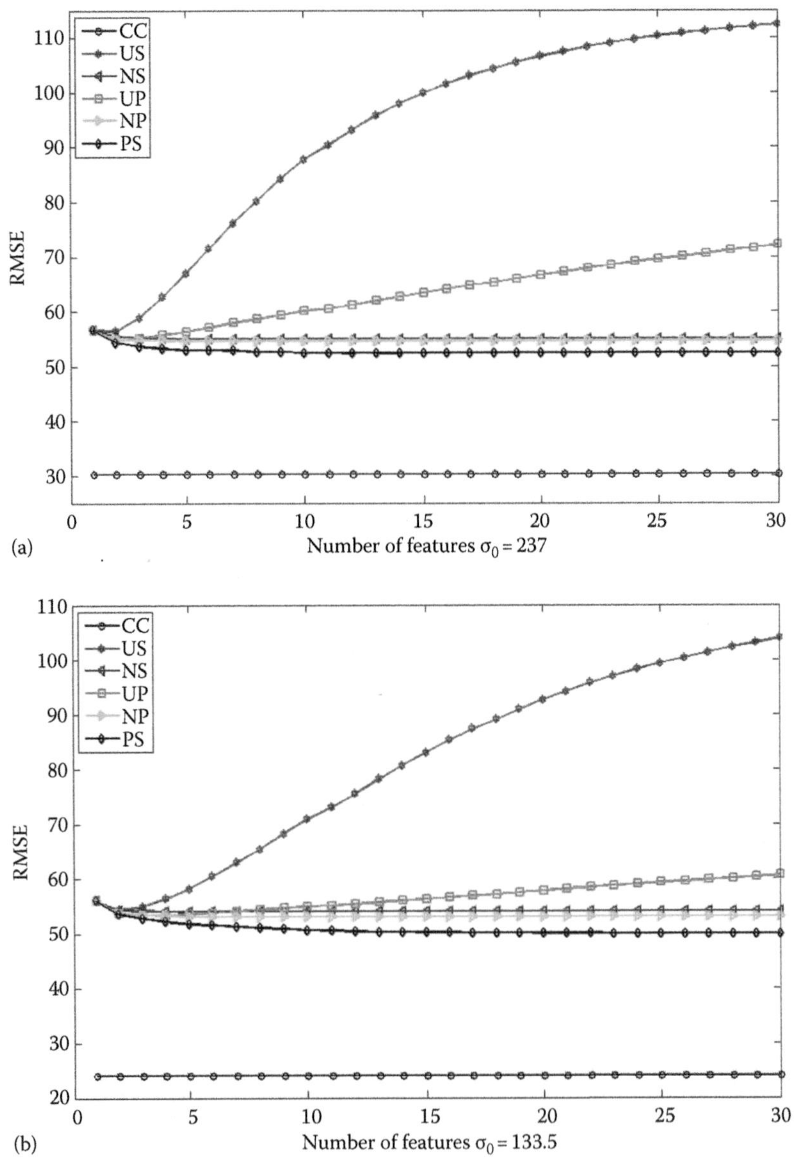

Figure 5.7 Reconstruction RMSE versus number of features for pseudorandom (PR) projections with K = 1600 and various noise strengths—(a) $\sigma_0 = 237$; (b) $\sigma_0 = 133.5$. *(Continued)*

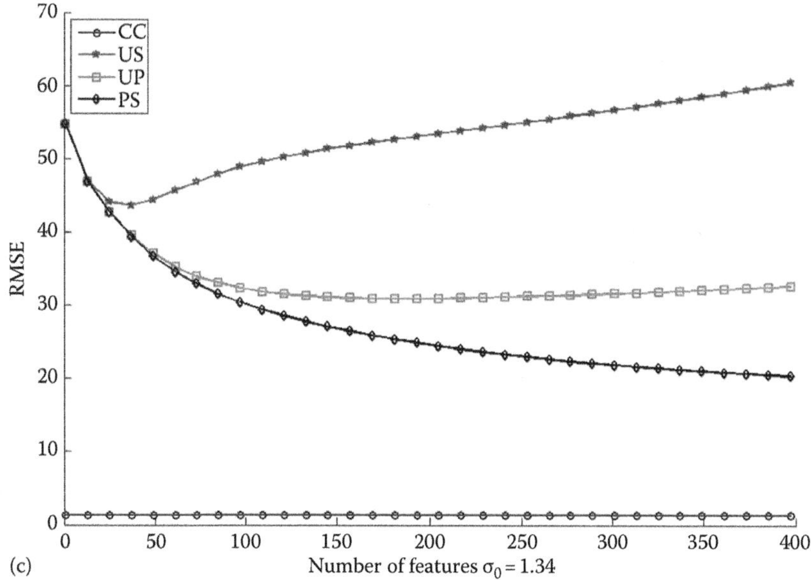

Figure 5.7 ***(Continued)*** Reconstruction RMSE versus number of features for pseudorandom (PR) projections with K = 1600 and various noise strengths—(c) $\sigma_0 = 1.34$.

the case of $\sigma_0 = 237$, whereas Figure 5.7b and c corresponds to noise levels of $\sigma_0 = 133.5$ and 1.34, respectively. Once again, optimal nonuniform photon allocations could not be found for the lowest noise case shown in Figure 5.7c. In Figure 5.8, two cases K1 = 1600 and K2 = 400 are presented together at a noise level $\sigma_0 = 237$. These results for PR projections are similar to those presented in Figure 5.3 for PC projections. Specifically, we note (1) an optimal number of measurements arising from the trade-off between M and σ and (2) the PS architecture providing the best performance among compressive imagers. However, an important difference between PR and PC projections is that for PR projections, we find that the CC always provides lower RMSE. This is simply because the compression achieved by the use of PR projections is weaker than that achieved using PC projections: PC projections exploit a stronger form of prior knowledge in order to *minimize* the number of measurements required to achieve a given level of truncation error. We also note that the performance of the NP and NS architectures are identical when noise is high, $\sigma_0 \geq 237$. This is because when the noise is very large, the total photon energy is allocated into a single feature, in which case the parallel and sequential architectures become identical. We also note from the data in Figure 5.8 that for a specific value of M, the performance of compressive imaging is slightly improved for smaller values of K.

Once again, it is instructive to extract the minimum RMSE values from each curve in Figures 5.7 and 5.8 and plot the resulting optimum performance as a function of σ_0. These data are shown in Figure 5.9. Once again, we see that the performance of conventional imaging is superior to that of PR compressive imaging for all meaningful values of σ_0.

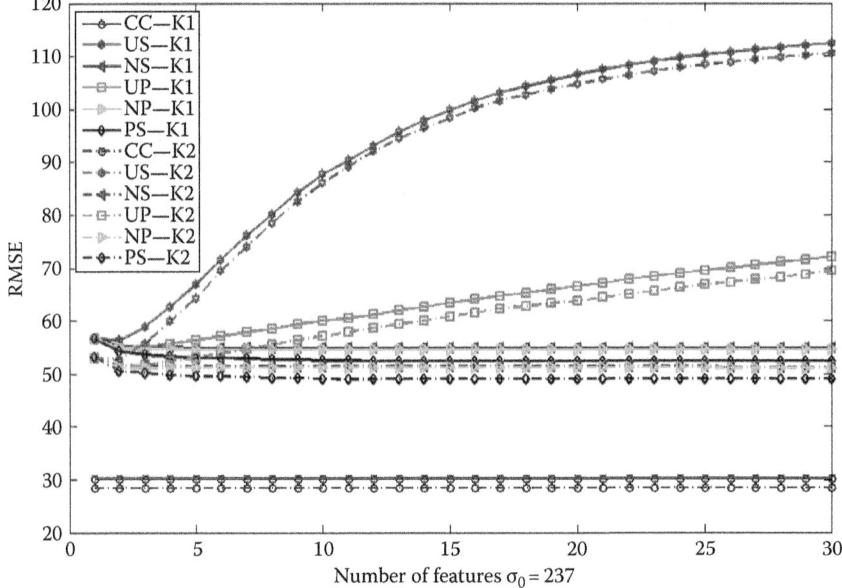

Figure 5.8 Reconstruction RMSE versus number of features for PR projections with noise strength $\sigma_0 = 237$ and two values of sparsity K1 = 1600 and K2 = 400.

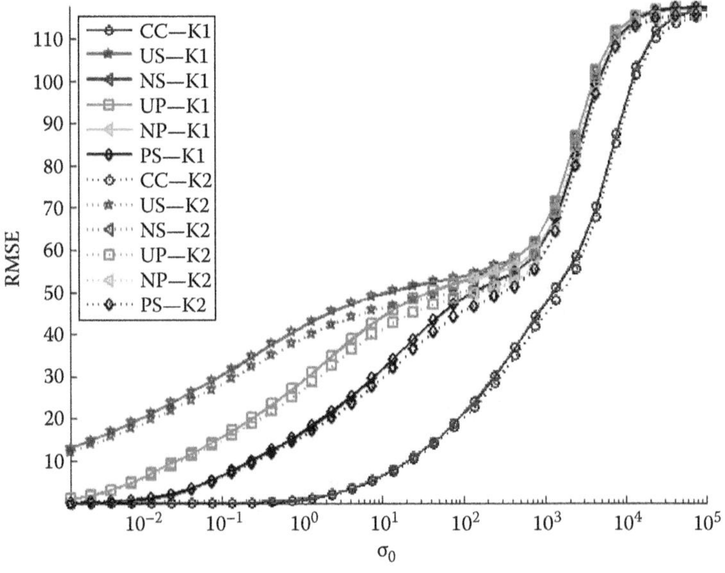

Figure 5.9 Minimum reconstruction RMSE versus σ_0 for compressive imaging using PR projections.

Results Using Nonlinear Reconstruction

We are interested in quantifying the extent to which nonlinear reconstruction might impact the architectural comparisons presented in the previous section. Note that this chapter is primarily concerned with a comparison among optical architectures and that our inclusion of nonlinear reconstruction is intended only as a simple probe of whether the architectural comparison remains valid when nonlinear processing is employed. In this exercise, we will focus on object class **X2** for which the images are known to be sparse in the wavelet domain. The measurement model is still given by Equation 5.1 with the projection matrix $\mathbf{F}$ formed from either PC or PR basis vectors. Nonlinear reconstruction of the object estimate $\hat{\mathbf{x}}$ from the measurement vector $\mathbf{m}$ exploits the prior knowledge of wavelet sparsity via one of the several candidate nonlinear algorithms. Figure 5.10a presents some results for compressive imaging using the PS architecture and PR projections with nonlinear reconstruction. These data are based on M = 2000 PR features and have been generated for a sparsity K = 400. Linear reconstruction performance is depicted by the solid curve. The dotted curves correspond to various nonlinear reconstruction methods and are labeled "Nowak" for the method reported in Haupt and Nowak [15], "MP" for matching pursuit [16], "L1qc" for minimum-l_1 norm with quadratic constraints, "TVqc" for minimum total variation with quadratic constraints [17], and "BPDN" for basic pursuit denoising [18]. We draw three conclusions from these data. (1) As expected, we see that the nonlinear reconstruction methods produce smaller RMSE than does the LMMSE solution when σ_0 is small. (2) We also see that the LMMSE solution can be superior when noise is large. Of course, improved starting points and/or stopping rules can be used to insure that these nonlinear methods do not cross over the LMMSE solutions in the high-noise domain [19]. (3) We note that among these five nonlinear reconstruction methods, the method reported by Nowak et.al. produces the smallest RMSE for our imagery. In Figure 5.10b, we show the RMSE results obtained from all three compressive architectures (US, UP, and PS) using both the LMMSE reconstruction (solid curves) and the Nowak algorithm (dotted curves). All other parameters are identical to those from Figure 5.10a. We note from the data in Figure 5.10b that the use of nonlinear reconstruction does not change the architectural performance trends that were observed in the previous section.

Figure 5.11 presents curves for RMSE versus number of features for the PS architecture using PR projections and both linear and nonlinear reconstruction methods. Two noise levels are shown $\sigma_0 = 0.37$ (solid curves) and 0.21 (dotted curves). The object sparsity associated with these data is K = 1600 in Figure 5.11a and K = 400 in Figure 5.11b. Once again, only the uniform photon allocation results were possible at these low noise levels. For this reason, we see the now-familiar trade-off between truncation error and measurement noise manifesting itself in the existence of minimum RMSE at specific values of M. We observe from Figure 5.11 that, as expected for these low levels of noise, the RMSE can be significantly reduced by the use of nonlinear reconstruction. For example, the minimum RMSE obtained using the LMMSE method is 11.06 for $\sigma_0 = 0.37$ and K = 1600, whereas the minimum RMSE obtained using the Nowak algorithm is 10.61 under these same conditions. This advantage derives from the explicit incorporation of sparsity within the reconstruction process. For this reason, we expect an ever greater benefit for K = 400. Figure 5.11b demonstrates that this is indeed the case. We see that the LMMSE method provides RMSE = 9.21 for $\sigma_0 = 0.21$ and K = 400, whereas the minimum RMSE obtained using the Nowak algorithm

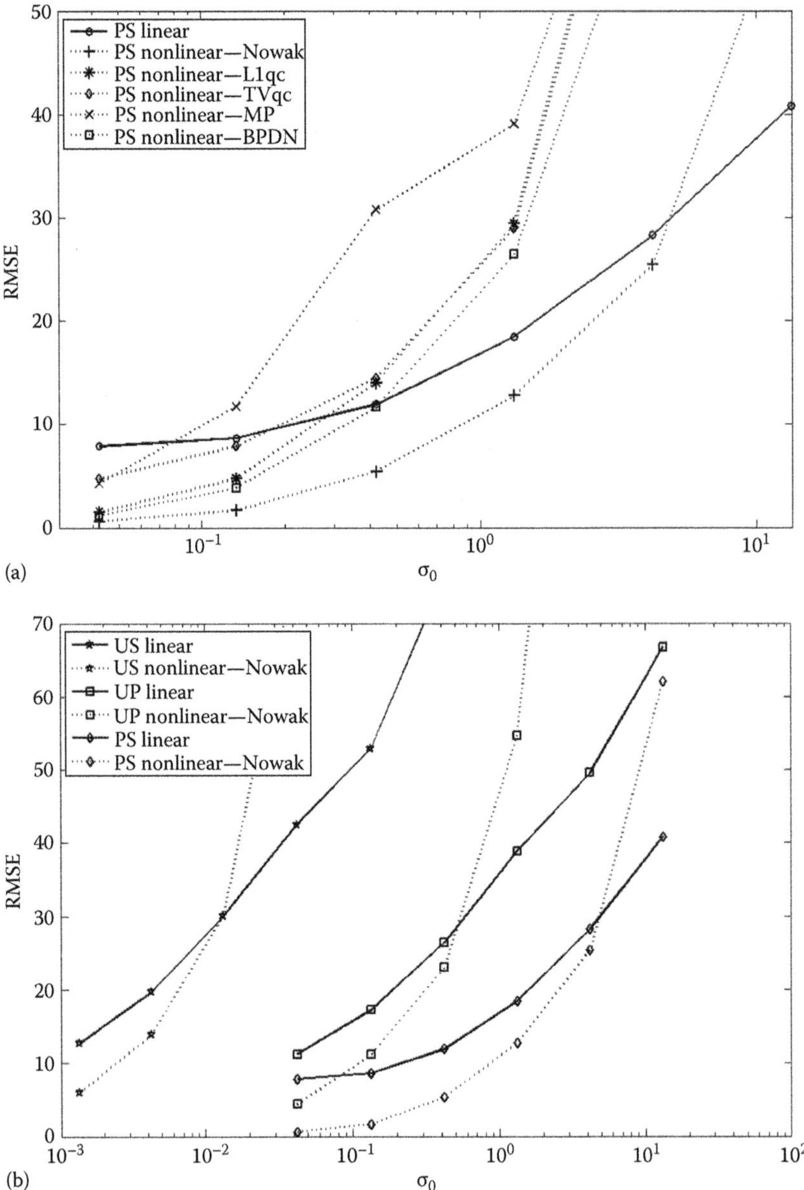

Figure 5.10 Reconstruction RMSE versus σ_0 obtained using PR projections and linear (solid) and nonlinear (dashed) reconstruction methods. (a) Various nonlinear algorithms compared using the photon-sharing (PS) architecture; (b) comparison among the uniform-sequential (US), uniform-parallel (UP), and PS architectures.

is 6.54 for this same set of parameters. Figure 5.12 presents some example reconstructions for the case $\sigma_0 = 0.21$ and two levels of sparsity. All of these images are obtained using the optimal number of projections. Original objects are shown in Figure 5.12a and d, LMMSE reconstructions in Figure 5.12b and e, and nonlinear reconstructions in Figure 5.12c and f for the cases K = 1600 and K = 400, respectively. We note that the nonlinear reconstructions tend to be visually superior to the linear reconstructions.

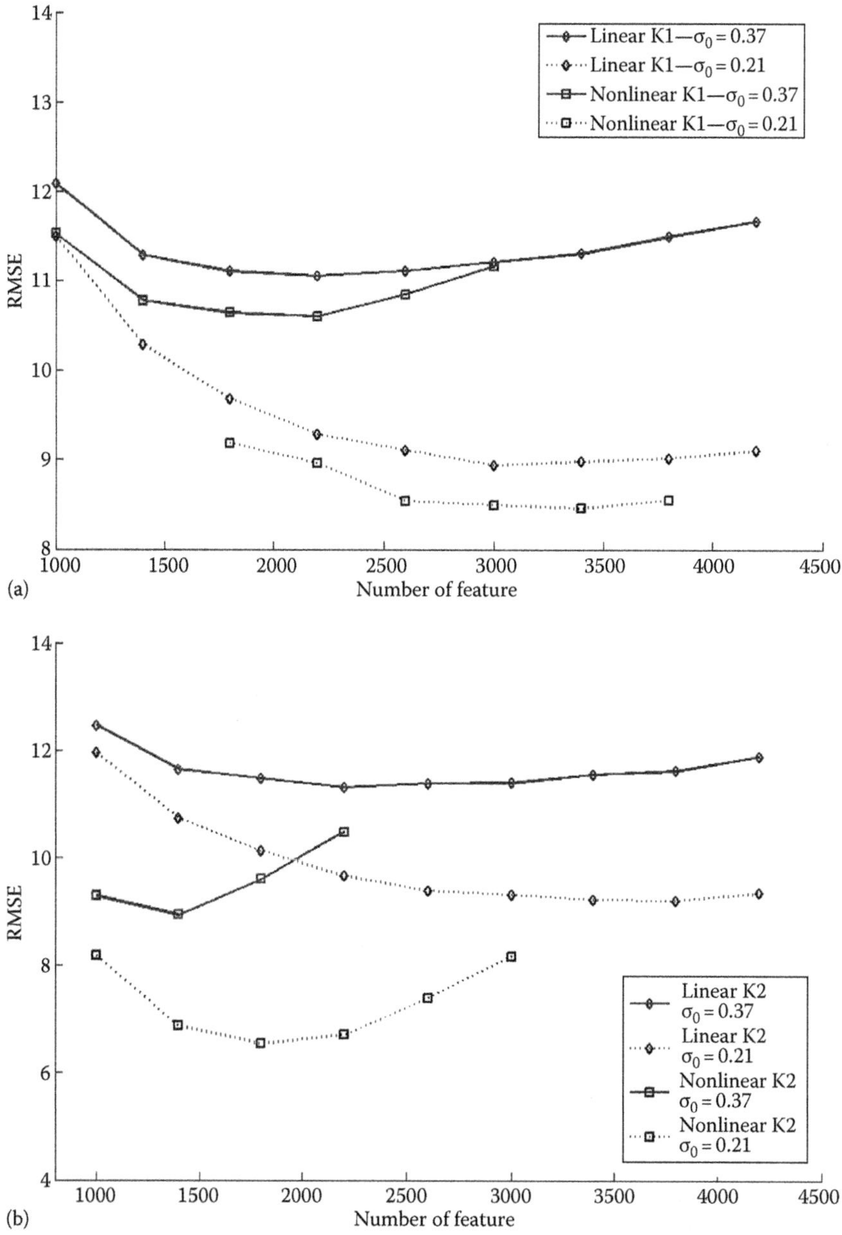

Figure 5.11 RMSE versus number of features obtained from the PS architecture using PR projections with linear (diamonds) and nonlinear (squares) reconstructions for (a) K = 1600; (b) K = 400.

We have also examined the use of nonlinear reconstruction for compressive imaging based on PC projections. We anticipate improved performance by virtue of incorporating object prior knowledge within both the basis vector design and the nonlinear reconstruction algorithm. We limit this study to the PS architecture and the two noise values $\sigma_0 = 0.37$ and $\sigma_0 = 0.21$. Figure 5.13 presents the resulting RMSE data as a function of number of features for K = 1600 (Figure 5.13a) and

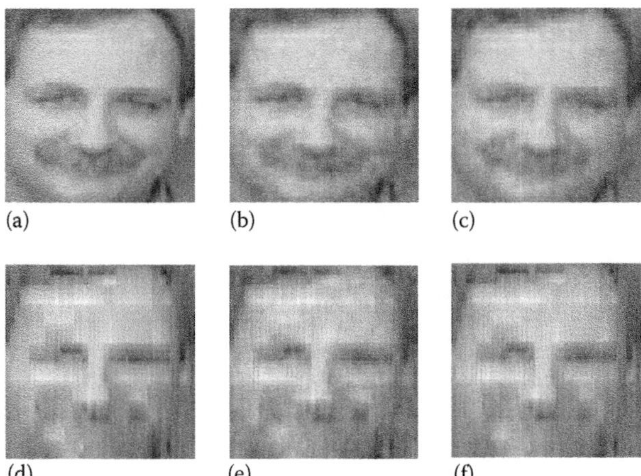

Figure 5.12 Example reconstructions for the PS architecture using PR projections and $\sigma_0 = 0.21$ from (a) through (c) K = 1600 and (d) through (f) K = 400; (a) and (d) are the original objects, (b) and (e) are the optimal linear reconstructions, and (c) and (f) are the optimal nonlinear reconstructions.

K = 400 (Figure 5.13b) using both linear and nonlinear reconstruction methods. As we observed for the PR projections, (a) minimum RMSE is reduced by the use of the Nowak nonlinear reconstruction method and (b) this improvement is more significant when the objects are more sparse. Comparing Figures 5.11 and 5.13, we observe that the additional prior knowledge embodied in the PC projections has indeed improved the performance as compared with the PR projections.

Shot-Noise-Limited Performance

Thus far we have dealt only with iid AWGN. The results from the previous sections therefore will be valid in the case of read-noise-limited sensors; however, many modern focal planes can now achieve shot-noise-limited performance. For this reason, we extend our analysis to this important domain. The key features of shot noise that distinguish it from AWGN are (a) it is governed by Poisson statistics and (b) its magnitude is correlated with the signal level. For the results reported here, we will assume that the number of photons incident on a pixel of the photodetector array in both the conventional and compressive systems will be large enough that Poisson statistics can be well approximated by Gaussian statistics (i.e., photon numbers > 40). This allows us to use our previous measurement models modified to capture the signal-dependent nature of the noise. Specifically, we consider the CC system as before wherein a single pixel measurement is given by $m_i = x_i + n_i$ and under the shot-noise model the signal-dependent noise variance is $\sigma_i^2 = x_i$. Similarly, our compressive measurement model becomes $m_i = \mathbf{f}_i \cdot \mathbf{x} + n$, where the variance of n is $\sigma_i^2 = \mathbf{f}_i \cdot \mathbf{x}$. We limit the scope of our shot-noise study to the case of LMMSE reconstruction and PC projections. We find that all other cases exhibit similar trends. In particular, we once again observe that there is an optimum number of measurements to balance noise against truncation error. Limiting our attention to these optimum values of RMSE, we obtain the data shown in Figure 5.14, which plots reconstruction quality versus total number of photons collected by the imaging apparatus. The most notable trends

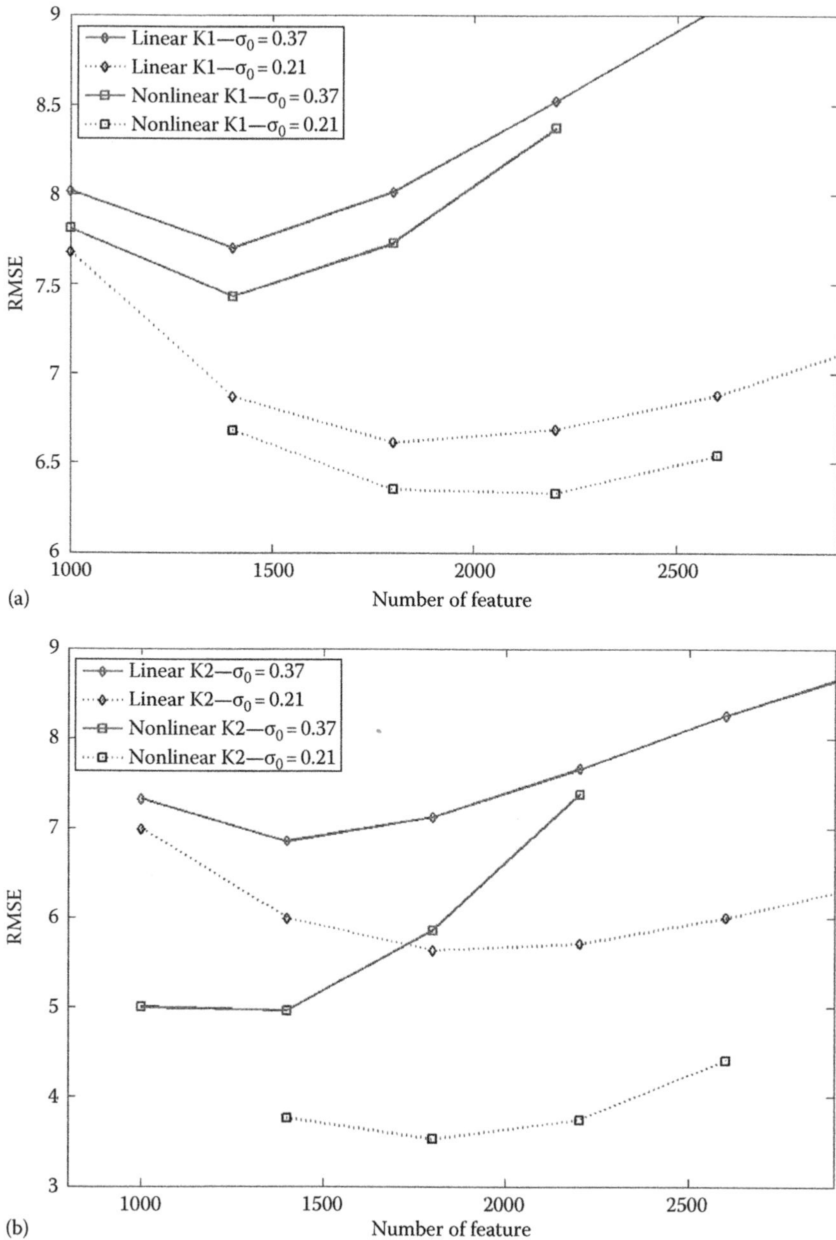

Figure 5.13 RMSE versus number of features using PC projection and the PS architecture with linear (diamonds) and nonlinear (squares) reconstruction methods when (a) K = 1600; (b) K = 400.

that emerge from these data are that (1) RMSE reduces with increasing numbers of collected photons as expected since the shot-noise-limited SNR is proportional to the square root of the total photon number, (2) the relative performance of the three CS architectures obeys the same trends that we saw in the case of read-noise-limited performance (i.e., in terms of RMSE we have PS < UP < US), and (3) we never see compressive imaging performance exceeding that of a CC. This

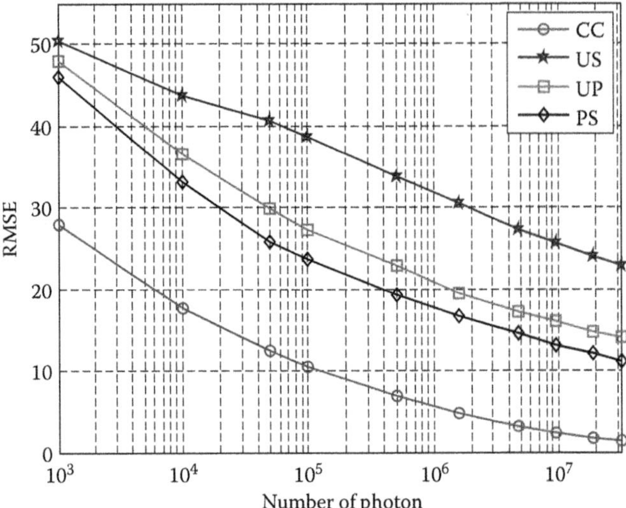

Figure 5.14 Shot-noise-limited minimum reconstruction RMSE versus total number of collected photons for all compressive imaging architectures using PC projections.

last observation is a significant departure from the trends we saw earlier in the chapter, and it arises from the absence of a multiplex advantage in the case of shot noise. Recall that in the case of read noise, we saw a linear increase in measurement SNR as additional photons were combined on a single measurement pixel. In the case of shot noise, this SNR is the *square root* of the photon number, and therefore when combined with truncation error, the shot-noise-limited performance never exceeds the performance of a CC for the same number of total collected photons. Of course, the reconstructions from these compressive measurements are still of very high quality and can be equivalent to the quality of a conventional image if sufficient photon resources are available.

Conclusions

As we discussed in the "Introduction" section, compressive imaging offers (1) reduced cost owing to a reduction in the number of photodetectors and the concomitant reduction in camera size, weight, power, etc. and (2) improved detector-noise-limited measurement fidelity. These observations suggest that compressive imaging is likely to be an important capability for applications/wavebands in which camera costs are high (e.g., IR imaging for which pixel counts are low and/or significant cryogenic resources are required or x-ray imaging for which a large number of tomographic projections are required) and/or photon counts are low (e.g., imaging at night and/or in severe environments). These same features may also provide an advantage for mobile platforms in which limited onboard processing and/or downlink bandwidth is available.

In this chapter, three candidate compressive imaging architectures (photon sharing, parallel, and sequential) have been compared using two types of projection (PC and PR) and two types of reconstruction algorithm (linear and nonlinear). From this comparison, we conclude that the PS architecture offers the best RMSE performance. When object energy is allocated equally into multiple feature measurements, compressive imaging architectures exhibit a trade-off between truncation error and measurement noise. We observe that this trade-off can be circumvented by the use of nonuniform energy allocation. In the case of linear reconstruction, we show that compressive imaging based on PC projections can be superior to conventional imaging in a high-noise environment, whereas PR projections were not able to offer an RMSE improvement compared with the CC. We also undertook this architectural comparison for several nonlinear reconstruction methods. We find that compressive imaging performance for both PC and PR projections can be improved by the use of nonlinear reconstruction in low-noise environments. Our analysis of compressive imaging in the case of shot-noise-limited detectors demonstrated that although high-quality images may be obtained, these images will be inferior to their conventional counterparts when total photon resources are held constant.

Since 2007, when this architectural study was originally reported, a great many optical demonstrations of compressive imaging have taken place [19–32]. Nearly all of these can be classified as some combination of the three architectures described here. For example, the use of a Toeplitz projection matrix, **F**, leads to simple, low-cost optical implementations via either PSF engineering (e.g., employing a simple mask in the pupil of a conventional imager) [31] or coded aperture imaging (e.g., a lens-free technique frequently used for imaging in the gamma- and x-ray bands) [28]. Both of these systems are examples of the parallel architecture albeit with some restriction on the allowed projections and so the performance limits described herein may continue to be viewed as upper bounds. Another important trend that has emerged since our original work relates to the use of compressive imaging in dimensionally mismatched applications. For example, compressive multispectral imaging, polarimetric imaging, and/or compressive video imaging inherently involves one or more object space dimensions that are not matched to the 2D spatial array of photodetectors available in a traditional image plane. Conventional approaches typically multiplex in space (i.e., pixel polarization filter arrays for polarimetric imaging) and/or time (e.g., push broom sensors for multispectral imaging), whereas recent compressive solutions have sought to generate joint projections across these dimensions. Once again, we find that these very powerful sensor concepts may be decomposed into some combination of sequential, parallel, and photon-sharing components. Continuing work on these architectural concepts and their various combinations will therefore be important as the very promising field of compressive imaging continues to evolve.

References

1. M.A. Neifeld and J. Ke, Optical architectures for compressive imaging, *Applied Optics*, 46(22), 5293–5303, 2007.
2. D. Donoho, Compressed sensing, *IEEE Transactions on Information Theory*, 52, 1289–1306, 2006.
3. E.J. Candès and M. Wakin, An introduction to compressive sampling, *IEEE Signal Processing Magazine*, 25(2), 21–30, 2008.
4. D. Takhar et al., A new compressive imaging camera using optical domain compression, *Proceedings of SPIE 6065, Computational Imaging IV*, 606509, February 2, 2006, doi: 10.1117/12.659602.
5. M.A. Neifeld and P. Shankar, Feature-specific imaging, *Applied Optics*, 42, 3379–3389, 2003.

6. H. Pal and M.A. Neifeld, Multispectral principal component imaging, *Optics Express*, 11, 2118–2125, 2003.
7. D.L. Marks, R. Stack, A.J. Johnson, D.J. Brady, and D.C. Munson, Cone-beam tomography with a digital camera, *Applied Optics*, 40, 1795–1805, 2001.
8. W.T. Cathey and E.R. Dowski, New paradigm for imaging systems, *Applied Optics*, 41(29), 6080–6092, 2002.
9. H.S. Pal, D. Ganotra, and M.A. Neifeld, Face recognition by using feature-specific imaging, *Applied Optics*, 44, 3784–3794, 2005.
10. I.T. Jolliffe, *Principal Component Analysis*, Springer-Verlag, New York, 2002.
11. E. Marszalec, B. Martinkauppi, M. Soriano, and M. Pietikäinen, A physics-based face database for color research, *Journal of Electronic Imaging*, 9(1), 32–38, 2000.
12. H.H. Barrett and K.J. Myers, *Foundations of Image Science*, John Wiley & Sons, Inc., Hoboken, NJ, 2004.
13. J. Ke, M.D. Stenner, and M.A. Neifeld, Minimum reconstruction error in feature-specific imaging, In Z. Rahman, R.A. Schowengerdt, and S.E. Reichenbach (Eds.), *Proceedings of SPIE-The International Society for Optical Engineering*, Vol. 5817, pp. 7–12, 2005.
14. W. Wenzel and K. Hamacher, A stochastic tunneling approach for global minimization of complex potential energy landscapes, *Physical Review Letters*, 82, 3003–3007, 1999.
15. J. Haupt and R. Nowak, Signal reconstruction from noisy random projections, *IEEE Transactions on Information Theory*, 52(9), 4036–4048, 2006.
16. M.F. Duarte, M.A. Davenport, M.B. Wakin, and R.G. Baraniuk, Sparse signal detection from incoherent projections, *Proceedings of the International Conference on Acoustics, Speech, and Signal Processing—ICASSP*, Toulouse, France, May 2006.
17. E. Candes, J. Romberg, and T. Tao, Stable signal recovery from incomplete and inaccurate measurements, *Communications on Pure and Applied Mathematics*, LIX, 1207–1223, 2006.
18. S.S. Chen, D.L. Donoho, and M.A. Saunders, Atomic decomposition by basis pursuit, *Society for Industrial and Applied Mathematics*, 43, 129–159, 2001.
19. H.H. Barrett, D.W. Wilson, and B.M.W. Tsui, Noise properties of the EM algorithm: I. theory, *Physics in Medicine and Biology*, 39, 833–845, 1994.
20. M.E. Gehm, R. John, D.J. Brady, R.M. Willett, and T.J. Schulz, Single-shot compressive spectral imaging with a dual-disperser architecture, *Optics Express*, 15(21), 14013–14027, 2007.
21. R. Horisaki and J. Tanida, Preconditioning for multiplexed imaging with spatially coded PSFs, *Optics Express*, 19(13), 12540–12550, 2011.
22. M. Shankar, N.P. Pitsianis, and D.J. Brady, Compressive video sensors using multichannel imagers, *Applied Optics*, 49(10), B9–B17, 2010.
23. T.-H. Tsai and D.J. Brady, Coded aperture snapshot spectral polarization imaging, *Applied Optics*, 52(10), 2153–2161, 2013.
24. P. Llull, X. Liao, X. Yuan, J. Yang, D. Kittle, L. Carin, G. Sapiro, and D.J. Brady, Coded aperture compressive temporal imaging, *Optics Express*, 21(9), 10526–10545, 2013.
25. M. Cho, A. Mahalanobis, and B. Javidi, 3D passive integral imaging using compressive sensing, *Optics Express*, 20(24), 26624–26635, 2012.

26. A. Mahalanobis, R. Shilling, R. Murphy, and R. Muise, Recent results of medium wave infrared compressive sensing, *Applied Optics*, 53(34), 8060–8070, 2014.
27. R.F. Marcia, Z.T. Harmany, and R.M. Willett, Compressive coded apertures for high-resolution imaging, *Proceedings of SPIE 7723, Optics, Photonics, and Digital Technologies for Multimedia Applications*, 772304, April 30, 2010.
28. R. Willett, R. Marcia, and J. Nichols. Compressed sensing for practical optical systems: A tutorial, *Optical Engineering*, 50(7), 072601, 1–13, 2011.
29. R. Marcia, Z. Harmany, and R. Willett, Compressive coded aperture imaging, *Proceedings of SPIE 7246, Computational Imaging VII*, 72460G, February 2, 2009, doi: 10.1117/12.803795.
30. A. Ashok and M.A. Neifeld, Compressive imaging: Hybrid measurement basis design, *Journal of the Optical Society of America A*, 28(6), 1041–1050, 2011.
31. V. Treeaporn, A. Ashok, and M.A. Neifeld, Space–time compressive imaging, *Applied Optics*, 51(4), A67–A79, 2012.
32. A. Ashok and M.A. Neifeld, Point spread function engineering for iris recognition system design, *Applied Optics*, 49(10), B26–B39, 2010.

6 Terahertz Imaging with Compressed Sensing

Yao-Chun Shen, Lu Gan, and Hao Shen

Contents

Introduction

Terahertz imaging is a newly developed imaging technique that utilizes the terahertz region of the electromagnetic spectrum (Saeedikia 2013). The definition of terahertz region varies widely, and generally speaking, terahertz region spans the frequency range between the mid-infrared and the millimeter/microwave. The center portion of terahertz radiation (300 GHz to 3 THz) has an unique combination of many useful properties (Zeitler and Shen 2013): (1) like infrared radiation, terahertz radiation gives rise to individual "fingerprints" spectra for many crystalline materials including explosives and active pharmaceutical ingredients; thus, terahertz imaging can be used for material characterization and identification; (2) like microwave radiation, terahertz radiation can penetrate a wide variety of nonconducting materials including clothing, paper, cardboard, wood, masonry, plastic, and ceramics; thus, terahertz imaging can be used for nondestructive and quantitative testing and evaluation; and (3) like infrared and microwave, terahertz radiation is nonionizing; thus, terahertz imaging does not give rise to safety concerns. At higher frequencies above a few terahertz into the infrared where there are even richer spectroscopic features, absorption and scattering limits penetration through barrier and packaging materials. At lower frequencies, such as the millimeter waves below, say, 300 GHz, there are no characteristic vibrational modes in solids, leading to featureless spectra that cannot be used for material identification.

Historically, measurement in terahertz region has been difficult owing to lack of efficient, coherent, and compact terahertz sources and detectors (Ferguson and Zhang 2002). These characteristics for the sources can be found in the common microwave-frequency sources such as transistors or RF/MW antennas and in devices working in the visible and infrared range like semiconductor laser diodes. However, it is not possible to adopt these technologies for operation in terahertz region without a significant reduction in power and efficiency. At the lower extreme of terahertz frequency range, the generated power by solid-state electronic devices, such as diodes, has roll-offs of $1/f^2$ due to reactive-resistive effects and long transit times (Armstrong 2012). On the other hand, optical devices, such as diode lasers, do not perform well at terahertz range limit because of the lack of materials with adequately small band gap energies (Williams 2007). Hence, the term "terahertz gap" is phrased to explain the infancy of this band compared to well-developed neighboring infrared and microwave spectral regions.

With the advances in semiconductor devices and femtosecond laser technologies, the past 20 years have seen a revolution in terahertz systems (Ferguson and Zhang 2002). Of particular significance is the development of terahertz time-domain imaging systems (Hu and Nuss 1995; Jepsen et al. 2011). For convenience and simplicity, the term "terahertz imaging" will be used throughout this chapter to represent "terahertz time-domain imaging," which is based on the coherent generation and detection of short pulses of broadband terahertz radiation by using ultrafast femtosecond laser. The use of pulsed radiation and the associated coherent detection scheme preserves the time-gated phase information, upon which terahertz imaging technology has been developed for characterizing the chemical compositions and the physical structures of a sample quantitatively and nondestructively.

The enormous inherent potential of the terahertz imaging technology led to a rapid development of terahertz systems, and the availability of commercial terahertz products has opened up many exciting opportunities in science and industry. A number of imaging application examples have been reported: metal contacts of a packaged integrated circuit chip, as well as the leaf structure of plants (Hu and Nuss 1995); polymer composites and flames (Mittleman et al. 1996); breakfast cereals and chocolate (Mittleman et al. 1999); tree-ring analysis (Jackson et al. 2009); medical tissues (Pickwell and Wallace 2006); analysis of polymer samples (Jansen et al. 2010); and pharmaceutical tablets (Zeitler et al. 2007; Zeitler and Gladden 2009; Shen 2011), among many other examples.

However, from the technology point of view, one considerable limitation of terahertz imaging is its long data acquisition time. Multielement detector schemes such as microbolometer arrays (Lee and Hu 2005) and electro-optic sampling with high-performance CCD cameras (Wu et al. 1996) provide accurate and real-time terahertz images. However, these imaging systems often require high-power sources and/or expensive complex detectors that lack the sensitivity of single-element detectors. Therefore, most terahertz imaging experiments have to be performed by mechanically raster scanning the object of interest and the image is mapped out in a pixel-by-pixel fashion using a single detector. This technique has the benefit of high spatial resolution and accurate imaging. However, a complete image may take minutes or even hours to acquire, depending on the total number of pixels and the required spectral range/resolution. This is a major limiting factor for real-time applications such as in vivo medical and security imaging or for online industrial process monitoring.

In this chapter, we will first briefly describe the terahertz imaging technology and then focus specifically on the experimental implementation of *compressive* terahertz imaging, aiming for high-speed and high-resolution terahertz imaging.

Terahertz Imaging

Coherent Generation and Detection of Terahertz Radiation

Figure 6.1 shows the schematic diagram of a typical experimental configuration for the coherent generation and detection of terahertz radiation using an ultrafast laser. The laser light from a Ti:sapphire femtosecond laser is split into two parts using a beam splitter: a pump laser beam for terahertz generation and a probe laser beam for terahertz detection. To date, the most commonly used device for generating ultrashort terahertz pulses is photoconductive antenna (Auston 1975). When the femtosecond laser light illuminates the biased GaAs photoconductive antenna, electron–hole pairs are generated in GaAs semiconductor crystal because the photon energy of the laser (1.55 eV at 800 nm) is higher than the band gap of GaAs (1.43 eV). These photo-generated carriers are then accelerated by the applied electric field, producing an ultrafast transient current, which generates a pulse of electromagnetic radiation in the terahertz frequency range. The terahertz pulses emitted from the GaAs photoconductive antenna are collimated and focused by a pair of off-axis parabolic mirror onto the sample. The transmitted terahertz pulses are then collected by another pair of off-axis parabolic mirror and focused onto a ZnTe crystal for electro-optic detection. The current

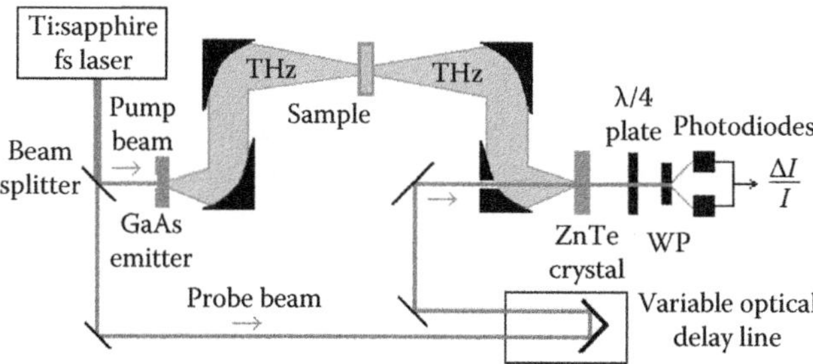

Figure 6.1 Experimental arrangements for coherent terahertz generation and detection. A biased GaAs photoconductive antenna is used for terahertz generation and a ZnTe crystal is used for terahertz detection via free-space electro-optic sampling. WP, Wollaston prism.

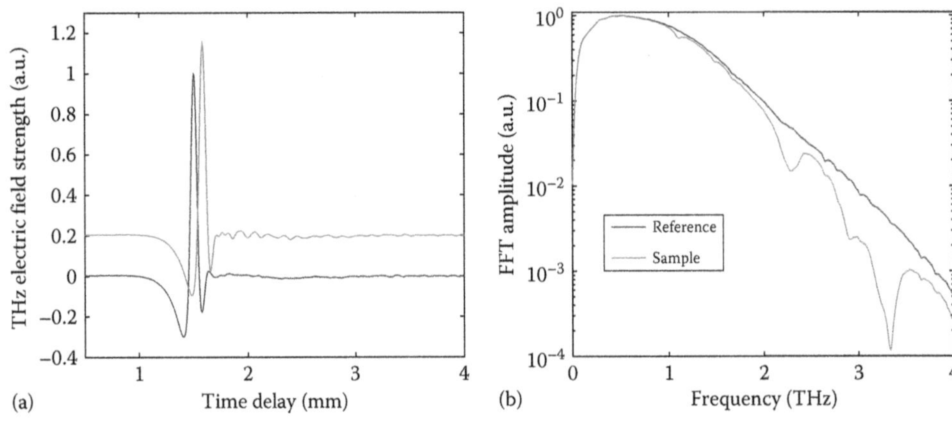

Figure 6.2 (a) Terahertz waveforms measured after propagating through a sample pellet (360 mg PE and 40 mg mannitol) and a reference pellet (360 mg PE); (b) their corresponding FFT amplitude.

difference from the two photodiodes ($\Delta I/I$) is directly proportional to the terahertz electric field strength (Wu and Zhang 1995).

In all measurements, the transient terahertz electric field is recorded as a function of the time delay between the terahertz pulse and the probe pulse using a variable delay stage. Usually, the sample chamber is either purged with dry nitrogen gas or evacuated throughout the measurement to reduce the effects of water vapor absorption. As an example, Figure 6.2a shows typical time-domain terahertz waveforms measured for a sample (a pellet made of 360 mg polyethylene and 40 mg mannitol) and a reference (a pellet made of 360 mg polyethylene), respectively, while Figure 6.2b shows their corresponding fast Fourier transform amplitude. The time-domain terahertz waveform can be used for the nondestructive analysis of pharmaceutical tablet coatings (Shen and Taday 2008) and for the clinical imaging of basal-cell skin carcinoma ex vivo and in vivo (Pickwell and Wallace 2006). The terahertz spectral data can be used for chemical mapping of heterogeneous mixtures and for detecting and identifying explosive (Shen et al. 2005a).

Terahertz Imaging Data and Sparsity

Most terahertz imaging measurements are usually performed by raster scanning the sample relative to a focused terahertz beam by using either a standard XY-stage or a six-axis robot system (Shen and Taday 2008). Time-domain terahertz waveform is taken at many points mapped over the surface of a sample, and at each pixel a terahertz waveform is recorded as a function of optical time delay. Thus, terahertz imaging experiment yields three-dimensional data $D(x, y, t)$ where the x- and y-axes describe vertical and horizontal dimensions of the sample and the t-axis represents the time-delay (depth) dimension. Applying a Fourier transform to the time-domain terahertz waveform yields frequency-domain terahertz spectral data for each pixel. This gives in total two set of three-dimensional data cubes, $D(x, y, t)$ and $D(x, y, \nu)$, providing vertical and horizontal information (spatial profile of a sample) and the time information (depth profile of a sample) as well as the frequency information (terahertz spectrum of a sample).

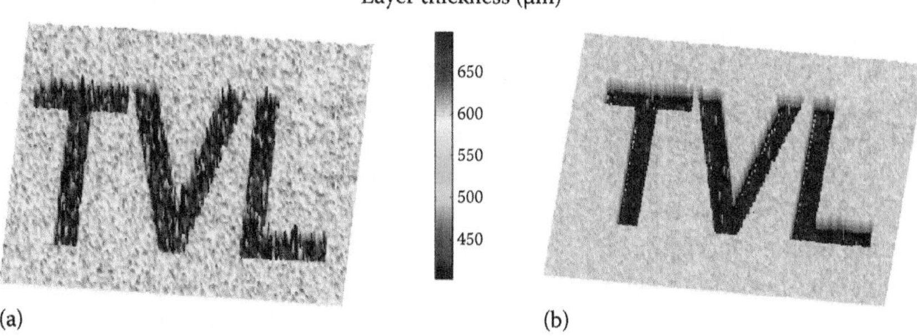

Figure 6.3 Layer thickness map extracted from the raw (a) and compressed (b) terahertz data set. After PCA compression, the size of the terahertz data has reduced from 160 MB to less than 10 MB, whereas all useful information including layer thickness information was retained.

It is well known that most natural images are sparse in spatial domain; thus, compressed sensing concept can be applied in spatial domain. Terahertz imaging data have an extra dimension in time domain, in addition to the spatial domain. Principal components analysis (PCA) is a classical technique to reduce the dimensionality of a data set by transforming to a new set of variables (the principal components) that summarize the features of the data (Jolliffe 1986 and see Chapter 5). Shen et al. (2005b) applied PCA to study the sparsity of terahertz image data, which has 200×200 pixels and each pixel has a time-domain terahertz waveform of 512 points. It was found that the time-domain terahertz data could be compressed dramatically while retaining all essential information on layer thickness and interface quality (Shen et al. 2005b). Figure 6.3 compares the layer thickness information extracted from the original terahertz data and the compressed terahertz data. The terahertz imaging data size has been reduced from 160 MB to less than 10 MB without losing any essential information on layer thickness. In fact, the signal-to-noise ratio of the terahertz waveforms after compression was improved over the original data. This is because the noise appears randomly at different depths for different pixels and therefore is not selected by the PCA algorithm, which only summarizes the common features of terahertz waveforms at all pixels. Therefore, three-dimensional terahertz data are compressible in both time domain and spatial domain.

Sampling Operator Design and Signal Recovery

Overview

Suppose that we want to sample an $I_r \times I_c$ image with $N = I_r\, I_c$ pixels in total. Let $f \in R^N$ denote the vector representation of the input image. Assume that we are allowed to take only $M(\ll N)$ linear measurements of f described as in Equation 6.1:

$$g_{M\times 1} = \Phi_{M\times N} f_{N\times 1} + n_{M\times 1} \tag{6.1}$$

where

Φ is an $M \times N$ sensing matrix and $g \in R^M$
n is the noise vector

Since $M \ll N$, the reconstruction of f from g is generally *ill-posed*. The compressed sensing theory is based on the fact that the original image signal f can be approximated with only $K \ll N$ coefficients in a known transformation domain Ψ (e.g., the wavelet and the discrete cosine transform). Under certain conditions, f can be reconstructed with only $M = O(K \log N)$ samples.

In compressive imaging, two fundamental questions include (1) the design and implementation of sampling operator Φ and (2) the development of fast reconstruction algorithms. The "Sampling Operator Φ" and "Signal Recovery/Reconstructions" sections will summarize existing work of compressive time-domain terahertz imaging on design of Φ and reconstruction algorithms, respectively.

Sampling Operator Φ

Some desirable properties for the sampling operator include (a) *near optimal performance* (the number of measurements should be as small as possible), (b) *fast computation* (due to the large data size in imaging applications, a fast computable Φ is desirable for both sensing and reconstruction algorithms), (c) *memory efficient* (the storage of Φ requires small memory size), and (d) *hardware friendly* (Φ can be easily implemented in the terahertz imaging systems).

In classical compressed sensing theory, Φ is usually chosen as a random binary matrix with Bernoulli entries (Chan et al. 2008a,b), in which the entries of Φ are random selected as 0 and 1 with equal probability. However, it has been noted that random projections do not work well at low signal-to-noise ratios or at low sampling rates (Haupt and Nowak 2006; Weiss et al. 2007). To address this issue, a binary valued Φ was optimized to approximate the Karhunen–Loeve transform (Shen et al. 2009), which effectively reduced the number of required samples.

In all aforementioned work (Chan et al. 2008a; Shen et al. 2009), a set of independent two-dimensional random binary masks are required, each of which corresponds to one row of Φ. As each row of Φ is independent, the imaging speed is limited by the slow mechanical translation of one mask to another (Chan et al. 2008a; Shen et al. 2009). An alternative approach is to use a spinning disk-based solution (Shen et al. 2012), where Φ is approximated as a block circulant matrix (Rauhut 2010). In this way, Φ can be implemented with a single rotating disk for fast compressive imaging. Details for physical implementation of the sampling operators can be found in the "Experimental Implementation of Compressive Terahertz Imaging" section.

Signal Recovery/Reconstructions

For reconstruction of terahertz images, a nonlinear optimization algorithm through the minimization of total variation (TV) was used in Chan et al. (2009). Unfortunately, TV minimizations require fairly heavy computations. To meet real-time imaging requirement, linear *minimum mean square error* (MMSE) estimation (Gan 2007) was employed to produce fast initial solutions (Shen et al. 2009). In fact, for practical compressive terahertz measurement, the linear estimator often offers comparable performance to the nonlinear image at much lower computational cost, as we will show in the next section. A more sophisticated algorithm (Xu and Lam 2010) made use of prior information about the phase and the correlation between the spatial distributions of the amplitude and phase in terahertz signal. More recently, a spatiotemporal dictionary learning algorithm was developed by Abolghasemi et al. (2015) for both terahertz signal reconstruction and estimation.

Experimental Implementation of Compressive Terahertz Imaging

Compressive Terahertz Imaging Using a Set of Independent Masks

Typical compressive terahertz imaging architectures (single-pixel terahertz camera) involve four key components: a terahertz light source, terahertz imaging optics, a single-element terahertz detector, and a terahertz spatial light modulator (SLM). The experiment operator (masks) discussed in previous sections is achieved by using a terahertz SLM, which, for binary encoding, selectively passes portions of the image to the detector and blocks others. The commercially available SLMs such as digital micromirror (Dudley et al. 2003) and liquid crystal (Johnson et al. 1993) systems do not operate at terahertz frequencies. Mechanical masks thus were used in the first experimental implementation of compressive terahertz imaging (Chan et al. 2008a,b). As shown in Figure 6.4, a collimated free-space terahertz wave travelling from an object to a single-point detector was spatially modulated by the insertion of a series of planar two-dimensional masks. Each binary mask comprised a random checkerboard pattern that could each either transmit or block the terahertz radiation. By recording the terahertz electric field in the presence of each mask whose pattern is known, a two-dimensional image of the object was reconstructed.

Chan et al. (2008a) used a set of 600 random patterns printed in copper on standard printed circuit boards (PCBs). Each pattern contains 32×32 pixels and the size of each pixel is 1×1 mm^2. A "copper" pixel corresponds to pixel value "0" on the random pattern, while a "noncopper" pixel corresponds to the value "1." Copper pixels block terahertz radiation while noncopper pixels transmit terahertz radiation since the PCB material is fairly transparent to the terahertz radiation. A motorized translation stage was used to change one mask to another, and for each random pattern one terahertz waveform consisting of the superposition of the terahertz radiation transmitted through all of the noncopper pixels was measured. It was shown that the system is capable of recovering a 32×32 image of a

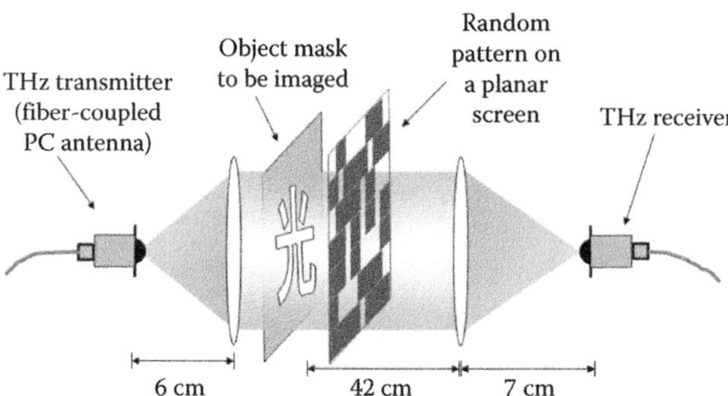

Figure 6.4 Schematic diagram of a typical compressive terahertz imaging system with a single-element detector. An approximately collimated beam from the terahertz transmitter illuminates an object mask and is partially (50%) transmitted through a random pattern of opaque pixels. The random patterns, the focusing lens and the receiver, are placed in order to most efficiently focus the terahertz beam onto the receiver antenna. (Reproduced from Chan, W.L. et al., *Appl. Phys. Lett.*, 93, 121105, 2008a. With permission.)

rather complicated object with only 300 measurements (e.g., a sampling rate of 30%) by using the compressed sensing theory. This represents a significant reduction in the number of measurements required for image reconstruction, which is advantageous for speeding up the image acquisition as compared to traditional raster scan terahertz imaging systems.

A major advantage of terahertz time-domain imaging is that the transient terahertz electric field rather than the terahertz radiation intensity is measured as a function of time. This coherent detection scheme yields terahertz signals with excellent signal-to-noise ratio and high dynamic range. Furthermore, an entire terahertz waveform is recorded at each pixel and Fourier transforming the measured terahertz time-domain waveform provides terahertz frequency-domain spectral information of the sample, thus enabling terahertz *spectroscopic* imaging for material identification. Shen et al. (2009) reported the first terahertz *spectroscopic* imaging based on compressed sensing concept. Figure 6.5 shows the experimental arrangement that is similar to that typically used for the coherent generation and detection of broadband terahertz radiation using a femtosecond laser. The terahertz emitter is a biased low-temperature-grown GaAs photoconductive emitter (Auston 1975), and the terahertz detection is achieved via electro-optic detection by using a ZnTe nonlinear crystal (Wu and Zhang 1995).

One of the distinct features of this compressive terahertz *spectroscopic* imaging system is the use of a set of optimized binary masks that were constructed from self-supported copper tape. The lack of a supporting substrate eliminates possible terahertz absorption/dispersion or phase delays in propagation through the supporting substrate materials of the transparent pixels. Therefore, the masks are truly binary in that the copper pixels are totally opaque to terahertz radiation while the noncopper pixels (air) are totally transparent to terahertz radiation, making this design ideal for broadband *spectroscopic* imaging applications. In addition, the set of 40 masks (Figure 6.6) were optimization based on a general isotropic two-dimensional model, rather than being based on a training set of images. Thus, the mask set is generically applicable to a wide range of samples.

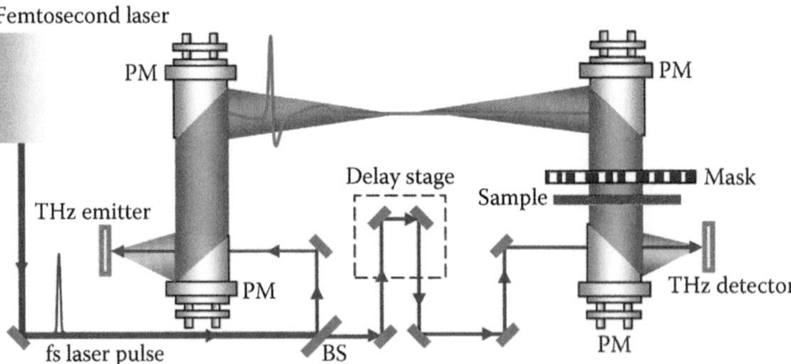

Figure 6.5 Experimental arrangements for using compressive sampling. Both the sample and the binary masks are placed in a collimated terahertz beam; thus, in principle, only one pair of parabolic mirrors is required for imaging using the compressive sampling technique. However, two pairs of parabolic mirrors are retained in the experimental arrangement so that both compressive and conventional THz spectroscopic imaging measurement can be performed without the need to modify the experimental configuration.

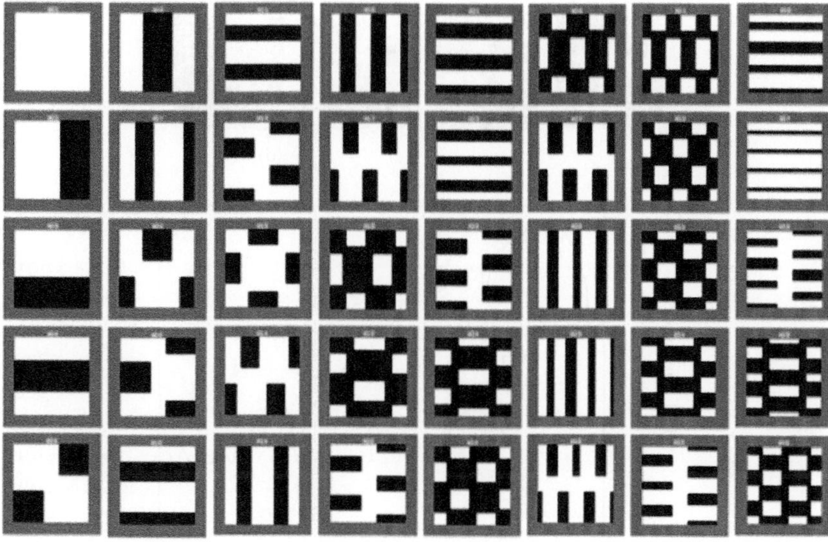

Figure 6.6 A set of 40 optimized mask set used for compressive terahertz spectroscopic imaging. Each of the 40 masks comprised 20×20 pixels and each pixel had a size of 2.0 mm×2.0 mm, thus providing a 40.0 mm×40.0 mm imaging area.

For compressive terahertz imaging measurement, a sample comprising regions of polyethylene, lactose, and copper tape (as shown in the inset of Figure 6.7) was placed in the collimated terahertz beam path together with one of the masks. The terahertz electric field was recorded as a function of time delay for each of the 40 masks. In all measurements, the variable delay stage, which provides the time delay between the terahertz pulse and the probe pulse, was scanned over a distance of 5 mm, providing a spectral resolution of 0.03 THz. A full set of compressive terahertz imaging measurements yield 40 waveforms and each terahertz waveform has 512 data points with a time interval between each data point of 10 μm.

For terahertz image reconstruction, all 40 THz waveforms were first Fourier transformed into the frequency domain. The resultant terahertz amplitude spectra were subsequently used for image reconstruction using the MMSE linear estimation (Haykin 1996). A good quality terahertz image could be reconstructed at any selected frequency in the range of 0.3–3.0 THz. The upper frequency is currently limited by the frequency response of the specific imaging system (the terahertz emitter and receiver, terahertz optics, and femtosecond laser) used, while the lower frequency range is mainly limited by the pixel size of the metal masks, which only allows efficient transmission of short wavelength (high frequency) terahertz radiations.

As an example, Figure 6.7a through c shows the terahertz images reconstructed at frequencies of 0.50, 0.54, and 1.38 THz, respectively. At 0.5 THz, the reconstructed image shows two bright areas corresponding to the polyethylene and lactose regions, indicating both materials are relatively transparent to terahertz radiation at this low frequency (Figure 6.7a). The dark areas correspond to the copper tape, which is opaque to terahertz radiation. Lactose monohydrate powder has two well-defined strong absorption features at 0.54 and 1.38 THz

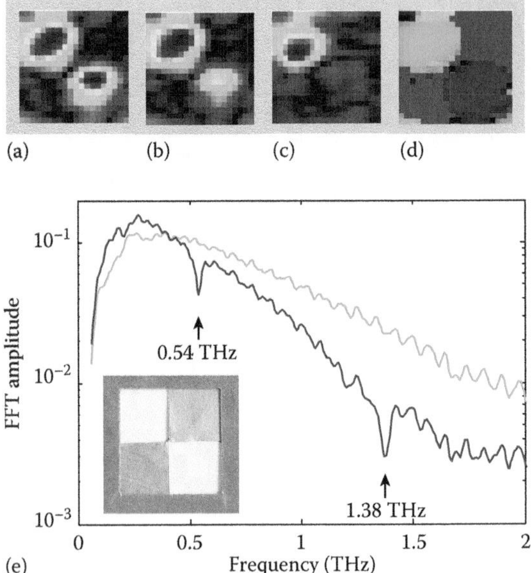

Figure 6.7 (a–c) Reconstructed terahertz images of the sample at 0.50, 0.54, and 1.38 THz, respectively. Each image is 40 mm×40 mm. (d) RGB chemical map of the sample where red is assigned to lactose, green to polythene, and blue to regions of no transmission (copper areas). (e) Terahertz spectra of polyethylene (upper trace) and lactose (lower trace). The inset shows a photograph of the sample that is made of copper tape with two square holes (each 20 mm×20 mm). A 3.0-mm-thick polyethylene pellet is placed at the top-left square while a 3.2-mm-thick lactose pellet is placed at the bottom-right square. (Reproduced from Shen, Y.C., *Appl. Phys. Lett.*, 95, 231112, 2009. With permission.)

(Brown et al. 2007). Consequently, at these two frequencies, the reconstructed sample image (Figure 6.7b and c) shows a much weaker transmission for the lactose (bottom-right quadrant) while there is reasonable terahertz transmission for the polyethylene (top-left quadrant).

At each pixel, a complete terahertz spectrum can be reconstructed. Figure 6.7e shows the reconstructed terahertz spectra of the lactose and polyethylene, determined by averaging over an area of 4×4 pixels at the center of the lactose (top-left square) and polyethylene (bottom-right square) region. Two well-defined absorption features are observed in the lactose spectrum at 0.54 and 1.38 THz, respectively, which agree well with the published data (Brown et al. 2007). The terahertz spectrum at each pixel of the image can be used to calculate spatially resolved chemical maps of the sample using cosine correlation mapping. For a better visualization of chemical distributions of the sample, the extracted chemical maps are displayed as an RGB map (Shen and Taday 2008). Figure 6.7d demonstrates that in this way the chemical distribution of the lactose and polyethylene can be clearly distinguished where pure R, G, B corresponds to pure lactose, polyethylene, and copper pixels, respectively (Shen et al. 2009).

In summary, the compressive terahertz imaging is advantageous in that it performs image compression simultaneously with image sampling by modulating the spatial profile of the terahertz beam with a set of known mask patterns. As demonstrated, a sampling rate of 10%–30% is sufficient to fully reconstruct an image of reasonable quality. This imaging scheme can speed up the acquisition process because the number of required measurement is significantly reduced. However, a major limitation of the compressive terahertz imaging setup reported (Chan et al. 2008a; Shen et al. 2009) is the slow translation of one mechanical mask pattern to another, which was done either manually or using a motorized stage. As will be shown in the following sections, this issue can be addressed by using a spinning disk (Shen et al. 2012) or terahertz SLMs that can be controlled either electronically or optically (Chan et al. 2009; Rahm et al. 2012; Shrekenhamer et al. 2013).

Compressive Terahertz Imaging Using a Spinning Disk

Most compressive imaging systems use a random Bernoulli operator, in which the entries of Φ are random selected as 0 and 1 with equal probability. In such systems, a set of independent two-dimensional random binary masks are required, each of which corresponds to one row of Φ. Despite their theoretical advantages, there are a couple of practical limitations. First, as each row of Φ is independent, the imaging speed is limited by the slow mechanical translation of one mask to another (Chan et al. 2008a; Shen et al. 2009). Second, a fully random binary operator incurs high computational complexity and huge memory, especially for high-resolution imaging.

To solve these problems, a single rotating disk (similar to the Nipkow disk used in confocal microscopy) (Xiao et al. 1988) is proposed for fast compressive terahertz imaging. This imaging scheme allows the measurement to be done automatically and continuously and thus is suitable for real-time terahertz imaging applications (Shen et al. 2012).

Figure 6.8a illustrates the schematic diagram of a spinning disk. This spinning disk has random binary patterns, where the black regions are stainless steel that block terahertz radiation while the white regions are through holes that transmit terahertz radiation. The spinning disk has a radius of 95 mm and the through holes have a radius of 2 mm. The use of circular-shaped holes minimizes potential terahertz scattering at the sharp edges/corners of square-shaped holes.

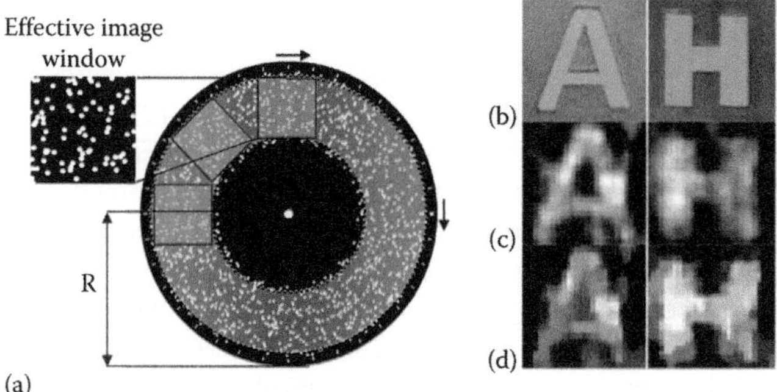

Figure 6.8 (a) Schematic diagram of a spinning disk. When spinning the disk, a set of masks is obtained from the rectangular window (in blue) with different binary patterns. The inset (top-left) shows one such a mask pattern, which also defines the effective image window of a size of 32 mm × 32 mm. (b) Photographs of the samples used in the terahertz imaging experiment, and their corresponding terahertz images reconstructed using MMSE linear estimation (c) and the TV-min nonlinear reconstruction algorithm (d). Each image has 32 × 32 pixels and was reconstructed from 160 measurements.

In addition, the mask pattern is truly binary because the stainless steel totally blocks terahertz radiation while the holes totally transmit terahertz radiation. The holes also transmit optical beam thus this spinning disk configuration is application to not only compressive terahertz imaging but also compressive optical and infrared imaging.

During the imaging process, the whole disk is covered except for a small fixed rectangular area, which also defines the effective imaging area. The effective image window has a dimension of 32 mm × 32 mm, with each pixel of size 1 mm × 1 mm. Hence, the reconstructed image is of size 32 × 32 pixels (1024 pixels in total). The spatial resolution of the image is mainly limited by the pixel size, which itself is related to the rotation angle steps size and the size of the through holes of the disk. Rotating the disk will generate a set of binary patterns on the fixed imaging window, and these known binary patterns are subsequently used for image reconstruction.

In order to validate this spinning disk approach and the robustness of the corresponding reconstruction algorithms, extensive simulations are performed using both MMSE linear estimation (Gan 2007; Ke and Neifeld 2007; Chapter 5) and the TV minimization (TV-min) nonlinear reconstruction algorithm (Candès et al. 2006). As shown in Figure 6.9, most of the reconstructed images can be easily recognized except ones using TV-min nonlinear reconstruction algorithm at relatively low signal-to-noise ratio. Note that only 120 measurements (each of these 120 effective masks is obtained by spinning the disk for each 3°) were necessary to obtain an image of 1024 pixels. This represents a nine times reduction in measurement number and thus measurement time. The quality of the reconstructed images becomes better when increasing the number of measurements or the signal to additive Gaussian white noise ratio. Compared with MMSE linear estimation, the TV-min nonlinear reconstruction algorithm provides much better reconstructed images in most of cases, especially in the presence of high signal to additive Gaussian noise ratio. However, in relatively low signal to additive Gaussian noise ratio, MMSE may work better. Another interesting finding is that in the realistic experiment where large noises might present, further increase in the measurement number may actually decrease the quality of the reconstructed images (Shen 2012).

With these numerical simulation results, a compressive terahertz imaging system using a spinning disk was developed (Shen et al. 2012). Figure 6.10 shows the schematic diagram of general experimental setup with Figure 6.11 showing a photograph of the compressive terahertz imaging system using a spinning disk. Terahertz radiation from a terahertz emitter was collimated using a parabolic mirror. After propagating through the sample and the spinning disk, the terahertz beam was then focused by using another parabolic mirror onto the terahertz receiver. The spinning disk is controlled by a motorized rotation stage. When rotating the disk, a new binary mask pattern is formed on the rectangular window at each rotation angle and in such a way the terahertz beam is spatially modulated. In the terahertz experiments, the maximum terahertz signal (peak position) was first found by varying the time delay between terahertz pulse and the optical probe pulse using a variable delay stage. The position of the variable delay stage was then fixed at the terahertz peak position, and the terahertz signal was recorded as a function of the rotation angle of the spinning disk, which was driven by a motorized rotation stage.

Owing to the limited signal-to-noise ratio of the specific terahertz imaging system used, the rotation speed of the spinning disk was 5°/s and it took about

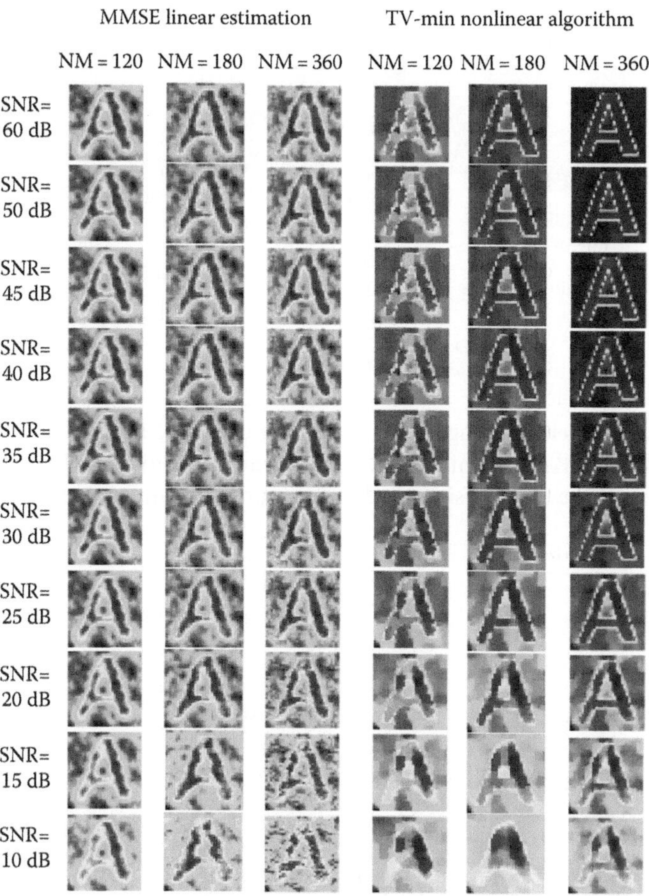

Figure 6.9 The simulation results for reconstructing an image of English character "A" using the spinning disk configuration. Both MMSE linear estimation and TV-min nonlinear reconstruction algorithm are used. All images have 32×32 pixels. NM, number of measurements; SNR, signal to additive Gaussian noise ratio.

80 s to measure 1 THz image. For the same reason, 160 measurements were for image reconstruction, as further increase in the measurement number decreases the quality of the reconstructed terahertz images. In fact, Chan et al. (2008a) also reported that more measurements did not necessarily improve the quality of the reconstructed terahertz images even with full random binary operators. Figure 6.8b through d shows the experimental results for terahertz images of "A," and "H" samples that are made from copper tape on a plastic substrate, which is transparent to terahertz radiation. Although the quality of reconstructed images is not as good as the simulated ones, all these characters are still recognizable. In addition, it was found that the quality of the reconstructed terahertz images obtained using classical MMSE reconstruction and the TV-min optimization is similar. A number of factors might contribute to the relatively poor quality of the reconstructed terahertz images. This could be caused by the imperfection of the operator Φ in practical implementations (e.g., fabrication errors and misalignment of the spinning disk patterns). The limited signal-to-noise ratio and the signal

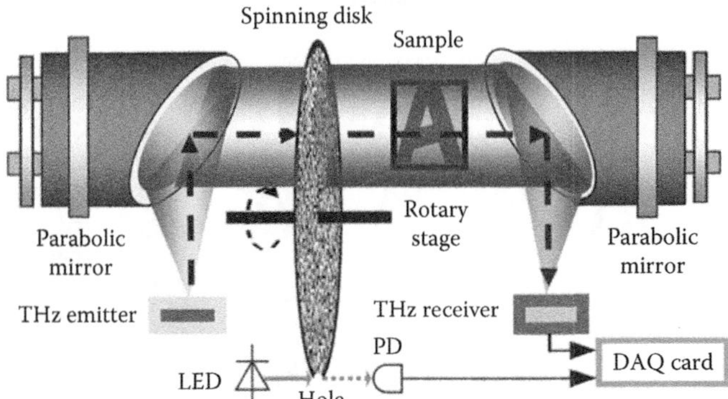

Figure 6.10 Schematic diagram of a spinning disk-based compressive terahertz imaging system. A light-emitting diode (LED) and a photodiode (PD) are used for synchronizing the mask position and the measured signal. DAQ, data acquisition.

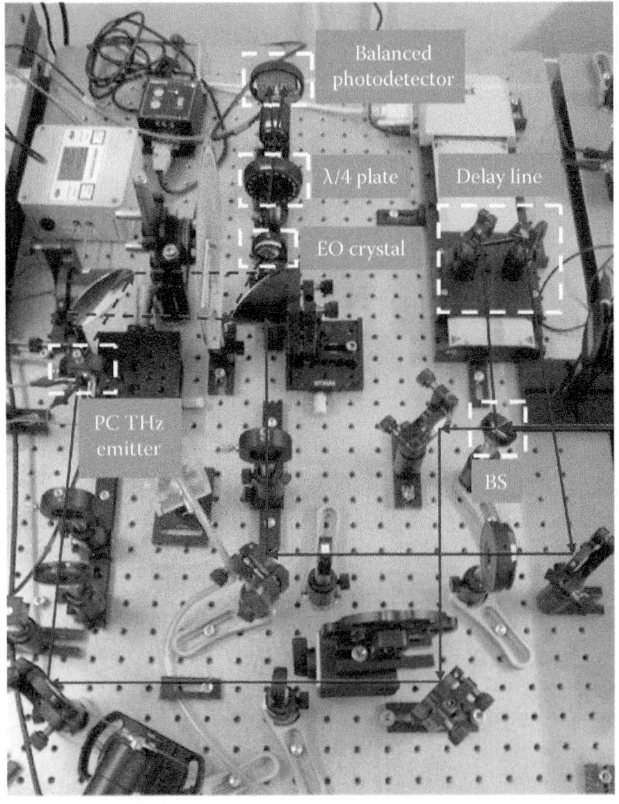

Figure 6.11 Photo of the compressive terahertz time-domain spectroscopic imaging system. BS, beam splitter; PC, photoconductive THz emitter; EO, electro-optic.

drift of the specific terahertz imaging system used may also play a role here. Nevertheless, the experimental results demonstrated the concept of using a continuously spinning disk approach for rapid compressive terahertz imaging.

Compressive Terahertz Imaging Using Reconfigurable SLMs

An SLM operating efficiently at terahertz frequencies is a crucial enabling component in any compressive terahertz imaging system. It allows the random spatial patterns to be encoded into the wave front of a terahertz beam. Mechanical masks such as a set of independent masks and a spinning disk have been proposed for compressive terahertz imaging. The mechanical mask approach is simple to implement and it has enabled the demonstration of the concept of compressive terahertz imaging (Chan et al. 2008a,b; Shen et al. 2009, 2012). However, the image acquisition rates are low because of the physical mask has to be changed mechanically either manually or by a motorized stage/motor. In order to achieve high-speed terahertz imaging, it would be advantageous to replace the slow mechanical mask with an electronically or optically reconfigurable SLM with fast switching speed.

Chan et al. (2009) reported the first electrically controllable SLM operating at terahertz frequencies. As shown in Figure 6.12, the device consists of a 4×4 pixel array, where each pixel is an array of subwavelength-sized split-ring resonator elements fabricated on a GaAs substrate. It was found that the spatial modulator has a uniform modulation depth of around 40% at the resonant frequency across all pixels. Each pixel can be independently controlled by applying an external voltage, thus the SLM patterns are electrically reconfigurable, making it an ideal SLM for fast compressive terahertz imaging.

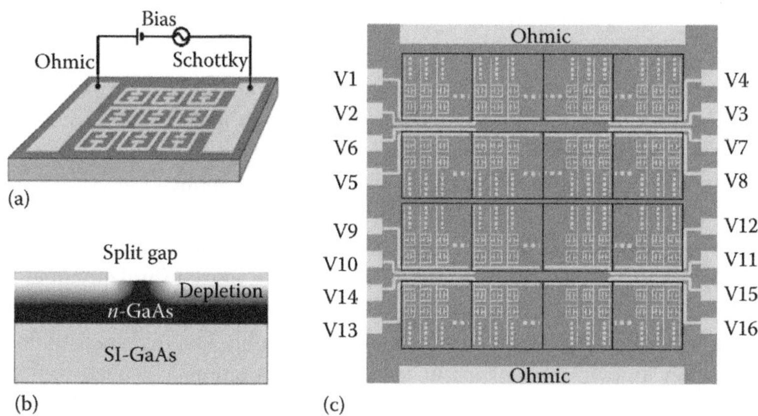

Figure 6.12 Schematic diagram of a terahertz SLM. (a) Each single pixel on the terahertz SLM contains a 4×4 mm^2, 2500 element array of metamaterial split-ring resonator. These elements are connected together with metal wires to serve as a metallic Schottky gate. An external voltage bias controls the substrate charge carrier density near the split gaps, tuning the strength of the resonance. (b) Diagram of the substrate and the depletion region near the split gap of a single split-ring resonator, where the gray scale indicates the free charge carrier density. (c) The terahertz SLM is a 4×4 array of individual pixels and each pixel is independently controlled by an external voltage. (Reproduced from Chan, W.L. et al., *Appl. Phys. Lett.*, 94, 213511, 2009. With permission.)

In a related work, Sensale-Rodriguez et al. (2013) proposed and experimentally demonstrated that arrays of graphene electro-absorption modulators can be employed as electrically reconfigurable patterns for compressive terahertz imaging applications. The active element of the proposed modulators consists of only single-atom-thick graphene but can achieve a modulation of the terahertz wave reflectance greater than 50% with a potential modulation depth approaching 100% (Sensale-Rodriguez et al. 2012). Since no mechanical parts are involved, these devices can potentially enable low-cost video-rate compressive terahertz imaging systems. In both works (Chan et al. 2009; Sensale-Rodriguez et al. 2013), the prototype devices presented contain only 4×4 pixels that are insufficient for most practical terahertz imaging applications. Nevertheless, it demonstrated the feasibility of developing reliable low-cost video-rate compressive terahertz imaging systems employing single detector.

In a further study, Watts et al. (2014) reported a new type of SLM that can be used for compressive terahertz imaging with only a single-pixel detector. The SLM device is based on dynamic metamaterial absorbers, which have demonstrated near-unity absorption across much of the electromagnetic spectrum (Watts et al. 2012). Figure 6.13 shows an electronically controlled 8×8 mask, where each pixel is composed of dynamic, polarization-sensitive metamaterial absorbers. The electromagnetic properties of the metamaterial can be tuned with an applied bias voltage, so each of the 64 pixels is individually and dynamically addressable, permitting operation as a real-time spatial mask for terahertz radiation. One of the distinct features of this new modulation technique is that it utilizes a lock-in detection scheme permitting imaging with [1, −1] mask values directly. Imaging with negative mask values is typically difficult to achieve with intensity-based SLM components. Using such a metamaterial SLM design and a lock-in detection scheme, high-quality terahertz images have been obtained at a sampling rate as low as 30%. The achieved frame rate is 1 image (8×8 pixels) per second with a terahertz source orders of magnitude lower in power than conventional terahertz imaging system using a focal-plane-array microbolometer detector (Lee and Hu 2005). The reported system is all solid state with no moving parts, allows the acquisition of high-frame-rate, high-fidelity images using a single terahertz detector, and thus is suitable for real-time terahertz imaging applications.

An alternative to the electrically controlled SLM is the optically reconfigurable SLM, which utilizes photo-generated carriers in semiconductors to spatially control the terahertz transmission of a semiconductor mask (Shrekenhamer et al. 2013). This was achieved by co-propagating a collimated optical laser beam and terahertz beam through a high-resistivity silicon wafer. The high-resistivity silicon wafer on its own is transparent to terahertz radiation. Upon absorption of above-the-band gap photons of an optical laser beam, photo-generated electron–hole pairs will block the terahertz radiation. Encoding a spatial pattern on the optical beam with a digital micromirror device could form a real-time terahertz SLM where pixels with laser beam illumination will block terahertz radiation, while pixels without laser beam illumination will transmit terahertz radiation. At terahertz frequencies, dynamically reconfigurable semiconducting device technology provides significant advantages over traditional mechanical masks due to the benefit of adaptability and real-time control. By using masks of varying complexities ranging from 63 to 1023 pixels, Shrekenhamer et al. (2013) were able to acquire terahertz images at a speed of up to 1/2 Hz. This result demonstrated the viability of obtaining real-time and high-fidelity terahertz images using an optically controlled SLM with a single-pixel detector. The drawback of this optically

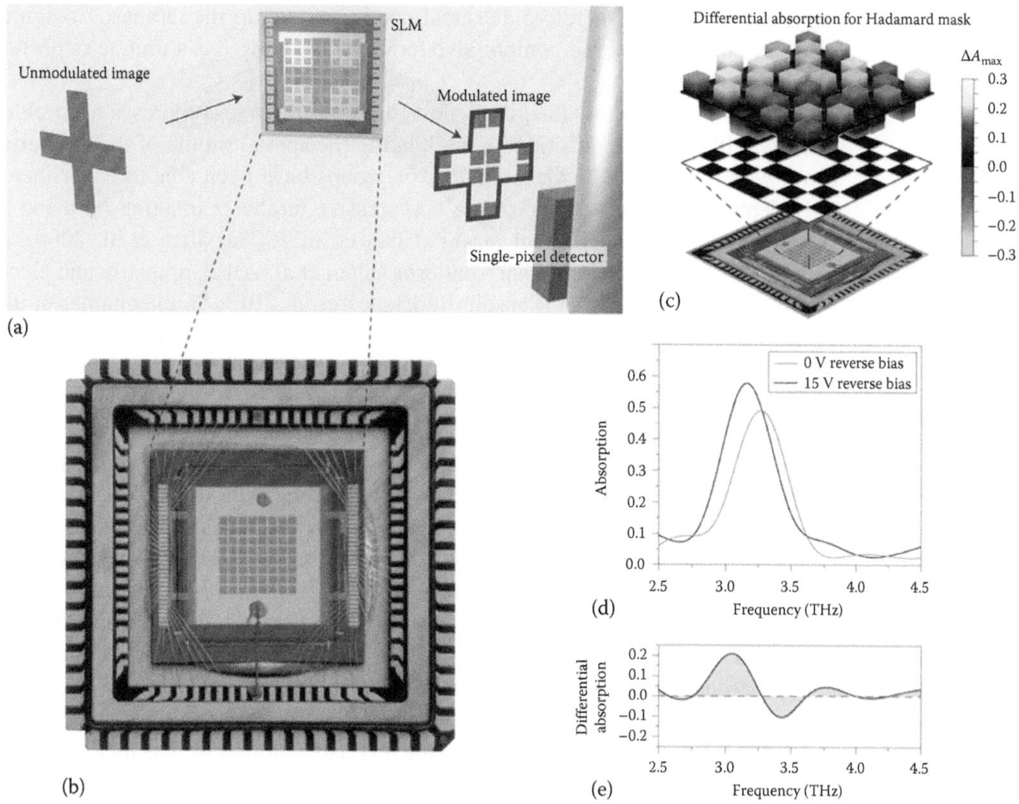

Figure 6.13 (a) Schematic of the single-pixel imaging process utilizing a terahertz SLM. (b) Photograph of the SLM. (Courtesy of K. Burke, Boston College Media Technology Services, Chestnut Hill, MA.) Total active area of the SLM is 4.8 mm^2. (c) Spatial map of maximum differential absorption for an example Hadamard mask over a photograph of the SLM device. (d) Frequency-dependent absorption of a single pixel for two bias voltages, 0 V reverse bias (blue curve) and 15 V reverse bias (red curve). (e) Differential absorption ($A15 - A0$ V) as a function of frequency. (Reproduced from Watts, C.M. et al., *Nat. Photon.*, 8, 605, 2014. With permission.)

controlled SLM approach is the need of an additional high-power laser and a digital micromirror device for writing in real-time terahertz masks, and this could add to the cost and complexity of compressive terahertz imaging system.

Summary and Future Work

In the past 20 years or so, we have seen a revolution in terahertz technology and the development of terahertz time-domain imaging technology in particular has enabled a number of applications in industry and science. Nevertheless, there is still a lack of sensitive focal-plane-array terahertz detectors and compact high-power terahertz sources. Consequently, most terahertz imaging experiments have to be performed by raster scanning the object of interest in a pixel-by-pixel fashion using a single detector. This single-pixel approach provides high signal-to-noise ratio and high spatial resolution terahertz images at the cost of long image acquisition time. A complete image data set of 256×256 pixels will take over an hour to acquire even at a reasonable measurement speed of 10 THz waveforms per second. There is therefore a critical need to speed up the terahertz imaging process, and in this sense compressive terahertz imaging has a unique opportunity here.

Compressive terahertz imaging is able to perform image compression simultaneously with image acquisition by modulating the spatial profile of the terahertz beam with a terahertz SLM. A number of groups have been able to experimentally demonstrate the concept of the compressive terahertz imaging by using a set of independent mechanical masks (Chan et al. 2008a; Shen et al. 2009), a spinning disk with random binary patterns (Shen et al. 2012), optically and electrically controllable SLMs (Sensale-Rodriguez et al. 2013; Shrekenhamer et al. 2013; Watts et al. 2014). This compressive terahertz imaging approach not only eliminates the need to raster scan the object or terahertz beam, but also reduces the number of required measurements. A sampling rate of 10%–30% has been reported to be sufficient for fully reconstructing a terahertz image with reasonable complexity. This is a significant improvement in speed compared with the traditional raster scanning used for conventional terahertz imaging, particularly when an electrically or optically reconfigurable SLM masks were used. However, relatively low-pixel density masks (20×20 or 32×32 for mechanical masks; 4×4 or 8×8 for reconfigurable masks) were used in these studies. For real-world applications, higher-resolution images are necessary; therefore, in my view, one of the future works is to develop high-pixel density reconfigurable SLMs that can operate efficiently at terahertz frequencies. For high-resolution terahertz imaging, a block-based compressed sensing scheme might be more appropriate since it reduces the computational burden of the reconstruction as well as measurement time (Gan 2007; Cho et al. 2011).

In addition, a typical terahertz imaging measurement involves the acquisition of a three-dimensional data set providing both spatial information and time (depth) information (Shen and Taday 2008). Up to now, all compressive terahertz imaging experiments have been performed in the spatial domain (by using an SLM to modulate the spatial profile of the terahertz beam). It has been shown that terahertz data are also compressible in time domain, and terahertz image data can be compressed to within 7% without losing essential information on coating layer thickness (Shen et al. 2005b; Shon et al. 2007). Therefore, an overall sampling rate of better than 1% might be plausible if a joint measurement operator in both spatial and time domains could be developed. Very recently, Abolghasemi et al. (2015) reported the image reconstruction from a subsampled three-dimensional terahertz data using spatiotemporal dictionary learning method, and it was shown that a terahertz image could be reconstructed from 5% observation. The design and experimental implementation of a measurement operator for true three-dimensional compressive terahertz imaging will, however, be challenging, and it remains to be a topic for future research focus.

References

Abolghasemi, V., H. Shen, Y. C. Shen et al. 2014. Subsampled terahertz data reconstruction based on spatio-temporal dictionary learning. *Digital Signal Process.* 43:1–7.

Armstrong, C. M. 2012. The truth about terahertz. *IEEE Spectrum* 49:36–41.

Auston, D. H. 1975. Picosecond optoelectronic switching and gating in silicon. *Appl. Phys. Lett.* 26:101–103.

Brown, E. R., J. E. Bjarnason, A. M. Fedor, and T. M. Korter. 2007. On the strong and narrow absorption signature in lactose at 0.53 THz. *Appl. Phys. Lett.* 90:061908.

Candès, E. J., J. Romberg, and T. Tao. 2006. Robust uncertainty principles: Exact signal reconstruction from highly incomplete frequency information. *IEEE Trans. Inf. Theory* 52(2):489–509.

Chan, W. L., K. Charan, D. Takhar et al. 2008a. A single-pixel terahertz imaging system based on compressed sensing. *Appl. Phys. Lett.* 93:121105.

Chan, W. L., H. T. Chen, A. J. Taylor et al. 2009. A spatial light modulator for terahertz beams. *Appl. Phys. Lett.* 94:213511.

Chan, W. L., M. L. Moravec, R. G. Baraniuk et al. 2008b. Terahertz imaging with compressed sensing and phase retrieval. *Opt. Lett.* 33(9):974–976.

Cho, S. H., S. H. Lee, C. N. Gung et al. 2011. Fast terahertz reflection tomography using block-based compressed sensing. *Opt. Express* 19:16401–16409.

Dudley, D., W. Duncan, and J. Slaughter. 2003. Emerging digital micromirror device (DMD) applications. *Proc. SPIE* 4985:14–25.

Ferguson, B. and X. Zhang. 2002. Materials for terahertz science and technology. *Nat. Mater.* 1:26–33.

Gan, L. 2007. Block compressed sensing of natural images. *In The 15th International Conference on Digital Signal Processing, IEEE*, Cardiff, U.K., pp. 403–406.

Haupt, J. and R. Nowak. 2006. Compressive sampling vs. conventional imaging. In *Proceedings of IEEE International Conference on Image Processing*, Atlanta, GA, pp. 1269–1272.

Haykin, S. 1996. *Adaptive Filter Theory*, 3rd edn. Prentice Hall Information and System Sciences Series, Prentice Hall, NJ.

Hu, B. B. and M. C. Nuss. 1995. Imaging with terahertz waves. *Opt. Lett.* 20:1716–1718.

Jackson, J. B., M. Mourou, J. Labaune et al. 2009. Terahertz pulse imaging for tree-ring analysis: A preliminary study for dendrochronology applications. *Meas. Sci. Technol.* 20(7):075502.

Jansen, C., S. Wietzke, O. Peters et al. 2010. Terahertz imaging: Applications and perspectives. *Appl. Opt.* 49(19):E48–E57.

Jepsen, P. U., D. G. Cooke, and M. Kock. 2011. Terahertz spectroscopy and imaging—Modern techniques and applications. *Laser Photon. Rev.* 5:124–166.

Johnson, K. M., D. J. McKnight, and I. Underwood. 1993. Smart spatial light modulators using liquid crystals on silicon. *IEEE J. Quant. Electron.* 29:699–714.

Jolliffe, I. T. 1986. *Principal Component Analysis*. New York: Springer.

Ke, J. and M. A. Neifeld. 2007. Optical architectures for compressive imaging. *Appl. Opt.* 46(22):5293–5303.

Lee, A. W. and Q. Hu. 2005. Real-time, continuous-wave terahertz imaging by use of a microbolometer focal-plane array. *Opt. Lett.* 30:2563–2565.

Mittleman, D. M., M. Gupta, R. Neelamani et al. 1999. Recent advances in terahertz imaging. *Appl. Phys. B* 68:1085–1094.

Mittleman, D. M., R. Jacobsen, and M. C. Nuss. 1996. T-ray imaging. *IEEE J. Sel. Top. Quant.* 2:679–692.

Pickwell, E. and V. P. Wallace. 2006. Biomedical applications of terahertz technology. *J. Phys. D—Appl. Phys.* 39:R301–R310.

Rahm, M., J. Li, and W. J. Padilla. 2012. THz wave modulators: A brief review on different modulation techniques. *J. Infrared Millim. Terahertz Waves* 34:1–27.

Rauhut, H. 2010. Compressive sensing and structured random matrices. In *Theoretical Foundations and Numerical Methods for Sparse Recovery*, Vol. 9, Radon Series on Computational and Applied Mathematics, ed. M. Fornasier, pp. 1–92. deGruyter. http://www.degruyter.com/view/product/43466.

Saeedikia, D. 2013. *Handbook of Terahertz Technology for Imaging, Sensing and Communications*. Woodhead Publishing Ltd, Cambridge, UK.

Sensale-Rodriguez, B., S. Rafique, R. Yan, M. Zhu, V. Protasenko, D. Jena, L. Liu, and H. G. Xing. 2013. Terahertz imaging employing graphene modulator arrays. *Opt. Express* 21:2324–2330.

Sensale-Rodriguez, B., R. Yan, S. Rafique, M. Zhu, W. Li, X. Liang, D. Gundlach et al. 2012. Extraordinary control of terahertz beam reflectance in graphene electro-absorption modulators. *Nano Lett.* 12(9):4518–4522.

Shen, H. 2012. Compressed sensing on terahertz imaging. PhD thesis, Liverpool University, Liverpool, U.K.

Shen, H., L. Gan, N. Newman et al. 2012. Spinning disk for compressed imaging. *Opt. Lett.* 37:46–48.

Shen, Y. C. 2011. Terahertz pulsed spectroscopy and imaging for pharmaceutical applications: A review. *Int. J. Pharm.* 417:48–60.

Shen, Y. C., L. Gan, M. Stringer et al. 2009. Terahertz pulsed spectroscopic imaging using optimized binary masks. *Appl. Phys. Lett.* 95:231112.

Shen, Y. C., T. Lo, P. F. Taday et al. 2005a. Detection and identification of explosives using terahertz pulsed spectroscopic imaging. *Appl. Phys. Lett.* 86:241116.

Shen, Y. C. and P. F. Taday. 2008. Development and application of terahertz pulsed imaging for nondestructive inspection of pharmaceutical tablet. *IEEE J. Sel. Top. Quant. Electron.* 14:407–415.

Shen, Y. C., P. F. Taday, D. A. Newnham et al. 2005b. 3D chemical mapping using terahertz pulsed imaging. *SPIE Proc.* 5727:24–31.

Shon, C. H., W. Y. Chong, G. J. Kim et al. 2007. Compression of pulsed terahertz image using discrete wavelet transform. *Jpn. J. Appl. Phys.* 46:7731.

Shrekenhamer, D., C. M. Watts, and W. J. Padilla. 2013. Terahertz single pixel imaging with an optically controlled dynamic spatial light modulator. *Opt. Express* 21:12507–12518.

Watts, C. M., D. Shrekenhamer, J. Montoya, G. Lipworth, J. Hunt, T. Sleasman, S. Krishna, D. R. Smith, and W. J. Padilla. 2014. Terahertz compressive imaging with metamaterial spatial light modulators. *Nat. Photon.* 8:605–609.

Watts, C. M., X. Liu, and W. J. Padilla. 2012. Metamaterial electromagnetic wave absorbers. *Adv. Opt. Mater.* 24:OP98–OP120.

Weiss, Y., H. S. Chang, and W. T. Freeman. 2007. Learning compressed sensing. In *Snowbird Learning Workshop*, Allerton, Monticello, IL.

Williams, B. S. 2007. Terahertz quantum-cascade lasers. *Nat. Photon.* 1:517–525.

Wu, Q., T. D. Hewitt, and X. C. Zhang. 1996. Two-dimensional electro-optic imaging of THz beams. *Appl. Phys. Lett.* 69:1026–1028.

Wu, Q. and X. C. Zhang. 1995. Free-space electro-optic sampling of terahertz beams. *Appl. Phys. Lett.* 67:3523.

Xiao, G. Q., T. R. Corle, and G. S. Kino. 1988. Real-time confocal scanning optical microscope. *Appl. Phys. Lett.* 53(8):716–718.

Xu, Z. and E. Y. Lam. 2010. Image reconstruction using spectroscopic and hyperspectral information for compressive terahertz imaging. *J. Opt. Soc. Am. A* 27(7):1638–1646.

Zeitler, A. and Y. C. Shen. 2013. Industrial applications of terahertz imaging. In *Terahertz Spectroscopy and Imaging: Theory and Applications*, Springer Series in Optical Sciences, eds. K. E. Peiponen, A. Zeitler, and M. Kuwata-Gonokami, Springer-Verlag Berlin Heidelberg, Germany, pp. 451–489.

Zeitler, J. A. and L. F. Gladden. 2009. In-vitro tomography and non-destructive imaging at depth of pharmaceutical solid dosage forms. *Eur. J. Pharm. Biopharm.* 71(1):2–22.

Zeitler, J. A., P. F. Taday, D. A. Newnham et al. 2007. Terahertz pulsed spectroscopy and imaging in the pharmaceutical setting—A review. *J. Pharm. Pharmacol.* 59(2):209–223.

III

Multidimensional Optical Compressive Sensing

Optical Design of Multichannel Data Acquisition System for Compressive Sensing

Ryoichi Horisaki

Contents

Introduction

Motivation

Image sensors are widely used in optical imaging systems. They consist of a two-dimensional detector array that measures a two-dimensional light intensity distribution. On the other hand, optical signals have more than two dimensions and include information about, for example, three-dimensional spatial position, time, spectrum (color or wavelength), and polarization. To observe such high-dimensional optical signals with a two-dimensional image sensor, most of the conventional approaches sacrifice the spatial resolution or temporal resolution in the imaging systems. The former and latter approaches are referred as space-division multiplexing and time-division multiplexing. Typical examples of space-division multiplexing and time-division multiplexing in spectral imaging, where the object optical signals are three-dimensional (x, y, λ), are the Bayer color filter and the *push broom* approach, respectively [1,2].

Compressive sensing (CS) is a powerful framework for observing a large amount of data with fewer measurements than those determined by the sampling theorem [3–5]. CS is especially useful for solving the issue mentioned earlier when observing a high-dimensional optical signal with a single shot using a two-dimensional image sensor. This chapter treats three CS-based generalized frameworks for single-shot multidimensional imaging.

Generalized Form

In this chapter, optical systems are assumed to be linear and objects multidimensional. The orthogonal spatial axes, x and y, are parallel to the image sensor. Each of the x–y planes in the object is defined as a channel, and this dimension is defined as the c-axis, which expresses depth, time, wavelength, polarization, and so on. One lateral spatial dimension (y-axis) in the system is omitted for simplicity in the following descriptions. The system model is written as

$$\mathbf{g} = \mathbf{\Phi f}, \tag{7.1}$$

where

- $\mathbf{g} \in \Re^{N_x \times 1}$ is the vector of the captured data
- $\mathbf{\Phi} \in \Re^{N_x \times (N_x \times N_c)}$ is the system matrix
- $\mathbf{f} \in \Re^{(N_x \times N_c) \times 1}$ is the vector of the object data
- N_x is the number of elements along the x-axis
- N_c is the number of channels

In this case, the size of the object data is N_c times larger than that of the captured data.

For single-shot multidimensional object acquisition with an image sensor shown in Figure 7.1, Equation 7.1 is modified to

$$\mathbf{g} = \mathbf{IEf}, \tag{7.2}$$

where

- $\mathbf{I} \in \Re^{N_x \times (N_x \times N_c)}$ is a matrix of an integration operator
- $\mathbf{E} \in \Re^{(N_x \times N_c) \times (N_x \times N_c)}$ is a matrix of an encoding operator

The matrix $\mathbf{I}$ is written as

$$\mathbf{I} = [\mathbf{I}_0 \ldots \mathbf{I}_0], \tag{7.3}$$

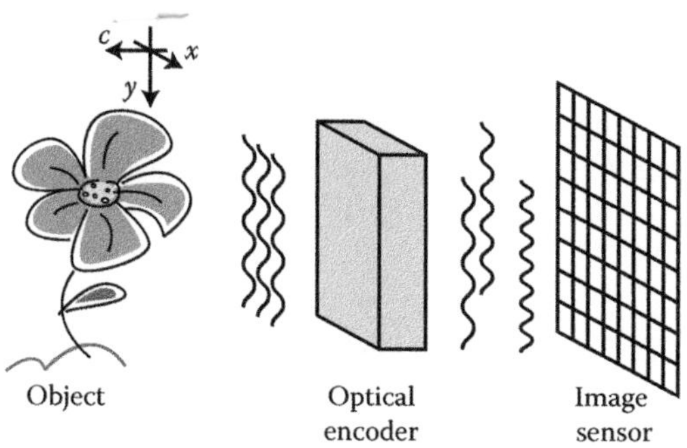

Figure 7.1 Single-shot multichannel object acquisition with an image sensor.

where $\mathbf{I}_0 \in \Re^{N_x \times N_x}$ is an identity matrix. The matrix $\mathbf{E}$ is written as

$$\mathbf{E} = \begin{bmatrix} \mathbf{E}_1 & 0 & \cdots & 0 \\ 0 & \mathbf{E}_2 & 0 & \vdots \\ \vdots & 0 & \ddots & 0 \\ 0 & \cdots & 0 & \mathbf{E}_{N_c} \end{bmatrix}, \tag{7.4}$$

where $\mathbf{E}_C \in \Re^{N_x \times N_x}$ is a matrix that represents an encoder of the object at the Cth channel. Equation 7.2 describes the following imaging process: (1) Each of the channels in the multidimensional object $\mathbf{f}$ is modulated with the encoder $\mathbf{E}_C$. (2) The modulated channels are integrated on the image sensor as the captured image $\mathbf{g}$. The system model in Equation 7.2 is the generalized form. A specific encoder $\mathbf{E}$ is given in each imaging modality in the following sections.

The inverse problem of Equation 7.2 is ill-posed. To solve the problem, a multiplexed imaging process with the optical encoder $\mathbf{E}$ and a sparsity-based reconstruction algorithm used in CS are employed. Various sparsity-based algorithms have been proposed [6–8]. A typical one is the two-step iterative shrinkage/thresholding (TwIST) algorithm. TwIST solves the following optimization problem iteratively:

$$\hat{\mathbf{f}} = \underset{\mathbf{f}}{\operatorname{argmin}} \left\| \mathbf{g} - \mathbf{IEf} \right\|_{\ell_2} + \tau\Theta(\mathbf{f}), \tag{7.5}$$

where

$\|\cdot\|_{\ell_2}$ is the ℓ_2 norm

τ is a regularization parameter

$\Theta(\cdot)$ is a regularizer to evaluate the sparsity in the domain of the regularizer

Point Spread Function Engineering

Point spread function (PSF) engineering has been used to enhance the performance of depth-from-defocus (DFD) based on a double-helix PSF or a coded aperture [9–11]. This imaging modality can be generalized to imaging for other dimensions [12].

Optics Design

To realize laterally space-invariant and channel-variant PSFs, an optical encoder for modulating channels is inserted in the pupil of the imaging optics, as shown in Figure 7.2. The PSF encoder should have a dependency for each channel.

System Model

In PSF engineering–based multichannel data acquisition, Equation 7.2 is rewritten as

$$\mathbf{g} = \mathbf{IPf}, \tag{7.6}$$

where $\mathbf{P} \in \Re^{(N_x \times N_c) \times (N_x \times N_c)}$ is a matrix of a convolution operator, which is the encoder $\mathbf{E}$ of Equation 7.2 in this imaging modality. The matrix $\mathbf{P}$ is written as

$$\mathbf{P} = \begin{bmatrix} \mathbf{P}_1 & 0 & \cdots & 0 \\ 0 & \mathbf{P}_2 & 0 & \vdots \\ \vdots & 0 & \ddots & 0 \\ 0 & \cdots & 0 & \mathbf{P}_{N_c} \end{bmatrix}, \tag{7.7}$$

where $\mathbf{P}_C \in \Re^{N_x \times N_x}$ is a Toeplitz matrix showing a convolution of the PSF at the Cth channel. Equation 7.6 describes the following imaging process: (1) Each of the channels in the multidimensional object $\mathbf{f}$ is modulated with the convolutions of different PSFs $\mathbf{P}_C$. (2) The modulated channels are integrated onto the image sensor as the captured image $\mathbf{g}$.

Simulation

A numerical demonstration of the PSF engineering–based approach is shown in Figure 7.3. The multidimensional object in Figure 7.3a is composed of six channels, where Shepp–Logan phantoms are located at two channels. The pixel count is $128 \times 128 \times 6$. The phantom is known to be sparse in the total variation (TV) domain [13]. The captured image, assuming randomly generated PSFs that are dependent on each channel for $\mathbf{P}_C$ in Equation 7.7, is shown in Figure 7.3b, where the pixel count is 128×128, and Gaussian noise with a signal-to-noise ratio (SNR) of 40 dB is added. For comparison, the reconstruction result obtained with the Richardson–Lucy (RL) method [14,15], which does not

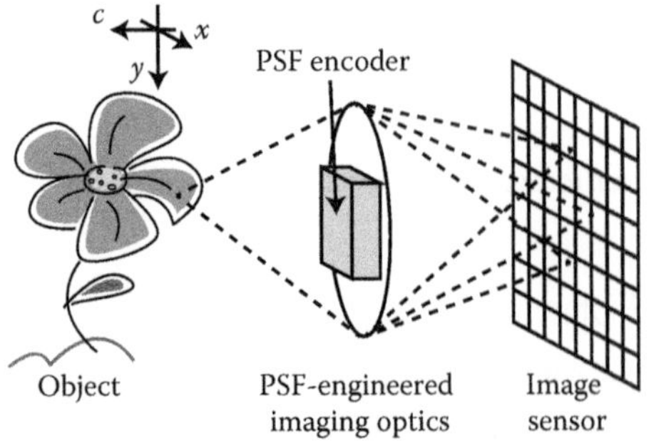

Figure 7.2 PSF engineering–based multichannel object acquisition.

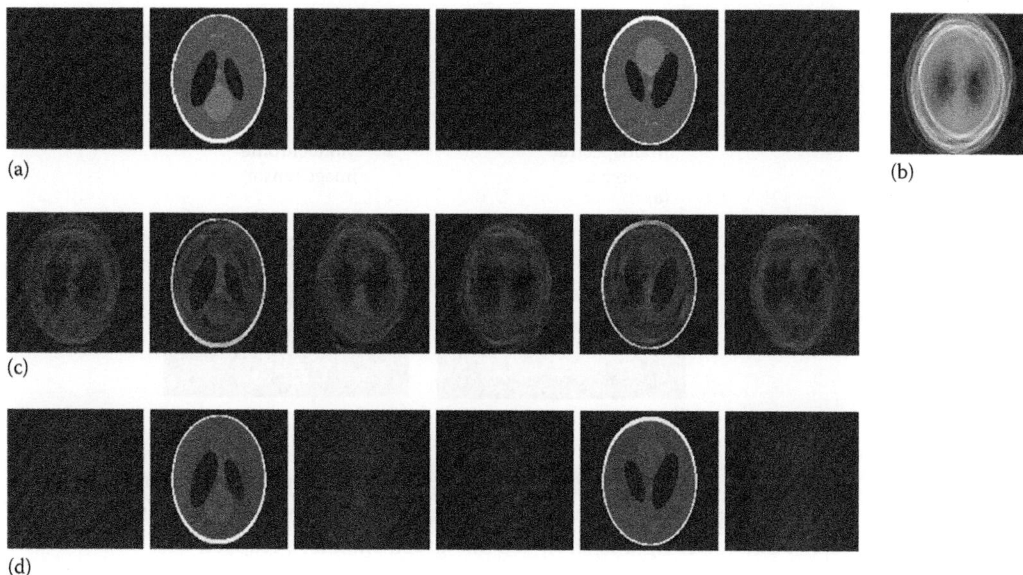

Figure 7.3 Simulation of the PSF engineering–based approach. (a) An object, (b) the captured image, and the reconstructions with (c) the RL method and (d) TwIST.

employ any sparsity-based regularization, is shown in Figure 7.3c. The result contains some artifacts, and the peak SNR (PSNR) between the object and the reconstruction is 22.6 dB. The reconstruction result obtained with TwIST, described in Equation 7.5, is shown in Figure 7.3c. TV is chosen as the regularizer. The multidimensional object is successfully reconstructed, and the TwIST-based reconstruction has fewer artifacts than the RL one. The PSNR of the TwIST-based reconstruction is 27.3 dB.

Applications

The approach based on PSF engineering is applicable to various multidimensional imaging applications. Table 7.1 shows examples of PSF encoders inserted in the pupil for several applications [12]. A coded aperture has been used for depth imaging to reduce the correlation between PSFs at different distances [11]. This approach also has been applied to wide field-of-view imaging [12]. Spectral imaging based on this approach is shown in Figure 7.4. A diffraction grating is inserted into the optical path of the imaging system, as shown in Figure 7.4a, where the image sensor is monochrome. High-order diffraction orders of the grating depend on the wavelength of light coming from the object. Thus, the PSF of this system is wavelength variant. The object

Table 7.1 Implementations of PSF Engineering–Based Multidimensional Imaging

Application	PSF Encoder
Depth imaging	Coded aperture
Spectral imaging	Grating
Polarimetric imaging	Birefringence element
Temporal imaging	Temporally variable coded aperture

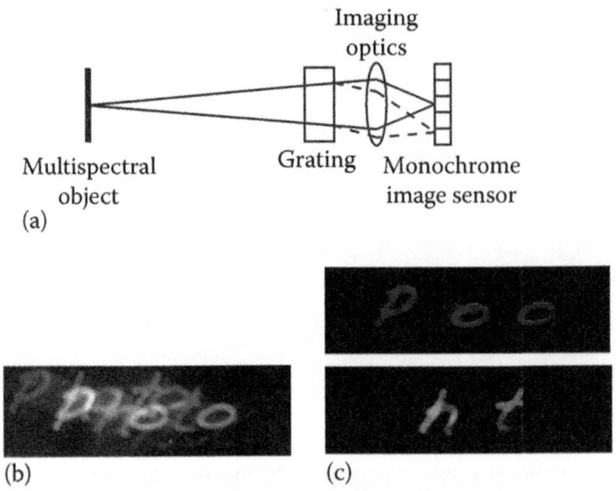

Imaging optics
Multispectral object
Grating
Monochrome image sensor
(a)
(b)
(c)

Figure 7.4 Spectral imaging with the PSF engineering–based approach. (a) The optical setup, (b) the captured image, and (c) the reconstruction result, where the upper and lower figures show red and green channels, respectively.

used in the demonstration is the printed word "Photo," where the "P" and two "o" are red and the "h" and "t" are green. The captured image is shown in Figure 7.4b. The two spectral channels are separated by the reconstruction process, as shown in Figure 7.4c.

Compound Eye

Compound eyes are the visual organs in many insects and crustaceans [16]. They are categorized into apposition and superposition types. Both of them consist of a number of small elemental optics. In the apposition type, one visual nerve (detector) is optically connected to one elemental optics, whereas in the superposition type, one visual nerve is connected to multiple elemental optics. Optics design based on apposition compound eyes has been used to reduce the size of imaging devices [17–19]. Apposition compound eyes are also used in integral imaging for three-dimensional object acquisition and display [20,21].

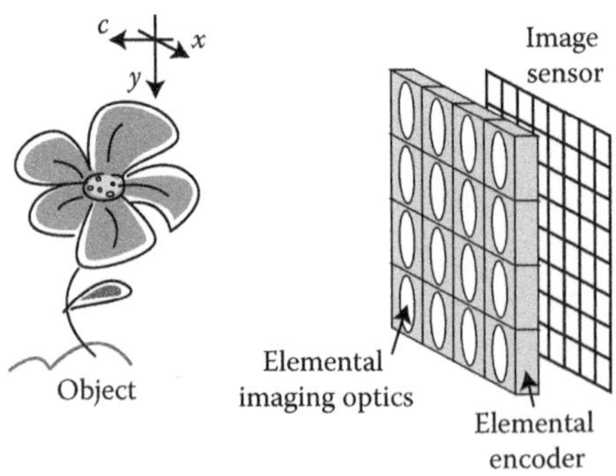

Figure 7.5 Compound-eye-based multichannel object acquisition.

Optics Design

Each of the elemental optics in compound-eye cameras implements different imaging processes. This flexibility is one of the advantages of the compound-eye-based multidimensional imaging approach. By inserting an optical encoder into the optical path of each of the elemental optics, various modulations for each channel of the multidimensional object can be realized, as shown in Figure 7.5 [22,23]. Two modulation schemes have been proposed: sheared integration and weighted integration. The former modulation laterally shifts each channel of the object, and the latter modulation multiplies the object by a weight distribution. Each elemental optics adds a different modulation to the object in this case.

System Model

The system model of Equation 7.2 in the compound-eye-based approach is rewritten as

$$\mathbf{g} = \mathbf{IMWSRf}, \tag{7.8}$$

where

$\mathbf{M} \in \Re^{(N_x \times N_c)\times(N_x \times N_c \times N_e)}$ is a matrix of a minification operator
$\mathbf{W} \in \Re^{(N_x \times N_e \times N_c)\times(N_x \times N_e \times N_c)}$ is a matrix of a weighting operator
$\mathbf{S} \in \Re^{(N_x \times N_e \times N_c)\times(N_x \times N_e \times N_c)}$ is a matrix of a shearing operator
$\mathbf{R} \in \Re^{(N_x \times N_e \times N_c)\times(N_x \times N_c)}$ is a matrix of a replication operator
N_e is the number of elemental optics

In this case, **MWSR** = **E**, which is the encoder in Equation 7.2. The matrix **M** is written as

$$\mathbf{M} = \begin{bmatrix} \mathbf{I}' & 0 & \cdots & 0 \\ 0 & \mathbf{I}' & 0 & \vdots \\ \vdots & 0 & \ddots & 0 \\ 0 & \cdots & 0 & \mathbf{I}' \end{bmatrix}, \tag{7.9}$$

where $\mathbf{I}' \in \Re^{1 \times N_e}$ is a matrix containing all ones. The matrix **W** is written as

$$\mathbf{W} = \begin{bmatrix} w_{1,1}\mathbf{I}_0 & 0 & \cdots & 0 \\ 0 & w_{2,1}\mathbf{I}_0 & 0 & \vdots \\ \vdots & 0 & \ddots & 0 \\ 0 & \cdots & 0 & w_{N_e,N_c}\mathbf{I}_0 \end{bmatrix}, \tag{7.10}$$

where $w_{E,C}$ is a weight for the *E*th elemental optics and the *C*th channel for the weighting operator. The matrix **S** is written as

$$\mathbf{S} = \begin{bmatrix} \mathbf{S}_{1,1} & 0 & \cdots & 0 \\ 0 & \mathbf{S}_{2,1} & 0 & \vdots \\ \vdots & 0 & \ddots & 0 \\ 0 & \cdots & 0 & \mathbf{S}_{N_e,N_c} \end{bmatrix}, \tag{7.11}$$

where $\mathbf{S}_{E,C} \in \Re^{N_x \times N_x}$ is an off-diagonal matrix to laterally shift the object signal at the Eth elemental optics and the Cth channel for the shearing operator. The matrix $\mathbf{R}$ is written as

$$\mathbf{R} = \begin{bmatrix} \mathbf{R}_0 & 0 & \cdots & 0 \\ 0 & \mathbf{R}_0 & 0 & \vdots \\ \vdots & 0 & \ddots & 0 \\ 0 & \cdots & 0 & \mathbf{R}_0 \end{bmatrix}, \tag{7.12}$$

where $\mathbf{R}_0 \in \Re^{(N_x \times N_e) \times N_x}$ is written as

$$\mathbf{R}_0 = \begin{bmatrix} \mathbf{I}_0 \\ \vdots \\ \mathbf{I}_0 \end{bmatrix}. \tag{7.13}$$

Equation 7.8 describes the following imaging process: (1) The multidimensional object $\mathbf{f}$ is replicated by individual elemental optics. (2) Channels of the replicated object are differently modulated with shearing and weighting in each elemental optics. (3) Each of the modulated channels is minified by pixelization. (4) The resultant channels are integrated on the image sensor as the captured image $\mathbf{g}$.

Simulation

The compound-eye-based approach is demonstrated numerically, as shown in Figure 7.6. The multidimensional object is composed of eight natural images, as shown in Figure 7.6a, where the pixel count is $256 \times 256 \times 8$. The simulated captured image is shown in Figure 7.6b. The size of each elemental image is 64×64, and the number of elemental optics is 4×4. Both the shearing and weighting encoders randomly modulate the object. The measurement SNR is 40 dB. The reconstruction with the pseudo-inversion is shown in Figure 7.6c [24]. It has strong artifacts, and the PSNR is 11.9 dB. The reconstruction result with TwIST is shown in Figure 7.6d. The ℓ_1 norm on the discrete wavelet domain is chosen as the regularizer in TwIST. The object is reconstructed well by TwIST compared with the pseudo-inversion, and its PSNR is 29.7 dB.

Applications

The compound-eye-based approach allows different modulations in individual elemental optics [22,23]. Moreover, these modulations can be integrated in compact hardware. Table 7.2 shows examples of the implementation of the modulations in the compound-eye-based approach. A stereo camera can be regarded as a kind of compound-eye camera, and its typical application is depth imaging by using parallax [25,26]. The compound-eye-based approach can also be used to extend the dynamic range and field of view [27].

A demonstration of depth and spectral imaging is shown in Figure 7.7 [23]. In this experiment, the object is composed of two depth channels (z_1 and z_2) and three color channels (red, green, and blue). Elementwise shearing for depth imaging and pixelwise weighting for spectral imaging are employed here. The captured image, where the number of elemental optics is 6×6, is shown in Figure 7.7a.

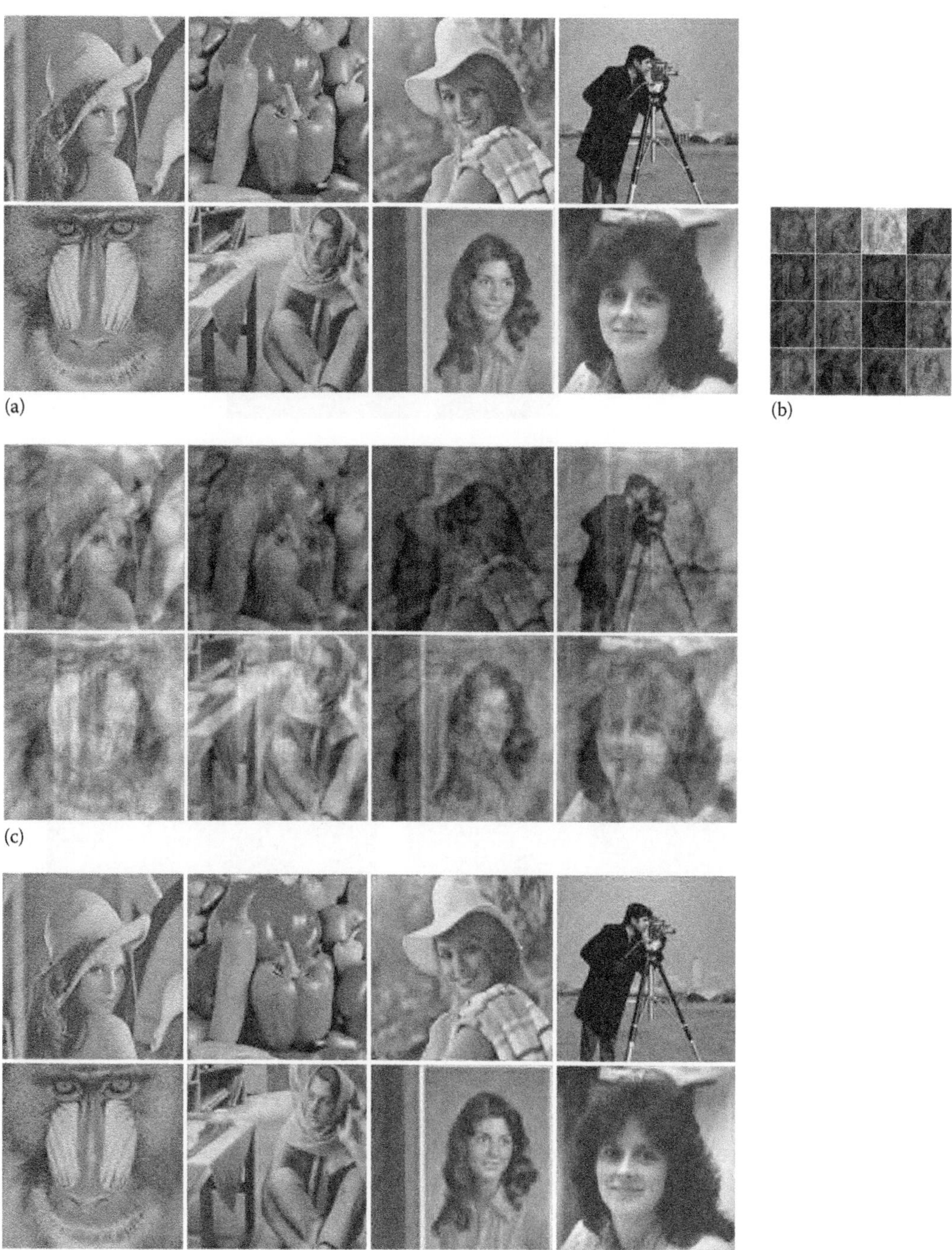

Figure 7.6 Simulation of the compound-eye-based multidimensional imaging. (a) A multidimensional object, (b) the captured image, and reconstructions with (c) the pseudo-inversion and (d) TwIST.

Table 7.2 Implementations of Compound-Eye-Based Multidimensional Imaging

Application	Shearing Encoder	Weighting Encoder
Depth imaging	Parallax	None
Spectral imaging	Disperser	Color filter
Polarimetric imaging	Birefringence element	Polarizer
Temporal imaging	Lens movement	Shutter

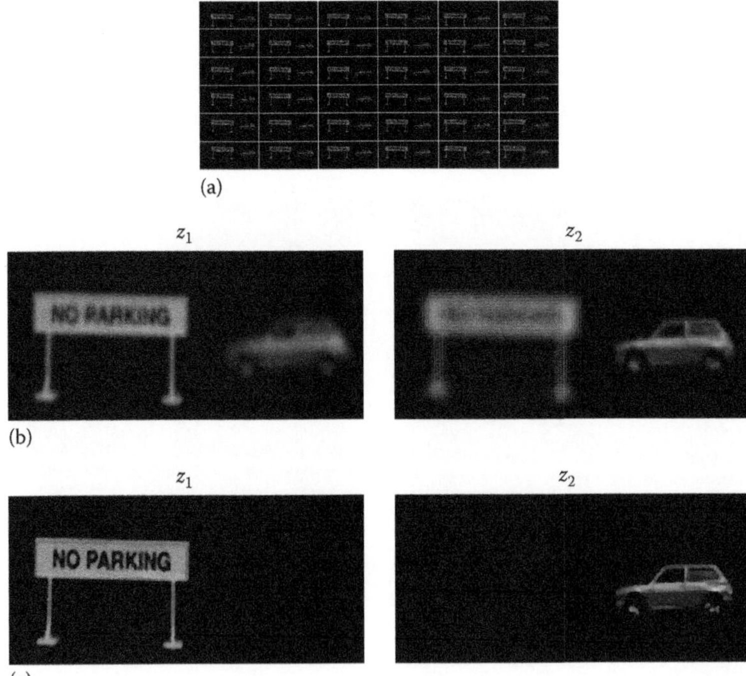

Figure 7.7 Depth and spectral imaging with the compound-eye-based approach. (a) A captured image and the reconstructions with (b) the back-projection and (c) TwIST.

The reconstruction with the conventional back-projection method is shown in Figure 7.7b. The result contains laterally and axially blurred objects. The reconstruction with TwIST is shown in Figure 7.7c. In the TwIST-based result, the object is axially well separated, and the image of each channel is sharper than those of the back-projection-based reconstruction.

Holography

Holography measures an optical complex field, which is composed of amplitude and phase. It allows us to image transparent objects, such as biomedical specimens, and to refocus three-dimensional objects digitally after capturing [28–30]. Thus, holography is especially promising in the field of biomedicine.

Optics Design

To observe multidimensional objects with holography, compressive Fresnel holography (CFH) is employed. CFH is a synthetic aperture technique, where the complex field of the object is reconstructed from sparsely sampled holographic data [31]. In multidimensional CFH, each of the channels is also sampled randomly and pixelwise by a filter array based on the sampling scheme of CFH shown in Figure 7.8 [32].

System Model

For multidimensional CFH, the system model of Equation 7.1 is rewritten as

$$\mathbf{g} = \mathbf{IAPf}, \tag{7.14}$$

where the vectors **g** and **f** are composed of complex numbers, unlike the real numbers used in the previous sections. Here, the matrix **P** in Equation 7.4 is also composed of complex numbers, and the submatrix $\mathbf{P}_C$ shows the convolution of the PSF by the Fresnel propagation at the *C*th channel [28]. In this imaging modality, **AP** = **E**, which is the encoder in Equation 7.2.

$$\mathbf{A} \in \Re^{(N_x \times N_c) \times (N_x \times N_c)}$$

is a diagonal matrix in which the diagonal components show the pixelwise filter at each of the pixels and the channels. Equation 7.14 describes the following imaging process: (1) Each of the channels in the multidimensional object **f** is convolved with different Fresnel PSFs $\mathbf{P}_C$. (2) The convolved channels are multiplied with the pixelwise filter array **A**. (3) The resultant channels are integrated on the image sensor as the captured image **g**.

Simulation

The CFH-based approach is simulated as shown in Figure 7.9. A multidimensional object is shown in Figure 7.9a. Its pixel count is 64×64×5, and it is composed of multiple Shepp–Logan phantoms. The spatial distribution of the transmission in each channel of the filter array is shown in Figure 7.9b. The filter patterns are complementary to each other. The captured image is shown in Figure 7.9c. The measurement SNR is 40 dB. For comparison, the reconstruction obtained with pseudo-inversion is shown in Figure 7.9d. In this case, the phantoms are not recovered, and the PSNR is 18.3 dB. The reconstruction with

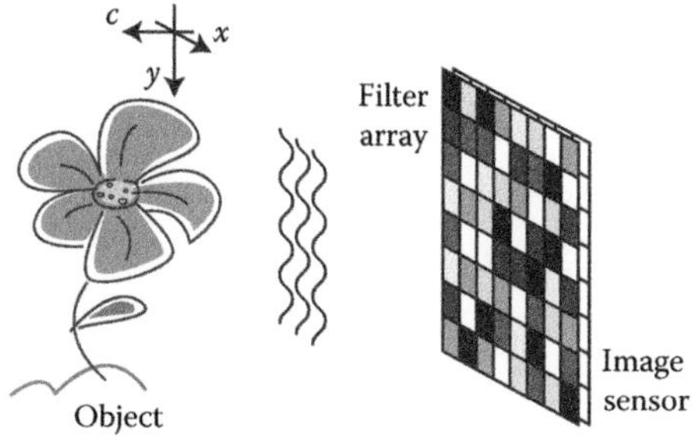

Figure 7.8 CFH-based multichannel object acquisition.

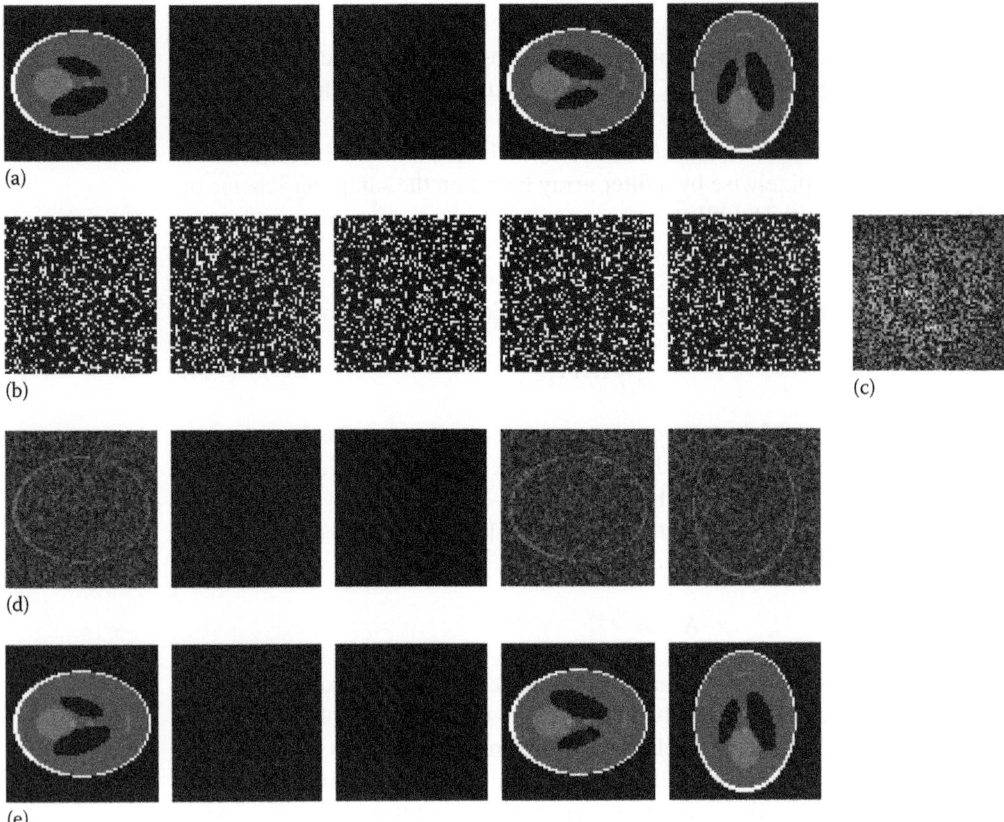

Figure 7.9 Simulation of the CFH-based multidimensional imaging. (a) A multidimensional object, (b) patterns of the filter array, (c) the captured image, and reconstructions with (d) the pseudo-inversion and (e) TwIST.

TwIST is shown in Figure 7.9e. The object is reconstructed successfully, and the PSNR is 54.6 dB.

Applications

Fresnel PSF is depth variant. Thus, the CFH-based approach can inherently be applied to depth imaging [28–30,33]. Table 7.3 shows other applications and their implementations. This imaging modality is also useful for wide dynamic-range imaging [32]. The CFH-based approach uses coherent illumination. Thus, it detects the phase of optical signals, and it is lensless and aberration-free.

Table 7.3 Implementations of CFH-Based Multidimensional Imaging

Application	Filtering Encoder
Depth imaging	None (Fresnel propagation)
Spectral imaging	Color filter
Polarimetric imaging	Polarizer
Temporal imaging	Pixelwise shutter

Conclusions

In this chapter, three imaging modalities for observing multidimensional objects have been introduced. These are based on PSF engineering, compound eyes, and holography. First, generalized forms of each imaging modality are presented, and then their specific matrix operators are derived. They are demonstrated numerically, and their applications are also described. Also other types of single-shot multidimensional imaging modality have been proposed based on coded spatial sampling although they are not referred here [34–37].

These imaging modalities are readily applicable to depth imaging, spectral imaging, polarimetric imaging, and temporal imaging (high-speed imaging). Some of these are mutually compatible, and this approach allows single-shot imaging of objects with a large number of dimensions.

References

1. Li, Q., He, X., Wang, Y., Liu, H., Xu, D., and Guo, F. (2013). Review of spectral imaging technology in biomedical engineering: Achievements and challenges. *J. Biomed. Opt.*, 18(10), 100901.
2. Hagen, N. and Kudenov, M. W. (2013). Review of snapshot spectral imaging technologies. *Opt. Eng.*, 52(9), 90901.
3. Donoho, D. L. (2006). Compressed sensing. *IEEE Trans. Inf. Theory*, 52(4), 1289–1306.
4. Baraniuk, R. (2007). Compressive sensing. *IEEE Signal Process. Mag.*, 24(4), 118–121.
5. Candes, E. J. and Wakin, M. B. (2008). An introduction to compressive sampling. *IEEE Signal Process. Mag.*, 25(2), 21–30.
6. Figueiredo, M. A. T., Nowak, R. D., and Wright, S. J. (2007). Gradient projection for sparse reconstruction: Application to compressed sensing and other inverse problems. *IEEE J. Sel. Top. Signal Process.*, 1(4), 586–597.
7. Bioucas-Dias, J. M. and Figueiredo, M. A. T. (2007). A new twist: Two-step iterative shrinkage/thresholding algorithms for image restoration. *IEEE Trans. Image Process.*, 16(12), 2992–3004.
8. Wright, S. J., Nowak, R. D., and Figueiredo, M. A. T. (2009). Sparse reconstruction by separable approximation. *IEEE Trans. Image Process.*, 57(7), 2479–2493.
9. Greengard, A., Schechner, Y. Y., and Piestun, R. (2006). Depth from diffracted rotation. *Opt. Lett.*, 31(2), 181–183.
10. Pavani, S. R. P., Thompson, M. A., Biteen, J. S., Lord, S. J., Liu, N., Twieg, R. J., Piestun, R., and Moerner, W. E. (2009). Three-dimensional, single-molecule fluorescence imaging beyond the diffraction limit by using a double-helix point spread function. *Proc. Natl. Acad. Sci. USA*, 106(9), 2995–2999.
11. Levin, A., Fergus, R., Durand, F., and Freeman, W. T. (2007). Image and depth from a conventional camera with a coded aperture. *ACM Trans. Graph.*, 26(3), 70:1–70:9.
12. Horisaki, R. and Tanida, J. (2010). Multi-channel data acquisition using multiplexed imaging with spatial encoding. *Opt. Express*, 18(22), 23041–23053.
13. Rudin, L. I., Osher, S., and Fatemi, E. (1992). Nonlinear total variation based noise removal algorithms. *Phys. D*, 60(1–4), 259–268.
14. Richardson, W. H. (1972). Bayesian-based iterative method of image restoration. *J. Opt. Soc. Am.*, 62(1), 55–59.

15. Lucy, L. B. (1974). An iterative technique for the rectification of observed distributions. *Astron. J.*, 79, 745–754.
16. Duparré, J. W. and Wippermann, F. C. (2006). Micro-optical artificial compound eyes. *Bioinspir. Biomim.*, 1(1), R1–R16. Retrieved from http://iopscience.iop.org/1748-3190/1/1/R01/pdf/1748-3190_1_1_R01.pdf.
17. Tanida, J., Kumagai, T., Yamada, K., Miyatake, S., Ishida, K., Morimoto, T., Kondou, N., Miyazaki, D., and Ichioka, Y. (2001). Thin observation module by bound optics (TOMBO): Concept and experimental verification. *Appl. Opt.*, 40(11), 1806–1813.
18. Duparré, J., Dannberg, P., Schreiber, P., Bräuer, A., and Tünnermann, A. (2005). Thin compound-eye camera. *Appl. Opt.*, 44(15), 2949–2956. Retrieved from http://ao.osa.org/abstract.cfm?URI=ao-44-15-2949.
19. Venkataraman, K., Lelescu, D., Duparr, J., Mcmahon, A., Molina, G., Chatterjee, P., Mullis, R., and Nayar, S. (2013). PiCam: An ultra-thin high performance monolithic camera array. *ACM Trans. Graph.*, 35(2), 1–13. Retrieved from http://www.pelicanimaging.com/technology/.
20. Stern, A. and Javidi, B. (2006). Three-dimensional image sensing, visualization, and processing using integral imaging. *Proc. IEEE*, 94(3), 591–607.
21. Xiao, X., Javidi, B., Martinez-Corral, M., and Stern, A. (2013). Advances in three-dimensional integral imaging: Sensing, display, and applications. *Appl. Opt.*, 52(4), 546–560.
22. Horisaki, R., Choi, K., Hahn, J., Tanida, J., and Brady, D. J. (2010). Generalized sampling using a compound-eye imaging system for multi-dimensional object acquisition. *Opt. Express*, 18(18), 19367–19378. Retrieved from http://www.opticsexpress.org/abstract.cfm?URI=oe-18-18-19367.
23. Horisaki, R., Xiao, X., Tanida, J., and Javidi, B. (2013). Feasibility study for compressive multi-dimensional integral imaging. *Opt. Express*, 21(4), 4263–4279. Retrieved from http://www.opticsexpress.org/abstract.cfm?URI=oe-21-4-4263.
24. Williams, D. B. and Madisetti, V. (Eds.). (1997). *Digital Signal Processing Handbook* (1st ed.). Boca Raton, FL, CRC Press.
25. Okutomi, M. and Kanade, T. (1993). A multiple-baseline stereo. *IEEE Trans. Pattern Anal. Mach. Intell.*, 15(4), 353–363.
26. Horisaki, R., Irie, S., Ogura, Y., and Tanida, J. (2007). Three-dimensional information acquisition using a compound imaging system. *Opt. Rev.*, 14(5), 347–350. Retrieved from http://link.springer.com/article/10.1007%2Fs10043-007-0347-z.
27. Horisaki, R. and Tanida, J. (2011). Multidimensional TOMBO imaging and its applications. *Proc. SPIE*, 8165, 816516.
28. Goodman, J. W. (1996). *Introduction to Fourier Optics*. New York, McGraw-Hill.
29. Nehmetallah, G. and Banerjee, P. P. (2012). Applications of digital and analog holography in three-dimensional imaging. *Adv. Opt. Photon.*, 4(4), 472–553.
30. Schnars, U. and Jüptner, W. P. O. (2002). Digital recording and numerical reconstruction of holograms. *Meas. Sci. Technol.*, 13(9), R85–R101. Retrieved from http://stacks.iop.org/0957-0233/13/i=9/a=201.
31. Rivenson, Y., Stern, A., and Javidi, B. (2010). Compressive Fresnel holography. *J. Disp. Technol.*, 6(10), 506–509. Retrieved from http://jdt.osa.org/abstract.cfm?URI=jdt-6-10-506.

32. Horisaki, R., Tanida, J., Stern, A., and Javidi, B. (2012). Multidimensional imaging using compressive Fresnel holography. *Opt. Lett.*, 37(11), 2013–2015.
33. Brady, D. J., Choi, K., Marks, D. L., Horisaki, R., and Lim, S. (2009). Compressive holography. *Opt. Express*, 17(15), 13040–13049. Retrieved from http://www.opticsexpress.org/abstract.cfm?URI=oe-17-15-13040.
34. Wagadarikar, A., John, R., Willett, R., and Brady, D. (2008). Single disperser design for coded aperture snapshot spectral imaging. *Appl. Opt.*, 47(10), B44–B51. Retrieved from http://ao.osa.org/abstract.cfm?URI=ao-47-10-B44.
35. Llull, P., Liao, X., Yuan, X., Yang, J., Kittle, D., Carin, L., Sapiro, G., and Brady, D. J. (2013). Coded aperture compressive temporal imaging. *Opt. Express*, 21(9), 10526–10545. Retrieved from http://www.opticsexpress.org/abstract.cfm?URI=oe-21-9-10526.
36. Tsai, T. H. and Brady, D. J. (2013). Coded aperture snapshot spectral polarization imaging. *Appl. Opt.*, 52(10), 2153–2161.
37. Gao, L., Liang, J., Li, C., and Wang, L. V. (2014). Single-shot compressed ultrafast photography at one hundred billion frames per second. *Nature*, 516(7529), 74–77.

8 Compressive Holography

Yair Rivenson and Adrian Stern

Contents

Introduction

In digital holography, a hologram is formed on a detector that converts the impinging illumination to charge carriers (such as charge coupled device [CCD] or complementary metal oxide semiconductor [CMOS] cameras in the visible optic spectrum). Digital holography became an important imaging modality and was embedded within diverse fields, such as in life sciences and materials engineering. One of the main advantages of digital holography is the ability to numerically reconstruct the object, using a standard computer, in contrast to standard holography that required film development procedures.

In this chapter, we demonstrate that digital holography is, in fact, a robust and highly effective optical compressive imaging modality. This claim is substantiated by the following points:

- The physical encoding of the object's field is performed by free space wave propagation, which is governed by the phenomena of diffraction (Goodman 1996). This means that, unlike other sensing modalities, no optical components are needed and, in many cases, a simple lensless optical setup can suffice. This also makes compressive holography a physically realizable compressive imaging application.
- The former statement also relaxes one of the most stringent requirements associated with computational imaging sensors, which is their *calibration*. Calibration issues are often disregarded in standard isomorphic imaging, but they play a crucial role in computational imaging applications, since recovery is based on a numerical model of the sensing process, which must be highly accurate. Incorrect calibration of the physical sensing operator could result in severe degradation in the performance of the computational imaging system (Chapter 4). However, since compressive holographic applications encode the object's field using free space diffraction, almost no physical elements are required, which means that related issues, such as misalignment, manufacturing errors, or inaccurate numerical models, can be avoided. Free space diffraction numerical models are well established (Schmidt 2010) and are dependent only on the illuminating wavelength, the distance to detector, and detector features (such as the pixel pitch and number of pixels).
- While sparsity-constrained signal reconstruction has been used for several decades and was also used for the reconstruction of objects from their digital holograms (Sotthivirat and Fessler 2004), compressive sensing (CS) has had a dramatic impact due to its successful demonstration that compressive measurements can be taken in a (signal) nonadaptive manner while providing sufficient *guarantees* for the reconstruction of classically considered subsampled signals (Candès and Wakin 2008). We shall discuss the reconstruction guarantees for compressive holography applications in the following sections and show that closed-form reconstruction guarantees can be obtained for several cases (summarized in Table 8.1), which are consistent with a physical intuitive perception of the system.

Hologram Forward Model

A general hologram recording setup is shown in Figure 8.1. Let us consider an object f that is illuminated with a collimated coherent light source in the form of a plane wave. The object's complex field amplitude (CFA) propagates in the $\hat{z}$ direction over a distance z, until it reaches the detector plane. A hologram is created by interfering the object's CFA, denoted by g, with the illuminating reference wave, E_r. The resultant interference on the detector plane is given by

$$I = |g + E_r|^2 = |g|^2 + |E_r|^2 + E_r g^* + E_r^* g, \tag{8.1}$$

where the $(\cdot)^*$ operator denotes the complex conjugate. Our primary interest lies in the object term, that is, $E_r^* g$, which contains all the information on the object

Table 8.1 Summary of the Different Modes, Applications, and Reconstruction of Compressive Digital Holographic Sensing

	Applications	Reconstruction Guarantees
Random subsampling of Fresnel transformed object	Limited pixel budget, suited for expensive detectors (THz, UV, etc.) Reduction of acquisition time, that is, faster acquisition Acquiring several signal modalities in a single shot (such as polarization, color) Removing noise/ reconstruction of objects under low illumination conditions Radar imaging Reconstruction of object from a single Fresnel transformed magnitude (phase retrieval)	Number of required compressive samples, M. Near-field approximation: $M \geq C'N_F^2 \frac{S}{N} \log N.$ Far-field approximation: $M \geq CS \log N.$ Optimal working distance is given by $z = \sqrt{N}\Delta x_0^2/\lambda.$ Using converging/diverging spherical illumination, the optimal working distance can be tailored.
General subsampling of the Fresnel transformed object	Optical super-resolution (diffraction limited) Geometrical super-resolution (high-precision particle localization and detection) Reconstruction of a partially occluded object with a single shot	Number of sparse object features that can be accurately reconstructed, S: $S \leq 0.5(1 + \theta^{-1})$, where $\theta^{FF} = \max_{m \neq l \cap u \neq v} \frac{\lvert \hat{O}(m-l,u-v) \otimes \hat{O}(m-l,u-v) \rvert}{\lVert o \rVert_2^2}$ where $o(x,y)$ is the subsampling function (pixel, finite aperture, occlusion) and $\hat{O}$ is its corresponding Fourier transform.
3D object inference from its 2D hologram	Single-shot 3D object tomography Axially super-resolved inference of sparse objects, surpassing classical limits Multiple aperture 3D object tomography, for enhanced axial resolution	Number of sparse object features that can be accurately reconstructed, S: $S \leq 0.5[1 + \lambda\Delta z/(\Delta x \Delta y)]$.

we wish to reconstruct. In the early 1960s, Leith and Upatnieks (1962) pioneered off-axis holography in order to reduce the influence of the bias and twin-image ($E_r g^*$) terms. However, with the advent of digital holography, other methods that enabled the complete removal of the unwanted terms became feasible, most notably phase-shifting holography (Yamaguchi et al. 2006). Assuming a paraxial optics regime, the object field g is related to the spatial object distribution f, illuminated by a plane wave with wavelength λ, by the Fresnel transform (Goodman 1996), given by (up to a constant multiplication factor)

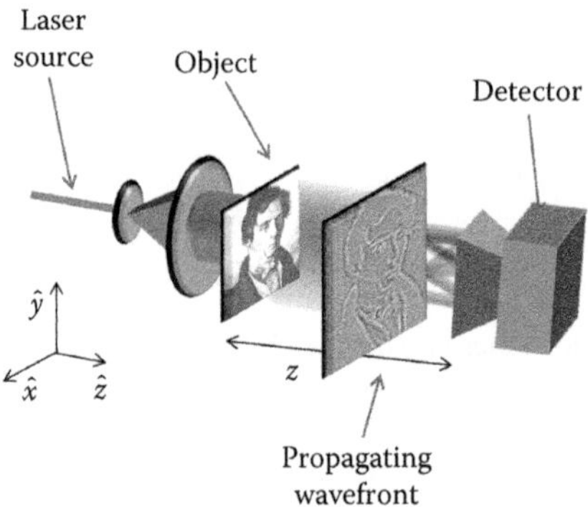

Figure 8.1 Schematic in-line (Gabor) holographic setup. An object (transparency of Augustin Fresnel) is illuminated by a coherent plane wave. The CFA propagates a distance z until it reaches the detector plane, where interference with a reference wave will occur to form the hologram.

$$\begin{aligned} g(x,y) &= f(x,y) * \exp\left\{\frac{j\pi}{\lambda z}(x^2+y^2)\right\} \\ &= \exp\left\{\frac{j\pi}{\lambda z}(x^2+y^2)\right\} \iint f(\xi,\eta)\exp\left\{\frac{j\pi}{\lambda z}(\xi^2+\eta^2)\right\} \\ &\times \exp\left\{\frac{-j2\pi}{\lambda z}(x\xi+y\eta)\right\} d\xi d\eta. \end{aligned} \tag{8.2}$$

While other types of holograms (such as Fourier holograms) also exist, in this chapter our focus is solely on Fresnel-type holograms. In subsequent sections, we shall derive the performance bounds of 2D and 3D compressive imaging applications that are based on the Fresnel transform as a sensing operator.

Reconstruction Guarantees for Compressive Fresnel Transform Imaging Systems

Reconstruction guarantees are one of the core products of CS (Candès and Wakin 2008) and probably what make it outshine previous sparsity-promoting optimization techniques. These guarantees are important in determining the operating conditions and performance of CS-based systems. For the sake of completeness, we briefly discuss the basics of CS and introduce the notation that will accompany us throughout the rest of this chapter.

Problem Statement for Compressive Sensing

CS is usually formulated in algebraic form, writing the forward operator (relating the object space to the measurement space) as a matrix–vector multiplication. Here, we shall define our measurement process as

$$\mathbf{g} = \mathbf{\Phi f}, \tag{8.3}$$

where **f** is an $N \times 1$ column vector that represents a lexicographic ordering of the 2D/3D sampled version of the object. We assume that N, the number of discrete object pixels, obeys the classic sampling theorem, that is, $N \geq 2LB$, where L is the object length and B is its bandwidth (Goodman 1996; Brady 2009). The measurement **g** is an $M \times 1$ vector, which is, again, the lexicographic representation of the measurements made by the detector. In the framework of CS, $M < N$. The *sensing operator* is defined as the operator that projects the object **f** to the measurement space **g** and is given by $\mathbf{\Phi}$, which is an $M \times N$ matrix.

In general, solving such an underdetermined problem seems impossible, but CS teaches us that it is, nevertheless, possible by using a priori knowledge about the object's sparsity (Candès and Wakin 2008). Formally, CS employs the accepted assumption that most natural signals can be sparsely represented using some mathematical operator, which we call the *sparsifying operator*, $\mathbf{\Psi}$, such that

$$\mathbf{f} = \Psi\alpha. \tag{8.4}$$

In this chapter, we take $\mathbf{\Psi}$ from Equation 8.4 to be an $N \times N$ matrix (mathematical basis), but, more generally, it can be described as a (overcomplete) dictionary (Elad and Aharon 2006). We denote α as an $N \times 1$ sparse vector with only $S \ll N$ meaningful entries. Such a vector is referred to as an S-sparse vector. CS then tells us that, out the signal we seek, **f** can be accurately recovered using the following norm-minimization problem:

$$\|\alpha\| \quad \text{subject to } \mathbf{g} = \Phi\Psi\alpha, \tag{8.5}$$

where $\|\cdot\|$ is the norm operator, which promotes the sparse structure of the signal. For compressive imaging applications, we will usually encounter the following:

- The ℓ_1-norm $\|\alpha\|_1 = \sum_i |\alpha_i|$, which promotes sparse solution and is convex; it is used to replace the combinatorial problem of finding the sparsest solution by the ℓ_0-norm, which is defined as counting the number of nonzero terms $\{\alpha_i \neq 0\}$ (see Chapters 1 and 2).

- The total variation (TV) norm, formulated using TV minimization (Rudin et al. 1992). For a 2D object, it is formulated as follows:

$$TV(\boldsymbol{\alpha}) = \sqrt{(\alpha_{i+1,j} - \alpha_{i,j})^2 + (\alpha_{i,j+1} - \alpha_{i,j})^2}\,. \tag{8.6}$$

The TV norm can be considered to be the minimization of the gradient of the object. Since this gradient is sparse in many cases, the sparsity operator is taken to be $\mathbf{\Psi} = \mathbf{I}$, that is, the canonical basis. In this case, $TV(\boldsymbol{\alpha}) = TV(\mathbf{f})$. Following reconstruction guarantees are not derived for the TV norm. However, practically, reconstruction using the TV norm follows the same trends.

Today, solving the aforementioned programs has become comfortably feasible both from aspects of hardware and software.

To solve the aforementioned programs, many algorithms are freely distributed on the Internet and can be applied using a standard computer. Although at the dawn of CS the results were derived for linear programming, many researchers, especially in the optics community (where the number of variables easily reaches ~10^6, corresponding to megapixel images), have abandoned them, due to the expense of their required computational resources. Today, the most popular solvers are based on the iterative shrinkage algorithms (Zibulevsky and Elad 2010).

The Fresnel Transform as a Sensing Operator

From this point to the end of this chapter, we will consider the Fresnel transform as our sensing operator, $\mathbf{\Phi}$ from Equation 8.3. As demonstrated in Equation 8.2, the diffraction of the object field is dependent on the illuminating wavelength, the working distance between the object and the detector, and the object's bandwidth. Let us fix the illuminating wavelength and object's bandwidth, and then the numerical Fresnel approximation only depends on the working distance, z. For numerical approximation of the Fresnel transform (8.2), it is convenient to distinguish between the *near-* and *far-*field regimes (Mas et al. 1999). The numerical near-field approximation is given by

$$g(p\Delta x_o, q\Delta x_o) = \mathcal{F}_{2D}^{-1} \exp\left\{-j\pi\lambda z_p\left(\frac{m^2}{N\Delta x_0^2} + \frac{n^2}{N\Delta y_0^2}\right)\right\} \mathcal{F}_{2D}\{f(l\Delta x_0, k\Delta y_0)\}, \quad (8.7)$$

where

Δx_0, Δy_0 are object and CCD resolution pixel sizes (inversely proportional to the signal's bandwidth), with $0 \le p, q, k, l \le \sqrt{N} - 1$

$\mathcal{F}_{2D}$ is the 2D Fourier transform

We assume that the object size and sensor size are $\sqrt{N}\Delta x_0 \times \sqrt{N}\Delta y_0$. The near-field model is valid for the regime where $z \le z_0 = \sqrt{N}\Delta x_0^2/\lambda$ (Mas et al. 1999). For the working regime of $z \ge z_0 = \sqrt{N}\Delta x_0^2/\lambda$, the far-field numerical approximation is given by

$$g(p\Delta x_z, q\Delta y_z) = \exp\left\{\frac{j\pi}{\lambda z}\left(p^2\Delta x_z^2 + q^2\Delta y_z^2\right)\right\}$$

$$\times \mathcal{F}_{2D}\left[f(k\Delta x_0, l\Delta y_0)\exp\left\{\frac{j\pi}{\lambda z}\left(k^2\Delta x_0^2 + l^2\Delta y_0^2\right)\right\}\right], \quad (8.8)$$

where $\Delta x_z = \lambda z/(\sqrt{N}\Delta x_0)$ is the output field's pixel size. From Equations 8.7 and 8.8, it is evident that the sensing operator is also very easy to implement from computational perspectives; it does not require the storage of a large sensing matrix, $\mathbf{\Phi}$, since it can be realized by fast Fourier transform operators.

Determining Reconstruction Guarantees for Subsampled Fresnel Fields

At the end of the previous section, we pointed out that the treatment of the numerical Fresnel transform should be divided into the near- and far-field regimes. We now further divide the analysis for reconstruction guarantees into two main categories. The first is where a low-budget measurement system should be designed, that is, where each pixel counts. This measurement scheme may not be justified in the visible optic regime, but it is suitable for IR, THz, UV, and x-ray regimes. We refer to this category as *random subsampling* of the Fresnel transformed signal (Figure 8.2). The second category is for *deterministic subsampling*. Such subsampling is usually encountered in modern optical experiments, where the object is

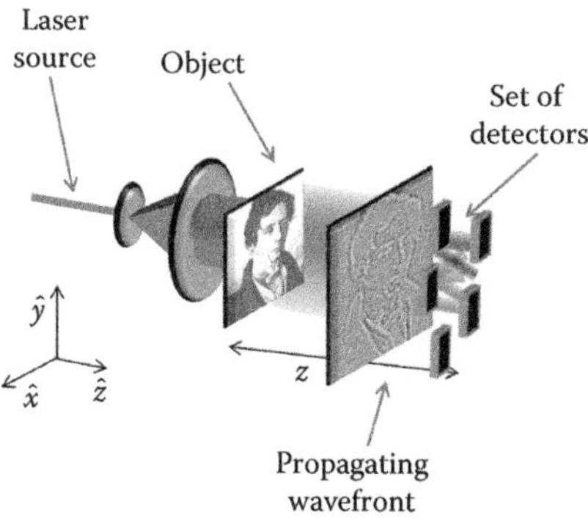

Figure 8.2 Illustrative example for sensing the Fresnel transformed object field using a sparse array of detectors.

measured by an array of detectors, such as CMOS or CCD devices. In this scenario, we have ~10^6 pixels in the detector (that costs only a few dollars; Gehm and Brady 2015) to perform the measurement with. Using the derived mathematical bounds, we can show that the signal can be reconstructed more faithfully than with classical sampling theories.

Random Subsampling of the Object's Wavefield

In order to gain an intuitive understanding of the aforementioned theory, we present the following example: consider a sparse set of detectors randomly spread across the measurement space, as shown in Figure 8.2. Referring to Equation 8.3, this setting is mathematically described as uniformly picking M out of N rows of $\mathbf{\Phi}$ at random. For this scenario, the reconstruction guarantees are determined by the coherence parameter (Candès and Plan 2011):

$$\mu = \max_{i,j} \left| \langle \phi_i, \psi_j \rangle \right|, \tag{8.9}$$

where

ϕ_i is a row vector of $\mathbf{\Phi}$
ψ_j is a column vector of $\mathbf{\Psi}$
$\langle \cdot,\cdot \rangle$ denotes the inner product

From a physical perspective, the coherence parameter, μ, measures the amount of the object's energy that has been spread between the detectors and is bounded by $1/\sqrt{N} \le \mu \le 1$. Ideally, we wish that each detector will sense an equal amount of energy from each object pixel. In this case, it can be shown that $\mu \to 1/\sqrt{N}$. On the other hand, if each object pixel is exactly mapped to a single detector pixel (which is the conventional goal with isomorphic imaging), one cannot discard any measurements, since this will most certainly result in information loss. In this case, $\mu \to 1$. The reasoning here is that if each detector pixel contains information about the entire object, then some measurements can be discarded. According to the CS theory, the signal can be reconstructed by taking M uniformly, at random projections, obeying (Candès and Plan 2011)

$$\frac{M}{N} \ge C\mu^2 S \log N, \tag{8.10}$$

where C is a small constant. It is clear from Equation 8.10 that the smaller the μ is, the lower the relative number of measurements that are required in order to enable accurate reconstruction of the object. Practically, for $\mu = 1/\sqrt{N}$, the number of samples required for accurate signal reconstruction is $M \approx S \log N$ measurements. A landmark example of this bound can be found in the earlier results in the CS theory, which have shown that subsampling of the Fourier-transformed signals is an extremely efficient CS mechanism (Candes et al. 2006; Candès and Plan 2011) and the signals can be accurately reconstructed from $M \approx S \log N$ samples.

Let us assume that the object is sparse in its native space, that is, $\mathbf{\Psi} = \mathbf{I}$ (we shall discuss more general conditions toward the end of this section). In this case, for the near-field numerical approximation, it can be shown that the coherence parameter can be approximated by (Rivenson and Stern 2011)

$$\mu_{\text{near field}} = \max_i |\phi_i| \approx \left[\frac{\Delta x_0 \Delta y_0}{\lambda z}\right], \quad z < z_0. \tag{8.11}$$

From a physical perspective, each ϕ_i represents the point spread response for each of the N object points. The function $|\phi_i|$ is shown in Figure 8.3 for a different working distance, z. As z is increased, the energy of each object point spreads across more measurements, thus reducing the coherence parameter. From Equation 8.11, we can derive the number of compressed samples, M, required for object recovery as

$$M \geq C' N_F^2 \frac{S}{N} \log N, \tag{8.12}$$

where

N_F denotes the recording device Fresnel number (Goodman 1996) given by $N_F = N((\Delta x_0 \Delta y_0)/(4\lambda z))$

C' is a small constant factor

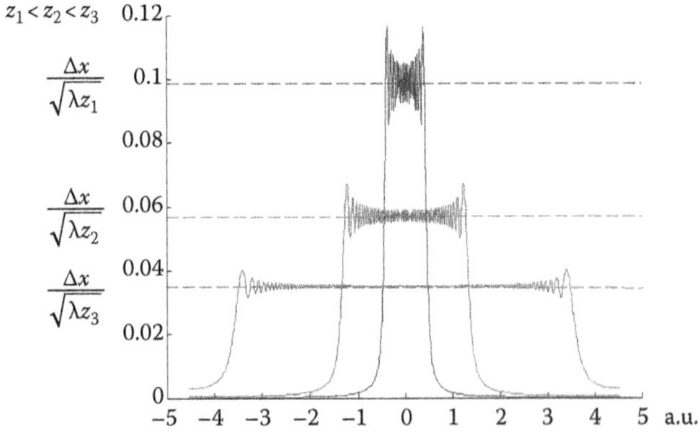

Figure 8.3 Magnitude of the one-dimensional near-field Fresnel point spread function. As the working distance, z, decreases, the magnitude increases, meaning that less energy is spread between the different pixels. The coherence parameter is inversely proportional to z. This, in turn, implies that more samples are necessary in order to retrieve the exact signal.

For the far-field numerical Fresnel approximation, the coherence parameter obeys

$$\mu_{\text{far field}} = \frac{1}{\sqrt{N}}, \tag{8.13}$$

which is exactly the lower bound that the coherence parameter can reach; therefore, the number of required Fresnel measurements of the Fresnel field is given by

$$M \geq CS \log N. \tag{8.14}$$

The trend reflected from Equations 8.12 and 8.14 has been demonstrated via a simulation (Rivenson and Stern 2011) and has also been demonstrated experimentally (Fan et al. 2014).

On this issue, we wish to make the following two practical comments:

- The optimal working distance should be at the point where $z = z_c = \sqrt{N}\Delta x^2/\lambda$. At this point, we can recover the field from a minimal number of measurements, while keeping the detector placement as compact as possible, since the support area for the diffraction pattern would be $\sqrt{N}\Delta x_0 \times \sqrt{N}\Delta y_0$, which is smaller than $\sqrt{N}\Delta x_z \times \sqrt{N}\Delta y_z$ for any $z > z_c$.
- In the earlier analysis, we have assumed that the object is sparse in the spatial domain. In the case where the object of interest is sparsely represented using a sparsifying operator, which is different from the canonical basis, for example, wavelet transform or the number of gradients, it is practically difficult to get an exact closed-form expression, as is demonstrated in the canonical case. However, a numerical investigation (Rivenson et al. 2013a) shows that the trend predicted from Equations 8.11 and 8.14 is valid for other sparsifying bases as well.

Reconstruction Guarantees for Spherical Wave Illumination

In many holographic applications, the object is illuminated by a spherical wavefront, especially in compact microscopy (lensless) systems (Coskun et al. 2010; Hahn et al. 2011) and, more recently, in x-ray digital holography (Rehman et al. 2014). Let us suppose that the illuminating wavefront originates from a point source located at distance z_i behind the object, as shown in Figure 8.4. In this case, the calculation of the coherence parameter needs to be changed accordingly. Plugging the diverging spherical illumination, instead of the planar illumination, into Equation 8.2, we obtain

$$\begin{aligned} g(x,y) &= \exp\left(j\pi\frac{x^2+y^2}{\lambda z_i}\right) f(x,y) * \exp\left(j\pi\frac{x^2+y^2}{\lambda z}\right) \\ &= \exp\left(j\pi\frac{x^2+y^2}{\lambda z}\right) \iint f(\xi,\eta) \exp\left(j\pi\frac{\xi^2+\eta^2}{\lambda}\left(\frac{1}{z}+\frac{1}{z_i}\right)\right) \\ &\quad \times \exp\left\{\frac{-j2\pi}{\lambda z}\left(x\xi+y\eta\right)\right\} d\xi d\eta. \end{aligned} \tag{8.15}$$

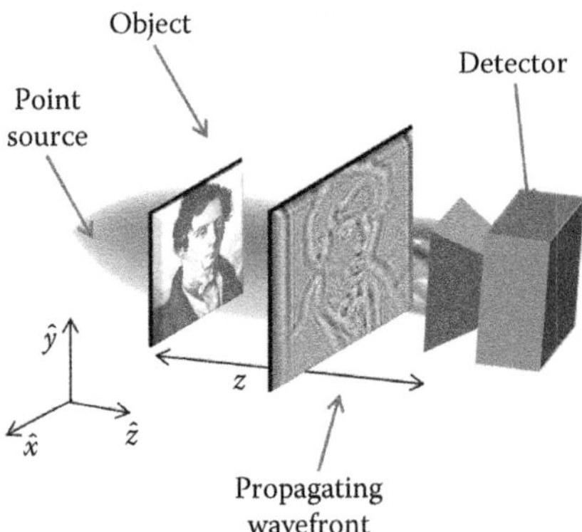

Figure 8.4 Illumination of the object using a spherical wave emanating from a point source.

Using similar arguments to those we have used for the planar illumination case, we can define the Fresnel kernel sampling condition (Mas et al. 1999), which, for simplicity, is expressed for the 1D case as follows:

$$\frac{\Delta x_0^2}{\lambda}\left(\frac{1}{z}+\frac{1}{z_i}\right)<\frac{1}{\sqrt{N}}. \tag{8.16}$$

Consequently, the working distance that defines the limit between the near- and far-field numerical approximations is given by (Stern and Rivenson 2014)

$$z_{c-\text{diverging}}=\frac{\sqrt{N}\Delta x_0^2}{\lambda-\sqrt{N}\Delta x_0^2/z_i}. \tag{8.17}$$

From Equation 8.15, we see that the diffraction is now dependent on the term $z_{eq} = z \cdot z_i/(z + z_i)$ rather than on z, as in the planar illumination case. Returning to the 2D model, we can deduce that the coherence parameter in the numerical near-field Fresnel approximation is given by

$$\mu_{\text{near field}}^{\text{diverging}}=\max_i\left|\phi_i^{\text{diverging}}\right|\approx\frac{\Delta x_0\Delta y_0}{\lambda\left[\dfrac{z\cdot z_i}{z+z_i}\right]}. \tag{8.18}$$

As in the plane wave illumination case, for the far-field approximation, $\mu_{\text{far field}}^{\text{diverging}}=1/\sqrt{N}$.

To align Equation 8.18 with a physical intuitive understanding, we notice that by taking the limit $z_i \rightarrow \infty$, $\mu_{\text{near field}}^{\text{diverging}}=\mu_{\text{near field}}^{\text{plane}}$, that is, the coherence parameter from the plane wave illumination. From Equations 8.10 and 8.18, the number of Fresnel field measurements required for accurate S-sparse object reconstruction is given by

$$M\geq C\left\{\frac{\sqrt{N}\Delta x_0\Delta y_0}{\lambda\left[\dfrac{z\cdot z_i}{z+z_i}\right]}\right\}^2 S\log N. \tag{8.19}$$

We can further extend the result to the case where the illuminating wavefront is a converging spherical illumination. Following the same analysis, the coherence parameter for this case is given by

$$\mu_{2D}^{\text{converging}} \approx \frac{\Delta x_0 \Delta y_0}{\lambda \left[\dfrac{z \cdot z_i}{z - z_i} \right]}. \tag{8.20}$$

It can be seen that by applying a convergent object illumination, the coherence parameter can be reduced, compared with converging or plane wave illumination for the same object to detector distance, z. The latter implies that for applications with a limited number of detector pixels that require proximity between the sample and the detector, illuminating the object with a converging wavefront should be preferred.

Applications of Random Subsampling of the Fresnel Field

Several applications that use the subsampling of the Fresnel field as a sensing mechanism have been suggested. Reconstruction of an object from its randomly subsampled Fresnel field has been demonstrated in Rivenson et al. (2010), where the sampling was adapted to put more emphasis on the center of the hologram. The same framework has also shown positive results for the reconstruction of subsampled frequency-shifting holography (Marim et al. 2010). As stated earlier, one of the immediate benefits of this development is the ability to use fewer pixels in spectral regimes where the pixel is relatively expensive, such as in the THz range (Cull et al. 2010). Returning to the visible optical regime, this framework was also implemented for the reconstruction of an object from a set of subsampled holograms in order to mitigate low object illumination conditions (Marim et al. 2011). In yet another interesting effort, a conceptual work (Horisaki et al. 2012) described how to use the compressive Fresnel holography framework by randomly dividing a standard detector array so that each group of randomly selected pixels will treat a different object modality, such as color channel (R, G, B), or state of polarization, using custom filters built upon the pixel sensors (see Chapter 7). This way, multidimensional imaging can be acquired in a single shot. Another interesting employment of this framework was demonstrated for the reconstruction of subsampled microwave radar holograms (Wilson and Narayanan 2014). The encoding efficiency of the Fresnel transform has also been found to be useful in the extraction of phase objects from their intensity measurements alone (i.e., non-interferometric recording), using a decoding element (Horisaki et al. 2014) or employing a conventional detector array (Rivenson et al. 2015).

We wish to conclude this section with an example, which is the compressive multiple view projection holography application (Rivenson et al. 2011). In the multiple view projection holography method (Shaked et al. 2009), a digital hologram is obtained using a simple optical setup that operates under spatially and temporarily "white" light illumination conditions. The method requires a conventional digital camera as a recording device. This method avoids many of the drawbacks related to coherent holographic acquisition systems, such as speckle noise, the need of high-coherence and high-intensity sources and the requirements of high mechanical and thermal stability of the optical setup. On the other hand, perhaps the major challenge with this method is the acquisition step. In this stage, multiple views of the scene are acquired by camera translation, which usually involves a tedious scanning effort, since in this method each view corresponds to a pixel in the hologram. Using the compressive Fresnel holography framework, we can dramatically reduce

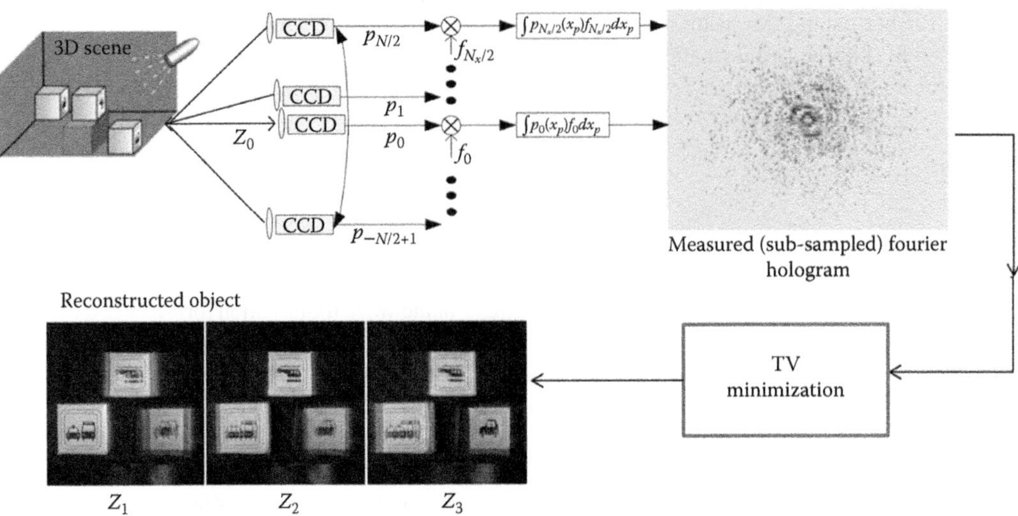

Figure 8.5 An illustration of compressive multiple view projection holography (Rivenson et al. 2011). Using a CCD camera located at a distance z_0 from a 3D scene, multiple 2D projections (denoted by p_i) are captured. Each acquired projection is digitally multiplied and summarized by a corresponding complex function to generate a subsampled hologram (Shaked et al. 2009). The number of captured projections in the compressive multiple view projection holography is $S \log N$, which is much smaller than the number of nominal projections, N. Despite this, the object is accurately reconstructed at the different depth planes, z_1, z_2, z_3.

the scanning effort of the acquisition step. This is done by taking only $M = S \log N$ ($S \ll N$) acquired samples of the object in the Fresnel domain instead of N sample views. Since each sample corresponds to a different view of the scene, an accurate reconstruction of the 3D scene can be obtained with only a fraction of the nominal number of exposures. This process is demonstrated in Figure 8.5. Following the reduced acquisition of the scene, the object can be exactly reconstructed by solving a TV minimization procedure (see the "Problem Statement for Compressive Sensing" section and "Fresnel Diffraction with Pixel Basis" section in Chapter 3).

Reconstruction Guarantees for Nonrandom Subsampling of the Fresnel Field

In the previous section, we dealt with reconstruction guarantees for the case of random subsampling of the Fresnel diffraction of the object; in this section, we discuss reconstruction guarantees for general subsampling of the Fresnel fields. For general (nonrandom) subsampling, it is a better practice to calculate the coherence parameter as follows:

$$\theta = \max_{i \neq j} \frac{\left|\langle \omega_i, \omega_j \rangle\right|}{\left\|\omega_i\right\|_2 \left\|\omega_j\right\|_2}, \tag{8.21}$$

where ω_i is the column vector of $\mathbf{\Omega} = \mathbf{\Phi\Psi}$, $\mathbf{\Omega} \in \mathbb{C}^{M \times N}$. It can be shown (Bruckstein et al. 2009) that $\sqrt{(N-M)/[M(N-1)]} \leq \theta \leq 1$. Using this definition, an S-sparse signal reconstruction guarantee is given by

$$S \le \frac{1}{2}\left\{1+\frac{1}{\theta}\right\}. \tag{8.22}$$

As θ gets smaller, we can accurately reconstruct higher dimensional S-sparse signals. As the number of measurements $M \rightarrow N$, the coherence parameter $\theta \rightarrow 0$. A simple way to calculate θ is by calculating the Gram matrix (Chapter 1), $\mathbf{G} = \left|\tilde{\Omega}^*\tilde{\Omega}\right|$, where $\tilde{\Omega}$ is the column normalized $\mathbf{\Omega}$ and $\tilde{\Omega}^*$ is the Hermitian conjugate of $\tilde{\Omega}$. The coherence parameter θ is evaluated as the largest off-diagonal element of this Gram matrix.

Nonrandom wavefield truncation is more relevant for the visible optical regime, such as subsampling induced due to the finite aperture of the detector or lens, occlusion, and finite pixel width. Basically, one can consider this problem as that of the extraction of more information from a conventionally measured hologram. Prior to deriving the guarantees, we formulate the problem in the spatially continuous domain as follows:

$$g(x,y) = \left[f(x,y) * \exp\left\{\frac{j\pi}{\lambda z_1}(x^2+y^2)\right\}\right] \times o(x,y). \tag{8.23}$$

Equation 8.23 describes the Fresnel diffracted field multiplied by a truncating function $o(x,y)$. The function $o(x,y)$ may represent, for instance, the finite size of the CCD, the finite pixel size of the CCD, or that of any aperture that blocks the field. In this sense, such a subsampling scheme also includes the random subsampling case. We also assume that after the diffracted field is truncated, no other information loss occurs. Several applications have used deterministic subsampling Fresnel encoded objects in order to demonstrate improved reconstruction, compared with standard imaging/holography. One of the earliest demonstrations was super resolution for incoherent fluorescent beads (Coskun et al. 2010). The recovery of partially occluded objects was proved possible with a single shot by capturing a truncated Fresnel field of the object (Rivenson et al. 2012). This framework was also employed for subpixel object localization, up to 1/45 of the pixel size in 1D localization (Liu et al. 2012) and $1/30^2$ of the pixel in 2D localization (Liu et al. 2014). Such a model was also successfully implemented for the measurement of the concentration of bubbles in a fluid (Chen et al. 2015).

Example: Reconstruction Guarantees for a Limited Aperture

From Equation 8.21, we can analytically calculate the coherence parameter for the numerical far-field approximation (Rivenson et al. 2012):

$$\theta^{FF} = \max_{m \ne l \cap u \ne v} \frac{\left|\hat{O}(m-l,u-v) \otimes \hat{O}(m-l,u-v)\right|}{\|o\|_2^2}, \tag{8.24}$$

where

$\otimes$ denotes the correlation operator
$\|\cdot\|_2$ is the ℓ_2-norm operator
$\hat{O}$ is the 2D Fourier transform of o, $\hat{O} = \mathcal{F}_{2D}\{o\}$

The indices $0 \le m,l,u,v \le \sqrt{N}-1$ denote the sensing matrix columns. The result in Equation 8.24 formulates the coherence parameter dependence on the structural properties of the truncating function. The number of S-sparse signal elements that can be accurately recovered is inversely proportional to θ, according to Equation 8.22. Equation 8.24 holds for the common case where $\Delta x_z = \lambda z/(\sqrt{N}\Delta x_0)$ and $\Delta y_z = \lambda z/(\sqrt{N}\Delta y_0)$ (Mas et al. 1999).

As reflected in Equation 8.21, the coherence parameter is actually defined by the similarity of two matrix columns of $\tilde{\Omega}$, where these columns are the lexicographic reordering of the point spread functions of the system (which are 2D functions). Thus, the coherence parameter measures the similarity between such 2D point spread functions originating from a 2D array of object point sources. That is the reason we formulated the coherence parameter in Equation 8.24 as the maximum of a 4D function, rather than in its standard 2D form.

In order to provide some intuitive understanding of the result in Equation 9.24, suppose now that the object's propagating wavefield is truncated by a square detector aperture of size $\sqrt{N}\Delta x_z \times \sqrt{N}\Delta y_z$, formally defined as

$$o(x,y) = \text{rect}\left(\frac{n}{\sqrt{N}\Delta x_z}\right)\text{rect}\left(\frac{l}{\sqrt{N}\Delta y_z}\right). \tag{8.25}$$

From Equation 8.24, it follows that the coherence parameter is given by

$$\theta^{FF} \approx \max_{\tau_x \neq 0 \cap \tau_y \neq 0}\left|\text{sinc}\left(\frac{\sqrt{N}\Delta x_o \Delta x_z \tau_x}{\lambda z}\right)\text{sinc}\left(\frac{\sqrt{N}\Delta y_o \Delta y_z \tau_y}{\lambda z}\right)\right|, \tag{8.26}$$

where $\tau_x = m - l$, $\tau_y = u - v$. A close inspection of this equation shows that the sinc() function argument is reduced to

$$\frac{\sqrt{N}\Delta x_o \Delta x_z}{\lambda z} = \frac{\sqrt{N}\Delta x_o \lambda z/(\sqrt{N}\Delta x_0)}{\lambda z} = 1, \tag{8.27}$$

which means that $\theta(\tau_x,\tau_y) = 0\ \forall\ \tau_x,\tau_y \neq 0$ (τ_x,τ_y are integer values) and the number of accurately reconstructed signal components $S \to \infty$ (Equation 8.22). Practically, all N signal components can be accurately reconstructed. This can also be interpreted as the Abbe resolution limit, since

$$\Delta x_o = \frac{\lambda z}{\sqrt{N}\Delta x_z} = \frac{\lambda}{2NA}, \tag{8.28}$$

where NA is the system's numerical aperture, defined as the sine of angle $((\sqrt{N}/2)(\Delta x_z))/z$. This means that Abbe's law is a specific setting, where N object pixels are mapped to N detector pixels with a proper pixel pitch (Equation 9.27) in a CS formulation of the object reconstruction problem. This also implies that for extremely sparse signals, the Abbe limit can be surpassed (which is well used in localization microscopy).

Reconstruction Guarantees for 3D Object Tomography from Its 2D Hologram

In the previous section, we have shown that holography can be used as an effective sensing mechanism to achieve superior transverse resolution. As

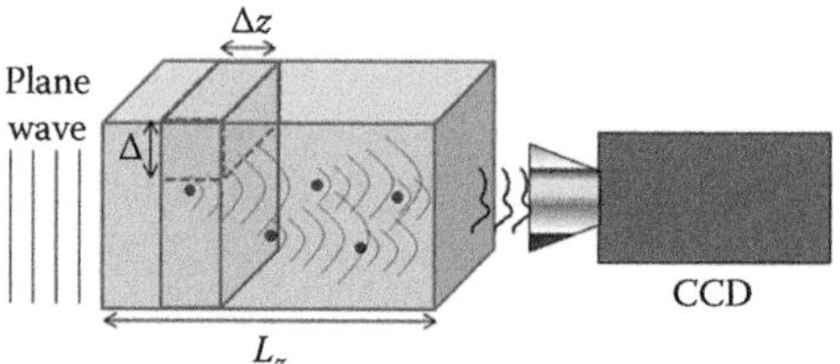

Figure 8.6 Schematic description of tomographic imaging with Fresnel in-line holography. A plane wave illuminating a volume of length L_z. The wavefront scattered from the different particles is holographically recorded on a CCD. (From Rivenson, Y. et al., *Opt. Lett.*, 38, 2509, 2013c. With permission.)

mentioned in the "Introduction" section, by recording the object's wavefield, holography encodes 3D information coming from the entire object planes into a 2D hologram. In this section, we demonstrate how 3D object tomography is indeed possible from 2D object holography. A schematic figure of the setup is shown in Figure 8.6. In order to reconstruct an object from its hologram, we use a numerical Fresnel propagation equation, such as in Equation 8.7. The numerical reconstruction obtained by digitally focusing on different object depth planes may be distorted due to out-of-focus object points located in other object planes, as seen in Figure 8.5, where in each reconstructed plane, z_1, z_2, z_3, out-of-focus object points that belong to different depth planes appear as a blurred object. These disturbances are the result of an incomplete model of the system, because the forward model of Equation 8.7 represents a 2D–2D model linking the hologram plane to a single depth plane, thus ignoring other object planes. This is a natural consequence, since, by applying reconstruction techniques based on a 2D–2D system model for 3D–2D object sensing, object points disobeying the model, that is, object points located in another depth plane, would be distorted.

In order to avoid these out-of-focus distortions, a more appropriate forward model should be formulated, that is, a 3D–2D forward model relating all the $N_{\text{object}} = N_x \times N_y \times N_z$ object voxels to $N_{\text{holo}} = N_x \times N_y$ hologram pixels should be used (Brady et al. 2009; Rivenson et al. 2013c). We formulate the model using the Fresnel numerical near-field approximation as follows:

$$g(k\Delta x_0, l\Delta y_0) = \sum_{r=1}^{Nz} \mathcal{F}_{2D}^{-1}\{e^{-j\pi\lambda r\Delta z[(\Delta\upsilon_x m)^2+(\Delta\upsilon_y n)^2]}\, e^{j(2\pi/\lambda)r\Delta z}$$
$$\times \mathcal{F}_{2D}[f(p\Delta x_0, q\Delta y_0; r\Delta z)]\}, \quad (8.29)$$

with $0 \le p, m \le N_x - 1$, $0 \le q, n \le N_y - 1$, and $1 \le r \le N_z - 1$. In Equation 8.29, the 3D object space is partitioned in a grid with $N_x \times N_y \times N_z$ voxels, each of size $\Delta x_0 \times \Delta y_0 \times \Delta z$. The $\mathcal{F}_{2D}$ operator denotes the 2D discrete Fourier transform. The spatial frequency variables are $\Delta\upsilon_x = 1/(N_x\Delta x_0)$ and $\Delta\upsilon_y = 1/(N_y\Delta y_0)$. This model assumes that the object depth dimension is sampled at constant sampling intervals Δz, and the entire object depth is $N_z\Delta z$. The problem of reconstructing the 3D object from its 2D hologram is *naturally underdetermined*, since there are N_z times more variables than equations.

Based on the knowledge that we have gained from the previous sections, we realize that we can formulate this problem as a CS problem:

- We have less equations ($M = N_x \times N_y$) than variables ($N = N_x \times N_y \times N_z$).
- We can assume that the object has an S-sparse representation, using an arbitrary mathematical transform, such that $S < M$, $S \ll N$.
- We have an efficient object encoding mechanism: the Fresnel transform.

Following this, we recast our problem as a CS problem. First, we formulate it as a standard matrix–vector multiplication:

$$\mathbf{g} = [\mathbf{H}_{\Delta z};\ldots;\mathbf{H}_{N_z\Delta z}][\mathbf{f}_{\Delta z};\ldots;\mathbf{f}_{N_Z\Delta z}] = \mathbf{\Phi f}, \tag{8.30}$$

where

$$\mathbf{H}_{r\Delta z} = \mathrm{e}^{j(2\pi/\lambda)r\Delta z}\mathcal{F}_{2D}^{-1}\mathbf{Q}_{r\Delta z}\mathcal{F}_{2D}. \tag{8.31}$$

The term $\mathbf{Q}_{r\Delta z}$ is a diagonal matrix that accounts for the quadratic phase elements of Equation 8.29, and $[\mathbf{f}_{\Delta z};\ldots;\mathbf{f}_{N_Z\Delta z}]$ is a lexicographical representation of the 3D object, where each $\mathbf{f}_{r\Delta z}$ is a column vector representing the 2D object distribution in depth plane $r\Delta z$. We note that the linear model (9.30), (9.31) assumes the Born approximation, that is, the object field in each plane does not interact with any other object plane (Brady et al. 2009).

Using the Gram matrix calculation, as in the previous section, we can derive mutual coherence as the intra-correlation between two propagation operators from two different planes (Rivenson et al. 2013c):

$$\theta_{3D} = \max_{k\neq l}\left|\tilde{\mathbf{H}}^*_{k\Delta z}\tilde{\mathbf{H}}_{l\Delta z}\right| = \max_{k\neq l}\left|\mathcal{F}_{2D}^{-1}\mathbf{Q}_{(k-l)\Delta z}\mathcal{F}_{2D}\right| \approx \max_{k\neq l}\left\{\frac{\Delta x\Delta y}{\lambda(l-k)\Delta z}\right\} = \frac{\Delta x\Delta y}{\lambda\Delta z}. \tag{8.32}$$

Thus, by combining Equations 8.32 and 8.22, we obtain the number of sparse object points that can be accurately reconstructed as

$$S \leq 0.5\left[1+\frac{\lambda\Delta z}{(\Delta x\Delta y)}\right], \tag{8.33}$$

Several conclusions can be drawn from Equations 8.32 and 8.33. The first is that, as expected, our ability to recover sparse objects is proportional to $\Delta z = L_z/N_z$ or inversely proportional to N_z. From a physical perspective, this means that the shorter the longitudinal separation between object points, the harder it becomes to differentiate them, which exactly agrees with the optical observations. From the mathematical perspective, it follows that as we divide our object into finer axial resolution, Δz, N_z increases, which means that the number of variables ($N_x \times N_y \times N_z$) we wish to resolve becomes larger, while there is no change in the number of equations ($N_x \times N_y$). The second conclusion we may draw considers the longitudinal resolution. The classical longitudinal resolution approximation is given by $\Delta z \approx 4\Delta x^2/\lambda$. This limit is generally accepted for differentiating two object point

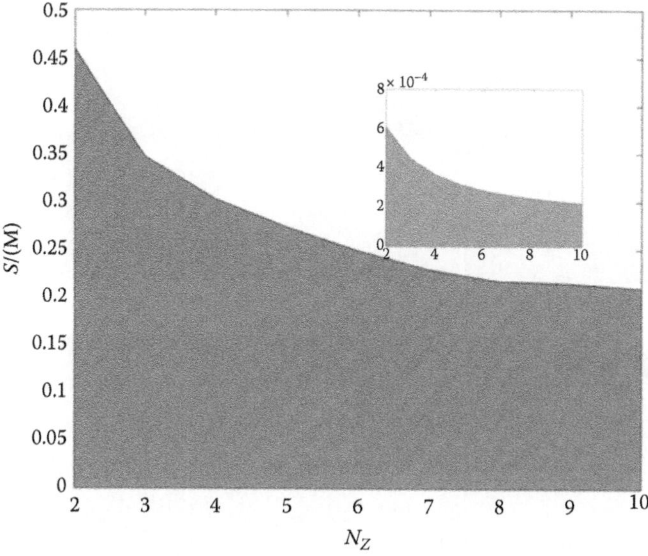

Figure 8.7 Simulation results showing the normalized number of reconstructed 3D object particles as a function of the number of object planes for a constant volume length, L_z, given a sensor with $M = N_x \times N_y$ pixels. The theoretical reconstruction guarantee according to Equation 8.33 is placed in the inset (Rivenson et al. 2013c). The ratio *S*/*M* indicates the ratio of the number of point particles that can be reconstructed (*S*) using *M* measurements. As predicted from Equation 8.33, as we divide our volume into finer *z*-sections (increasing of N_z), the number of reconstructed samples decreases. Exact reconstruction was assumed when the reconstruction mean squared error (MSE) was $<10^{-4}$.

sets on the optical axis, Δz apart. In our case, if we substitute $S = 2$ (points) to Equation 8.33, we find that $\Delta z = 3\Delta^2/\lambda$, which is ~33% superior to the classical resolution limitation. Simulation results, which are shown in Figure 8.7, support this claim and show that the possible number of particles that can be resolved is in fact much higher than predicted by Equation 8.33. We will further discuss this difference toward the end of this chapter.

In the simulation results shown in Figure 8.7, we have constrained the object to be sparse using the ℓ_1-norm; however, for natural, piecewise objects, such as the ones shown in the multiple view projection holography example (Figure 8.5), we may use other sparsity constraints. For example, we define the quasi-3D TV norm:

$$\|f\|_{TV} = \sum_r \sum_{p,q} \sqrt{(f_{p+1,q,r} - f_{p,q,r})^2 + (f_{p,q+1,r} - f_{p,q,r})^2}. \tag{8.34}$$

The use of this constraint has been demonstrated to be more effective for piecewise objects and promotes high-quality object reconstruction for 3D–2D forward holographic models (Brady et al. 2009). The result of applying the minimization procedure based on Equation 8.34 on the multiple view projection hologram that is shown in Figure 8.5 is shown in Figure 8.8. As demonstrated in Figure 8.8, we are able to get an almost accurate 3D object tomographic reconstruction at three

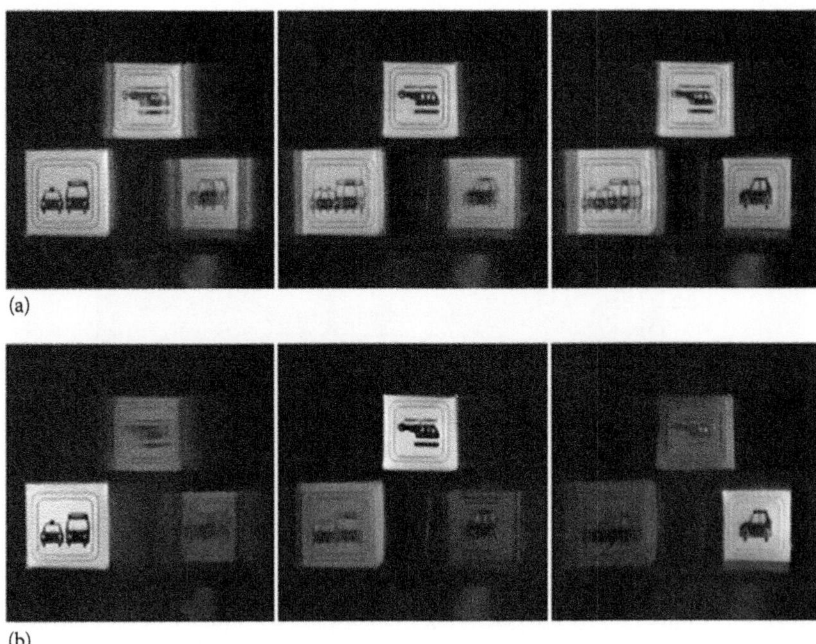

Figure 8.8 Reconstruction of 3D object tomography from their 2D holographic projection. (a) Reconstruction using the conventional 2D–2D Fresnel back-propagation model. (b) Reconstruction using the compressive holographic tomography approach. (From Rivenson, Y. et al., *Opt. Express*, 19, 6109, 2011. With permission.)

different depth planes (the out-of-focus terms are significantly reduced for each corresponding depth plane).

The arguments described earlier show that a *2D digital hologram* can indeed "compress" a *3D object* with *S* degrees of freedom, where the maximum number of degrees of freedom depends on the sensor's resolution, the axial resolution of the object, and the illuminating wavelength.

Several successful applications have already been implemented for the 3D–2D forward model framework. The pioneering work of Brady et al. (2009) used such a framework for the reconstruction of a 3D object from a single recorded Gabor hologram (Gabor 1948). At about the same time, another work (Denis et al. 2009) used related principles to demonstrate the reconstruction of an object from its in-line hologram. These works were followed by a reconstruction of a 3D object in the THz regime (Cull et al. 2010) and incoherent optical scanning holography (Zhang and Lam 2010). Using this framework, it was also demonstrated that it is possible to obtain superior axial resolution for multiple view projection holography when compared with standard multi-aperture acquisition systems (Rivenson et al. 2011). This framework was also applied to the reconstruction of a 3D object from its formed holographic projections at different angles (Nehmetallah and Banerjee 2012).

We conclude this section by discussing the combination of the 3D–2D forward model with the single-exposure in-line (SEOL) holographic recording setup (Javidi et al. 2005). The SEOL holography method uses a Mach–Zehnder interferometer setup to record the Fresnel diffraction field of the 3D object in a way similar to phase-shifting digital holography. However, in contrast to phase-shifting in-line digital holography techniques, SEOL digital holography uses only a *single exposure*. It was shown that the bias and twin-image interferences are substantially reduced or can be neglected in SEOL digital holographic microscopy under certain conditions (Stern and Javidi 2007), but there are reconstruction artifacts due to the inherent ill-posed nature of the 3D object reconstruction from its 2D hologram. The combination of the SEOL and 3D–2D compressive holography frameworks can yield an almost ideal setup for practicing compressive holography (Rivenson et al. 2013b). This follows from three properties of the SEOL holographic recording setup. First, the resolution of the system is the same as that of the in-line holography (i.e., Gabor) setup, which is at least twice the achievable resolution from an off-axis recording setup. The second property is that SEOL uses only a single shot, as opposed to the multiple (typically four) shots that are needed in order to apply the phase-shifting holography technique. The first and second properties are also common for Gabor holography. The third property, in which regard SEOL differs from in-line holography, is that SEOL holography behaves like a heterodyne system. This enables the recording of digital holograms with an improved signal-to-noise ratio (SNR) by proper control of the amplitude partition of the reference and object arms. Generally, an improvement in the SNR yields enhanced lateral object resolution details. This property of the SEOL setup, combined with CS reconstruction, has yielded improved axial resolution in comparison to compressive reconstruction obtained with Gabor holographic recording, as demonstrated in Figure 8.9. Similar results were demonstrated with biological (microorganisms) samples in Rivenson et al. (2013b).

Discussion and Conclusion

In this chapter, we demonstrated the effectiveness of the Fresnel transform, which naturally emerges through the process of diffraction and can be used as a physically realizable sensing modality. The Fresnel transform can be acquired using well-established digital holographic techniques. Using the Fresnel transform as a sensing mechanism relaxes stringent requirements on hardware implementation and calibration (Stern et al. 2013; Gehm and Brady 2015). The Fresnel transform has a very efficient realization and also provides a very efficient sensing mechanism, allowing the reconstruction of highly subsampled 2D and 3D object fields. The formulated reconstruction guarantees stand alongside other popular sensing mechanisms, such as the Fourier transform and random Gaussian measurements (Candès and Wakin 2008). We summarize the compressive digital holographic sensing reconstruction guarantees along with their corresponding applications in Table 8.1.

Though the reconstruction guarantees are, in general, pessimistic, they are nonetheless the only available guideline for the *accurate* reconstruction of *any* signal. They certainly provide an indication of performance trends. The guarantees summarized in this chapter provide technical guidelines that relate the physical attributes of the optical system (including illumination and detector features) to the overall performance of the system and predict what performance gain one might expect when modifying the system parameters or equipment.

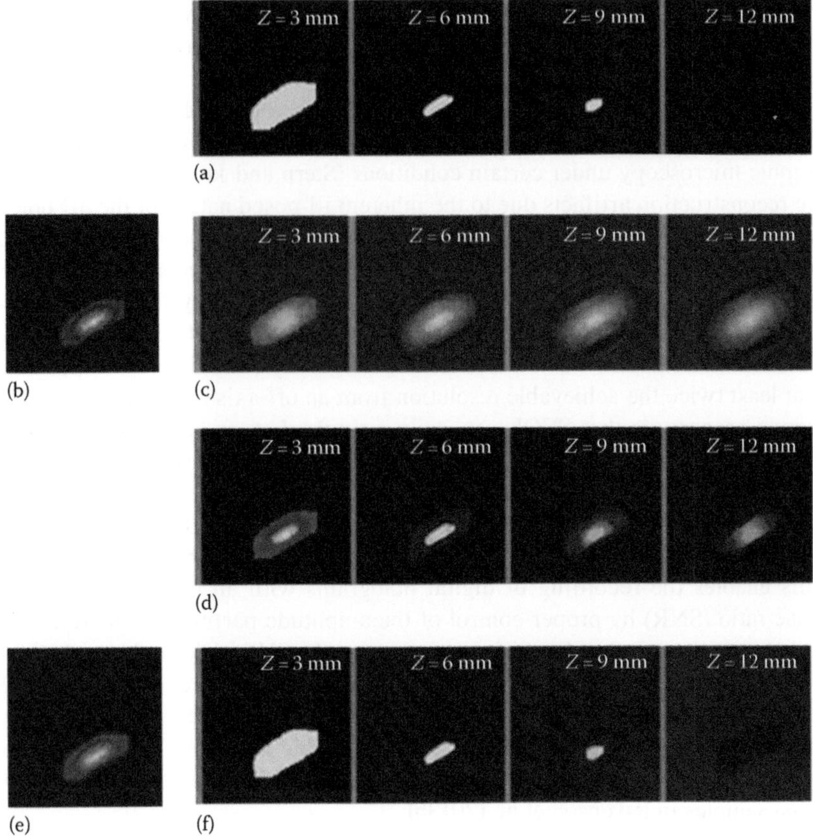

Figure 8.9 (a) Three depth planes of a 3D object. (b) Acquired, noisy in-line hologram. (c) Depth plane reconstruction using conventional field back-propagation. (d) 3D–2D object reconstruction using a CS approach to the problem, where the hologram is given by (b). (e) SEOL noisy hologram with an amplified object to reference wave intensity ratio when compared to (b). (f) 3D–2D object reconstruction from the SEOL hologram (e). This qualitative evaluation demonstrates that the out-of-focus object points are less dominant when compared with (d) and resemble the original 3D object distribution shown in (a). ([a–c,e,f] From Rivenson, Y. et al., *Appl. Opt.*, 52, A423, 2013b. With permission.)

References

Brady, D.J., *Optical Imaging and Spectroscopy* (Wiley, Hoboken, NJ, 2009).

Brady, D.J., Choi, K., Marks, D.L., Horisaki, R., and Lim, S., Compressive holography, *Opt. Express* 17 (2009): 13040–13049.

Bruckstein, A.M., Donoho, D.L., and Elad, M., From sparse solutions of systems of equations to sparse modeling of signals and images, *SIAM Rev.* 51 (2009): 34–81.

Candès, E.J. and Plan, Y., A probabilistic and RIPless theory of compressive sensing, *IEEE Trans. Inf. Theory* 57 (2011): 7235–7254.

Candès, E.J., Romberg, J.K., and Tao, T., Stable signal recovery from incomplete and inaccurate measurements, *Commun. Pure Appl. Math.* 59 (2006): 1207–1223.

Candès, E.J. and Wakin, M., An introduction to compressive sampling, *IEEE Signal Process. Mag.* 2 (2008): 21–30.

Chen, W., Tian, L., Rehman, S., Zhang, Z., Lee, H., and Barbastathis, G., Empirical concentration bounds for compressive holographic bubble imaging based on a Mie scattering model, *Opt. Express* 23 (2015): 4715–4725.

Coskun, A.F., Sencan, I., Su, T.-W., and Ozcan, A., Lensless wide-field fluorescent imaging on a chip using compressive decoding of sparse objects, *Opt. Express* 18 (2010): 10510–10523.

Cull, C.F., Wikner, D.A., Mait, J.N., Mattheiss, M., and Brady, D.J., Millimeter-wave compressive holography, *Appl. Opt.* 49 (2010): E67–E82.

Denis, L., Lorenz, D., Thiébaut, E., Fournier, C., and Trede, D., Inline hologram reconstruction with sparsity constraints, *Opt. Lett.* 34 (2009): 3475–3477.

Elad, M. and Aharon, M., Image denoising via sparse and redundant representations over learned dictionaries, *IEEE Trans. Image Process.* 15 (2006): 3736–3745.

Fan, W., Yuhong, W., Tianlong, M., and Xiaole, G., Experimental investigation on reconstruction guarantees in compressive Fresnel holography, *Proceedings of SPIE 9271, Holography, Diffractive Optics, and Applications VI*, Beijing, Hebei, China, November 11, 2014.

Gabor, D., A new microscopic principle, *Nature* 161 (1948): 777–778.

Gehm, M. and Brady, D., Compressive sensing in the EO/IR, *Appl. Opt.* 54 (2015): C14–C22.

Goodman, J.W. *Introduction to Fourier Optics* (McGraw-Hill, New York, 1996).

Hahn, J., Lim, S., Choi, K., Horisaki, R., and Brady, D.J., Video-rate compressive holographic microscopic tomography, *Opt. Express* 19 (2011): 7289–7298.

Horisaki, R., Ogura, Y., Aino, M., and Tanida, J., Single-shot phase imaging with a coded aperture, *Opt. Lett.* 39 (2014): 6466–6469.

Horisaki, R., Tanida, J., Stern, A., and Javidi, B, Multidimensional imaging using compressive Fresnel holography, *Opt. Lett.* 37 (2012): 2013–2015.

Javidi, B., Moon, I., Yeom, S., and Carapezza, E., Three-dimensional imaging and recognition of microorganism using single-exposure on-line (SEOL) digital holography, *Opt. Express* 13 (2005): 4402–4506.

Leith, E.N. and Upatnieks, J., Reconstructed wavefronts and communication theory, *J. Opt. Soc. Am.* 52 (1962): 1123–1130.

Liu, Y., Tian, L., Hsieh, C., and Barbastathis, G., Compressive holographic two-dimensional localization with $1/30^2$ subpixel accuracy, *Opt. Express* 22 (2014): 9774–9782.

Liu, Y., Tian, L., Lee, J., Huang, H., Triantafyllou, M., and Barbastathis, G., Scanning-free compressive holography for object localization with subpixel accuracy, *Opt. Lett.* 37 (2012): 3357–3359.

Marim, M., Angelini, E., Olivo-Marin, J.-C., and Atlan, M., Off-axis compressed holographic microscopy in low-light conditions, *Opt. Lett.* 36 (2011): 79–81.

Marim, M., Atlan, M., Angelini, E., and Olivo-Marin, J.C., Compressive sensing with off-axis frequency-shifting holography, *Opt. Lett.* 35 (2010): 871–873.

Mas, D., Garcia, J., Ferreira, C., Bernardo, L.M., and Marinho, F., Fast algorithms for free-space diffraction patterns calculation, *Opt. Commun.* 164 (1999): 233–245.

Nehmetallah, G. and Banerjee, P., Applications of digital and analog holography in three-dimensional imaging, *Adv. Opt. Photon.* 4 (2012): 472–553.

Rehman, S., Duan, Y., Chen, W., Matsuda, K., and Barbastathis, G., Phase imaging with X-ray digital holography and compressive sensing approach, *Imaging and Applied Optics 2014, OSA Technical Digest,* July 13–17, 2014, Seattle, Washington, DTh2B.1: Optical Society of America.

Rivenson, Y., Aviv (Shalev), M., Weiss, A., Panet, H., and Zalevsky, Z., Digital resampling diversity sparsity constrained-wavefield reconstruction using single-magnitude image, *Opt. Lett.* 40 (2015): 1842–1845.

Rivenson, Y., Rot, A., Balber, S., Stern, A., and Rosen, J., Recovery of partially occluded objects by applying compressive Fresnel holography, *Opt. Lett.* 37 (2012): 1757–1759.

Rivenson, Y. and Stern, A., Conditions for practicing compressive Fresnel holography, *Opt. Lett.* 36 (2011): 3365–3367.

Rivenson, Y., Stern, A., and Javidi, B., Compressive Fresnel holography, *J. Display Technol.* 6 (2010): 506–509.

Rivenson, Y., Stern, A., and Javidi, B., Improved three-dimensional resolution by single exposure in-line compressive holography, *Appl. Opt.* 52 (2013a): A223–A231.

Rivenson, Y., Stern, A., and Javidi, B., Overview of compressive sensing techniques applied in holography, *Appl. Opt.* 52 (2013b): A423–A432.

Rivenson, Y., Stern, A., and Rosen, J., Compressive multiple view projection incoherent holography, *Opt. Express* 19 (2011): 6109–6118.

Rivenson, Y., Stern, A., and Rosen, J., Reconstruction guarantees for compressive tomographic holography, *Opt. Lett.* 38 (2013c): 2509–2511.

Rudin, L., Osher, S., and Fatemi, E., Nonlinear total variation based noise removal algorithm, *Physica D* 60 (1992): 259–268.

Schmidt, J.D., *Numerical Simulation of Optical Wave Propagation with Examples in MATLAB* (Washington, DC: SPIE, 2010).

Shaked, N.T., Katz, B., and Rosen, J., Review of three-dimensional holographic imaging by multiple-viewpoint-projection based methods, *Appl. Opt.* 48 (2009): H120–H136.

Sotthivirat, S. and Fessler, J.A., Penalized-likelihood image reconstruction for digital holography, *J. Opt. Soc. Am. A* 21 (2004): 737–750.

Stern, A., August, Y., and Rivenson, Y., Challenges in optical compressive imaging and some solutions, *Proceedings of 10th International Conference on Sampling Theory and Applications, SampTA 2013*, Bremen, Germany, July 2013.

Stern, A. and Javidi, B., Theoretical analysis of three-dimensional imaging and recognition of micro-organisms with a single-exposure on-line holographic microscope, *J. Opt. Soc. Am. A* 24 (2007): 163–168.

Stern, A. and Rivenson, Y., Theoretical bounds on Fresnel compressive holography performance (Invited Paper), *Chin. Opt. Lett.* 12 (2014): 060022–060025.

Wilson, S.A. and Narayanan, R.M., Compressive wideband microwave radar holography, *Proceedings of SPIE 9077, Radar Sensor Technology XVIII*, Baltimore, MD, May 29, 2014, p. 907707.

Yamaguchi, I., Yamamoto, K., Mills, G.A., and Yokota, M., Image reconstruction only by phase data in phase-shifting digital holography, *Appl. Opt.* 45 (2006): 975–983.

Zhang, X. and Lam, E.Y., Edge-preserving sectional image reconstruction in optical scanning holography, *J. Opt. Soc. Am. A* 27 (2010): 1630–1637.

Zibulevsky, M. and Elad, M., L1-L2 Optimization in signal and image processing, *IEEE Signal Process. Mag.* 27 (2010): 76–88.

9 Spectral and Hyperspectral Compressive Imaging

Isaac Y. August and Adrian Stern

Contents

Introduction to Spectroscopic Imaging Sensing

Spectroscopic analysis has played an important role in the development of the most fundamental theories in physics (Born and Wolf 1999). Spectroscopy and, more generally, spectroscopic imaging (SI) such as hyperspectral imaging or ultraspectral imaging methods are used in a wide field of real-life applications. In many cases, SI systems operate by acquiring data in a large number of very narrow spectral bands throughout the visible and near-infrared spectra. Some examples of the extensive use of spectroscopic methods may be found in the fields of quality control, the food and agriculture industries (Thenkabail et al. 2012; Wu and Sun 2013, 1–14), medical imaging and spectroscopic microscopy (Martin et al. 2006, 1061–1068; Akbari et al. 2012, Li et al. 2013), remote sensing (Warner et al. 2009; Eismann 2012), art conservation (Dupuis et al. 2002, 1329–1336; Martinez et al. 2002, 28–41; Delaney et al. 2010, 584–594; Ribés 2013, 449–483), gas identification, homeland security applications (Manolakis and Shaw 2002, 29–43; Manolakis et al. 2003, 79–116), and many more.

Despite the many advantages of using spectral information, the implementation of spectroscopic sensing systems and the subsequent acquiring and processing of data pose significant challenges. Spectral imaging datacubes are extremely large, which imposes many challenges on the storage, transmission, and processing of the captured data. Other challenges may also arise due to the spectroscopic system design and the imaging time requirements. We shall now show that compressive sensing (CS) methods implemented in spectral imaging systems can handle these challenges effectively.

Notation Conventions

In this chapter, we frequently use SI cubes and related tensor data. We have tried to remain as consistent as possible with terminology that would be familiar to the community of optical and spectroscopic engineers and with the terminology of previous publications in the area. We denote scalars by regular lowercase letters, for example, f, and vectors (tensors of order one) are denoted by boldface lowercase letters, for example, **f**. Second-order tensors (matrices or *2D* images) are denoted by boldface uppercase letters, for example, **F** or **Φ**, and third-order tensors (*3D* cube) are denoted by boldface Euler script letters, for example, $\mathcal{F}, \mathcal{G}$, or $\mathcal{T}$. The ith entry of a vector **f** is denoted by f_i, the element (i, j) of a matrix **F** is denoted by f_{ij}, and element (i, j, k) of a third-order tensor $\mathcal{F}$ is denoted by f_{ijk}. Generally, in this chapter, indices of the input signal **f** typically range from 1 to N, for example, $i = 1, \ldots, N$. For a third-order tensor $\mathcal{F}$, the indices are $i = 1, \ldots, N_x$; $j = 1, \ldots, N_y$; $k = 1, \ldots, N_\lambda$. Indices of the measurement signal **g** typically range from 1 to M, for example, $i = 1, \ldots, M$. For a third-order measured data tensor $\mathcal{G}$, the indices are $i = 1, \ldots, M_x$; $j = 1, \ldots, M_y$; $k = 1, \ldots, M_\lambda$. The sparsity of a signal is denoted by S_x, S_y, and S_λ in the spatial and the spectral domains, respectively. The nth element in a sequence of vectors, matrices, or cubes is denoted by a superscript in parentheses, for example, $\mathbf{f}^{(n)}$, $\mathbf{F}^{(n)}$, or $\mathcal{F}^{(n)}$, respectively. The jth column of image **F** is denoted by $\mathbf{f}_{:j}$, and the ith row of image **F** is denoted by $\mathbf{f}_{i:}$. In an additional format, the jth column $\mathbf{f}_{:j}$ of image **F** may be denoted more compactly as $\mathbf{f}_j$. Third-order tensors have column, row, and tube fibers, denoted by $\mathbf{f}_{:jk}$, $\mathbf{f}_{i:k}$, and $\mathbf{f}_{ij:}$, respectively, and they are assumed to be oriented as column vectors. *Fiber* vectors are the analogue of rows and columns in matrices, but with the additional third dimension of a tensor cube (Kiers 2000). *Slices* are 2D sections of a tensor that are defined by fixing a single index and are denoted by $\mathbf{F}_{i::}$, $\mathbf{F}_{:j:}$, and $\mathbf{F}_{::k}$. Alternatively, the kth spectral slice $\mathbf{F}_{::k}$ of a third-order tensor, $\mathcal{F}$, may be denoted more compactly as $\mathbf{F}_k$.

Classical Spectroscopic Measurements

An SI system generally consists of an illumination source and a spectral light measurement system. In most common SI systems, the scene is uniformly illuminated with a wide spectral range source and the reflected light is detected with narrow spectral window filters. Spectral separation is achieved by means of optical prisms or gratings (Wolfe 1997; Eismann 2012) that geometrically separate the spectral components into different angular directions. Figure 9.1b and c shows the common scanning schemes used to capture the *3D* data of a spectral image $\mathcal{F}$, as shown in Figure 9.1a. Figure 9.1b shows the spatial scanning mechanism based on dispersive/diffractive elements, called "push broom" scanning (aka, "along track" scanning). According to this mechanism, in each time slot, a slice $\mathbf{F}_i$ is captured. Figure 9.1c shows the spectral scanning mechanism based on a tunable filter. With this mechanism, in each time slot, a slice $\mathbf{F}_k$ is captured. Figure 9.1d shows the scanning mechanism based on spatial scanning with a spectrometer called "whisk broom" (aka "across track scanner"). With this mechanism, in each time slot, a spectral vector $\mathbf{f}_{ij}$ is captured.

Despite the clear differences in the various technical and physical implementations (Figure 9.1b through d), all of these methods are similar in the sense that they attempt to perform a direct measurement. Such systems can be modeled as a linear mapping, where each point from the object f_{ijk} is mapped to a single point g_{ijk} in the *3D* digital array $\mathcal{G}$. We express this mathematically by a matrix–vector multiplication:

$$\mathbf{g} = \mathbf{\Phi}\mathbf{f} + \mathbf{n}, \tag{9.1}$$

where the object $\mathbf{f}$ and the image $\mathbf{g}$ are "vectorized" forms of $\mathcal{F}$ and $\mathcal{G}$, arranged in a lexicographical order (see the "Compressive Spectral Imaging with Spectral Encoding in the Spatial Domain" section). The matrix $\mathbf{\Phi}$ represents the system's spatial–spectral optical transmission, and $\mathbf{n}$ is additive noise. For the direct imaging of $\mathbf{f}$, ideally, the sensing matrix is a huge diagonal square matrix $\mathbf{\Phi} \in \mathbb{R}^{N_x N_y N_\lambda \times N_x N_y N_\lambda}$.

The use of indirect measurements in spectroscopy (Sloane 1979, 71–80) is sometimes called *coded spectroscopy* or *multiplexed spectroscopy*. We will use both terms; "multiplexed spectroscopy" is usually associated with continuous spectral encoding, while the term "coded spectroscopy" is used for the discrete spectral encoding form. With coded SI, each measurement provides information carried by multiple wavelengths and spatial points; therefore, each measurement g cannot be associated with a specific wavelength or position. The sensing matrix of an indirect measurement system $\mathbf{\Phi}$ has many nonzero off-diagonal elements. For coded spectroscopy data, the captured data need to be digitally processed to return it to the standard form of $\mathbf{\Phi}$ (Figure 9.1a). In order to transform the captured measurements $\mathbf{g}$ to the form of $\mathbf{f}$, the inverse of the sensing matrix, $\mathbf{\Phi}^{-1}$, is required. For this, $\mathbf{\Phi}$ has to be an orthogonal, or at least a nonsingular, matrix. Generally, encoded spectroscopy requires more processing than direct spectroscopy systems, but the use of multiplex measurement for spectroscopy provides several advantages over that of direct measurements, such as Fellgett's advantage, which states that since all the spectral wavelengths are measured simultaneously,

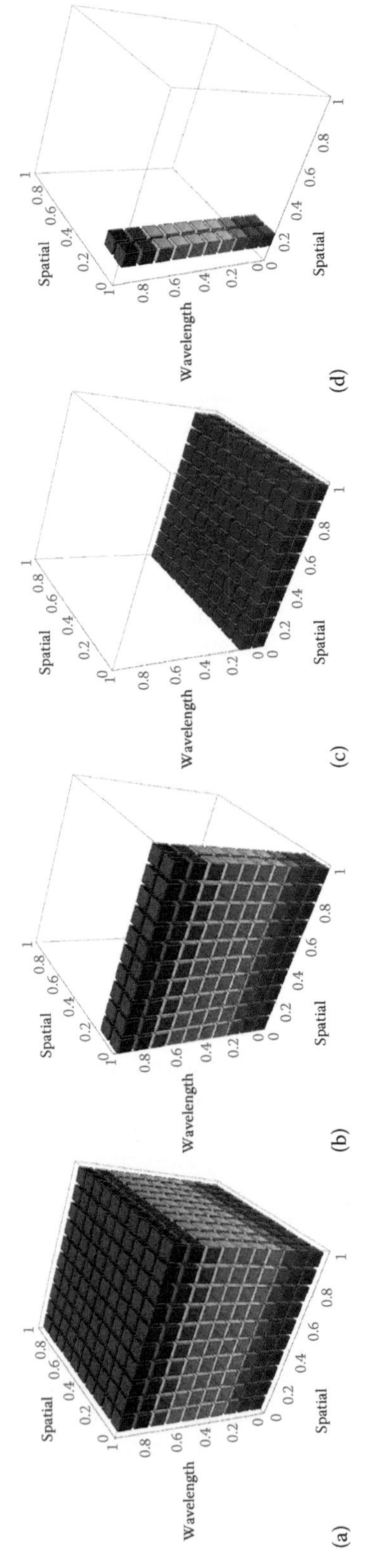

Figure 9.1 (a) The spectroscopic datacube array, $\mathcal{F}$. One way for capturing the datacube is by spatial scanning (b); in each exposure time slot, a single slice $\mathbf{F}_i$ (shown) or $\mathbf{F}_j$ (not shown) is captured. Another way for capturing the datacube is by spectral scanning (c); in each exposure time slot, a single slice $\mathbf{F}_k$ is captured. A third way for capturing the datacube is by spectral scanning where, in each time slot, a single point spectrum $\mathbf{f}_{ij}$ is captured (d).

an improvement in signal-to-noise ratio is gained (Fellgett 1967, C2-165–C2-171; Hirschfeld 1976, 68–69; Vickers et al. 1991, 42–49).

Properties of Spectroscopic Data

The amount of data typically captured with SI systems is huge. The problems associated with these large dimensions can be alleviated by utilizing the fact that SI cubes are highly compressible, or even sparse (Ryan and Arnold 1997a, 546–550, 1997b, 419–436; Lim et al. 2001a, 97–99, 2001b, 109–111; Qian et al. 2001, 1459–1470; Keshava and Mustard 2002, 44–57; Shaw and Hsiao-hua 2003; Lukin et al. 2008, 375–377; Iordache et al. 2011, 2014–2039). The redundancy of information $\mathbf{\Phi}$ is evident if we consider each slice of the datacube $\mathbf{F}_k$ from a single narrow spectral window k. Such a slice is a regular 2D image, which is well known to be compressible in the discrete wavelet or block discrete cosine domain (Vetterli and Kovaseviss 1995; Mallat 2009).

The redundancy is evident also if we consider the spectral information $\mathbf{f}_{ij}$. The spectral redundancy can be demonstrated, for example, by applying endmember decomposition (Manolakis and Shaw 2002, 29–43; Manolakis et al. 2003, 79–116). The endmember model is commonly used to represent the spectral information $\mathbf{f}_{ij}$ as a linear mixture of only few S_λ spectral components. Each spectral component $\boldsymbol{\psi}_k$ has its own weight α_{ijk} in the mixture model. Equation 9.2 presents the hyperspectral vector $\mathbf{f}_{ij}$ as a mixture of only a few components, and $\mathbf{n}_{ij}$ is the model error and noise:

$$\mathbf{f}_{ij} \cong \sum_{k=1}^{S_\lambda} \alpha_{ijk}\psi_k + \mathbf{n}_{ij}. \tag{9.2}$$

The spectral sparsity of the signal $\mathbf{f}_{ij}$ in the endmember domain is less than S_λ and for real scenes $\sim 3 \leq S_\lambda \leq \sim 7$ (Keshava and Mustard 2002, 44–57; Iordache et al. 2009, 2011, 2014–2039). Locally, each pixel in the cube contains a mixture of only a few endmembers, but globally the total number of endmembers may be higher. Other ways for sparse representation of the spectral data can be achieved by using other transforms, such as wavelets, the block Fourier transform, vector quantization (Ryan and Arnold 1997b, 419–436; Qian et al. 2001, 1459–1470), or specialized dictionaries. In some cases, such as in lines spectroscopy (Smith 1952; Meggers 1961; Sansonetti et al. 1996, 74–77), the signal is sparse in the canonical (λ) itself. For hyperspectral remote sensing, it is assumed that the spectral information has higher compressibility than the spatial information. This assumption is supported empirically; the number of different materials in the scene is small relative to the spatial information (Iordache et al. 2011, 2014–2039). The high compressibility feature of SI information, with no meaningful information loss incurred, is the key to justifying the effective operation of CS techniques for spectroscopy or for SI. In the following section, we assume that the high compressibility assumption holds.

Compressive Sensing Spectroscopy and Spectroscopic Imaging

From Theoretical to Applied Compressive Spectral Imaging

CS systems that are based on random and other sensing structures were presented in Chapter 1. However, the optical implementation of those descriptions is not fully achievable and various compromises may need to be made.

In SI with random multiplex measurements, rather than using optical elements and devices to filter out a single physical value f_{ijk} in a narrow spectral band k at a specific position i, j, each scanning step captures a multiplex measurement g_ν obtained by summing a combination of randomly weighted object spatial–spectral components. This can be described by spatial spectral filtering of the cube $\mathcal{F}$ with a (*3D* volume) virtual mask $\mathcal{T}$ and a summation to a single measurement g_ν. For simplicity, we start by describing binary sensing. The binary sensing pattern can be implemented, for example, by using a partial Hadamard ensemble or a Bernoulli binary ensemble. In these cases, the volume masks contain only two values that are associated with the optical "open/close" apertures. Figure 9.2 illustrates binary random *3D* volume masks $\mathcal{T}$ in the standard spatial spectral domain (x,y,λ). In this figure, four different random masks are shown, $\mathcal{T}^{(1)}$, $\mathcal{T}^{(2)}$, $\mathcal{T}^{(3)}$, and $\mathcal{T}^{(4)}$, and the probability of each entry t_{ijk} being 0 or 1 is half. The volume mask enables the light to pass through the open aperture; the color of the boxes indicates their wavelengths. In CS spectroscopy, light from all the marked spectral–spatial components is summed together to yield a single measurement. The process is repeated, each time i with a different mask $\mathcal{T}^{(i)}$, yielding a vector of compressive measurements $\mathbf{g}$ that is later used for the recovery process.

One possible optical implementation, according to coded volume masks (Figure 9.2a through d), can be made as follows. In the first step, the spectral cube $\mathcal{F}$ is split into spectral slices $\mathbf{F}^k$. The second step comprises a spatial encoding of each different spectral slice $\mathbf{F}^k$ with a different random mask $\mathcal{T}^{(i)}$. The last step is implemented by collecting all the encoded spectral slices into a single optical power measurement device. By repeating the process with different masks $\mathcal{T}^{(i)}$, each time i, a set of measurements, $\mathbf{g}$, is built and a recovery process can be applied to estimate the spectral cube $\mathcal{F}$.

Although random sampling as shown in Figure 9.2 is highly desirable, because of its applicability to virtually any type of SI data, its implementation may pose one or more challenges and difficulties that prevent a random sampling scheme. Compressive SI systems may be limited due to *fundamental issues*, various *technical and physical constraints*, and also *information constraints*:

1. From a technical point of view, the signal $\mathbf{f}$ and measurements $\mathbf{g}$ are typically of very high dimension, which implies serious computational challenges. Other major difficulties arise due to the size of the system sensing matrix $\mathbf{\Phi}$, which may have as many as $O(10^{17})$ elements. It is obvious that storing and processing such huge amounts of data raise severe computational difficulties and storage limitation issues.
   An additional difficulty arises at the calibration stage of the SI system. In the calibration process, typically, there is a need for many measurements in order to specify the exact system operation state. In order to calibrate $\mathbf{\Phi}$, one may need to measure $N_x \cdot N_y \cdot N_\lambda$ impulse responses (each determining one column of $\mathbf{\Phi}$), each having $M_x \cdot M_y \cdot M_\lambda$ samples.
   The inability to map the optical system precisely produces errors, which may result in poor performance. In some cases, this could lead to system resolution degradation, an appearance of false patterns, or the disappearance of the low-energy signal components in the reconstructed signal.

2. There are also some fundamental limitations that prevent the realization of ideal compressive SI schemes. A fundamental limitation in optics arises

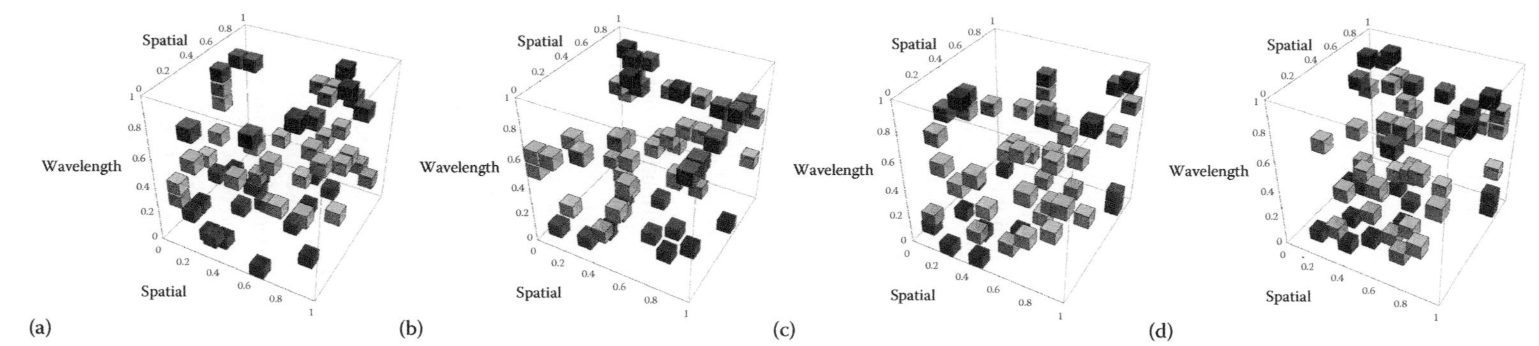

Figure 9.2 (a)–(d) Examples of spectral–spatial random masks for CS measurement in the standard (x,y,λ) *3D* cube representation. Ideally, universal CS SI should sample randomly in both the spectral and spatial domains.

due to the fact that, in general, the sensing matrices, $\mathbf{\Phi}$, and the measurements cannot have negative values (see Chapter 4). This means that $\mathbf{\Phi}$ spans only the positive orthant. As a result, the coherence parameter μ defined in Equation 1.7 of $\mathbf{\Phi}$ is high, indicating limited system performance.

3. Physical implementation limitations may occur when the optical system model is only an approximation of the physical behavior, or when physical principles restrict some optical implementations. An example of a physical restriction arises when an effort is made to minimize the coded aperture (CA) elemental feature. Minimization of the CA elemental feature is desired to improve the systems' spatial resolution, but this, in turn, induces limitations related to diffraction, fabrication, and optical efficiency. Other physical limitations may appear when an optical linear model is used in order to describe nonlinear physical behavior.

4. From the information point of view, there are additional aspects that need to be taken into account. In some applied CS SI systems, the realizable i.i.d. random sampling matrices required are not possible. For example, in the systems to be presented in the "Compressive Sensing System with Spectral Encoding in the Spectral Domain" section, the rows of the sensing matrix are determined by a set of partially randomly chosen physical states (such as voltage, distance). Given such a sensing matrix $\mathbf{\Phi}$, a sparsifying operator $\mathbf{\Psi}$ should be selected such that it is both efficient (i.e., it concentrates the signal in a small number of S coefficients) and has a low coherence parameter μ.

Accordingly, we summarize that the performance of CS SI systems is bounded by theoretical information limitations, as well as by restrictions due to the optical implementation.

Figure 9.3 demonstrates spectral distributions obtained in a virtual experiment that simulated the ideal compressive spectroscopic process. The spectrum reconstructions in Figure 9.3 are obtained by simulating a random coded CS spectrometer. The spectrum reconstruction is presented versus the real spectral density signal. Here, the term *ideal* is used in order to emphasize that the only constraint on the implementation was due to the optical sensor's ability to measure only positive optical power values. In order to give a quantitative measure of the ideal spectrometer performance, a peak signal-to-noise ratio (PSNR) measurement was made.

Compressive Spectral Imaging with Spectral Encoding in the Spatial Domain

This section introduces compressive spectroscopic and SI systems that are based on spatial encodings. Spatial encoding is a technical method that can be used to encode either the spatial or the spectral information or the joint spatial–spectral information. A common way of spatial encoding is by encoding the image plane, which can be achieved, for example, with a static/moving CA. CAs are built of arrays of "open" or "closed" cells that transmit or block light, thus performing Bernoulli (1/0) encoding. Figure 9.4d shows a CA that can be used in order to encode spatial data. Other ways to encode the spatial information is by employing spatial light modulators (SLMs), digital micromirror devices (DMDs), or liquid crystal (LC) SLMs. DMDs also perform Bernoulli encoding of the incoming light by employing an array of tiny mirrors that can be directed or undirected in specific two angular directions. The advantages

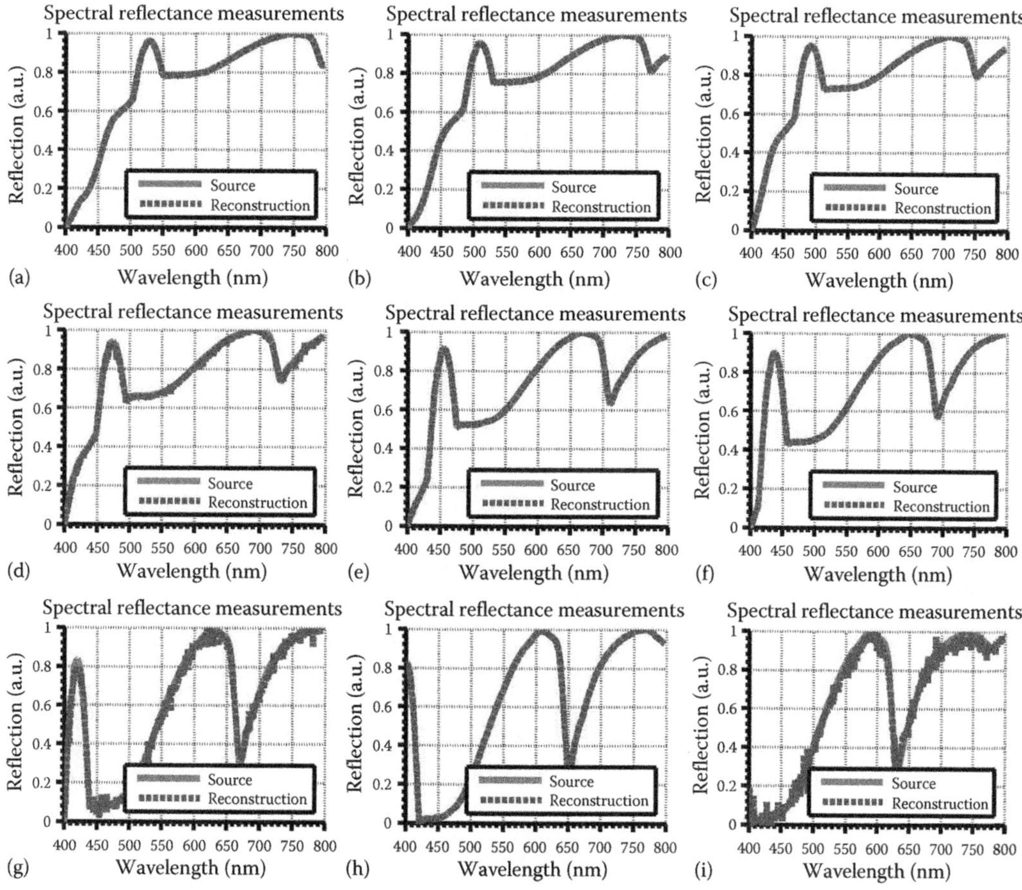

Figure 9.3 Simulation results of compressive spectral sensing by random sampling; $N = 1024$ spectral bands (continuous line) of nine different biogenic crust-like spectra. The reconstructions (dashed lines) are from $M = 256$ Bernoulli (0/1) i.i.d. random measurements; Daubechies wavelets (db3) were used as the sparsifying operator. The reconstruction PSNR values are (a)–(i) 50.7, 52.0, 51.9, 39.1, 51.3, 50.6, 32.1, 49.8, and 29.5 (dB).

of using DMD lie in the flexibility of choosing the spatial encoding pattern and the ability to dynamically change the encoding pattern. Nonbinary encoding patterns can be achieved by using partially transparent CA cells (Figure 9.4c), or even DMDs, by using appropriate temporal–spatial modulation for a single exposure, thus introducing encoding levels between 0 and 1. An alternative to DMD is the LC SLM; it can be embedded in SI systems due to its ability to form partially transparent patterns in a single pattern for a single exposure. The drawback of using this kind of element is that it can only be implemented for a single polarization mode.

Spectral domain encoding can be achieved with spatial encoding devices after performing spectral-to-spatial optical conversion. Spectral-to-spatial optical transformation is mainly achieved with dispersive or diffractive elements, such as prisms and diffractive gratings. These components separate the incident light into its spectral components by reflection and transmission of different wavelengths to different angles or locations, thereby enabling the spectral light encoding operation effectively in the spatial domain.

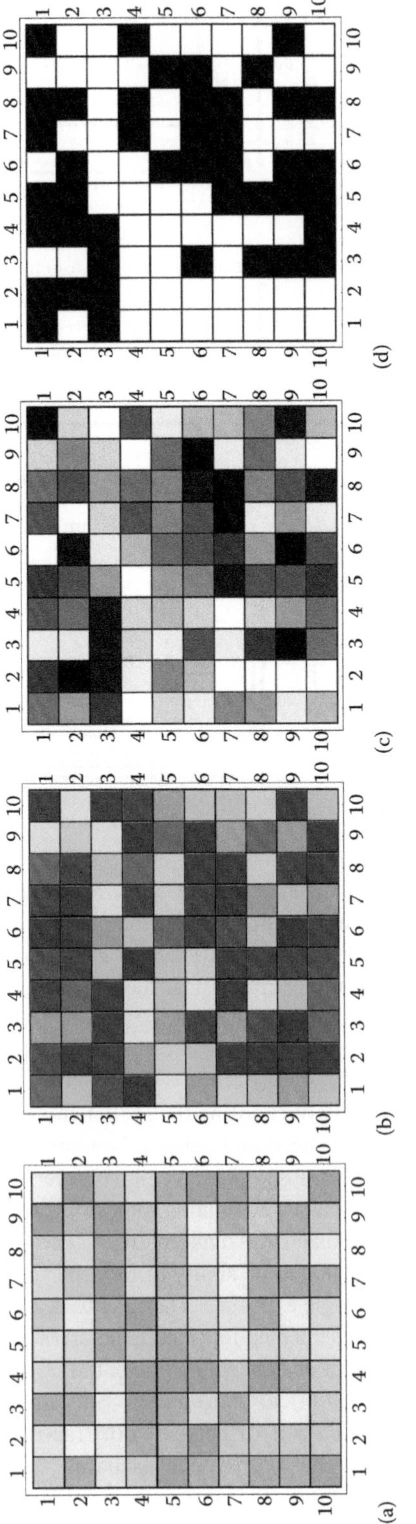

Figure 9.4 Generalized amplitude coded apertures: (a) wide spectral band coded aperture, (b) color-coded aperture with narrow spectral bands, (c) gray-level coded aperture with partially transparent patterns, and (d) binary coded aperture.

Recently, a more generalized form of amplitude CA has been made available for use. Such devices or elements may contain masks with cell elements characterized by a single narrow spectral band (Figure 9.4b) or a wide spectral band (Figure 9.4a). In some cases, the fabricated element may also use a coded polarizer in order to encode the polarization state.

A colored CA (Figure 9.4b) was used in Arguello and Arce (2014, 1896–1908) in a system dubbed colored CASSI, which is an extension of the coded aperture snapshot spectral imaging (CASSI) method described in the next section. The use of a color-coded (filtered) aperture offers an additional degree of freedom to encode the spectral–spatial datacube that can be used to achieve better performance; in Arguello and Arce (2014, 1896–1908) and Rueda et al. (2014, 7799–7803), a 6 dB gain in spatial PSNR was demonstrated.

The spatial–spectral encoding of the 3D SI cube resulting from the application of the four mentioned types of apertures is illustrated in Figure 9.5. A comparison between the ideal volume masks for CS (Figure 9.2a through d) and the volume mask in Figure 9.5 shows two properties. The first one is that, as mentioned earlier, without dispersive or diffractive elements spectrum encoding cannot be achieved (Figure 9.5c and d). The second notable property is that the multicolored CA in Figure 9.4a behaves most similarly to an ideal spatial–spectral filter.

Spectral imaging systems that are based on an SLM or a CA have many benefits, yet these systems also present different implementation limitations. One of the main limitations is attributed to the size and length of these systems. Other difficulties may arise with imaging in the infrared region, due to the nonneglectable diffraction and scattering effects.

Additional spatial encoders are the phase CAs, which are phase encoders, unlike spectral amplitude encoders that were discussed thus far. Phase CAs can be implemented with diffusers, such as the "RIP diffuser" in Golub et al. (2015) or with an LC SLM. Such masks introduce different phase transmission values at different positions over the mask. In the common use, in order to obtain phase multiplexing in the sensor plane the phase CA is placed in the aperture plane. The phase transmission is spectrally dependent; therefore, if the mask is placed at the entrance pupil, it acts as a randomly spectrally encoding generalized pupil.

Coded Aperture Snapshot Spectral Imaging

Probably, the best known compressive SI scheme is the CASSI scheme (Arguello and Arce 2011, 2400–2413; Wu et al. 2011, 2692–2694; Arce et al. 2014, 105–115; Lin et al. 2014, 233). In principle, CASSI systems are based on a CA, a dispersive prism, and a photosensor array. In Figure 9.6, CASSI optical information processing is illustrated. CASSI processing starts by projecting the object image onto a spatial CA. The CA encodes the image in the same way for all its wavelengths. This yields an encoding of the SI cube, as shown in Figure 9.6a and b. Next, a lateral shearing transformation is made using dispersive elements. The dispersive element shifts the images that are associated with different wavelengths to different lateral positions, as shown in Figure 9.6c. Long-wavelength image slices undergo a relatively small diversion while short-wavelength image slices undergo a relatively large diversion. A focusing system creates a spectrally blurred image of all the spectrally shifted images over the sensor array. Each measurement across the optical sensor contains power contributions from multiple locations and multiple wavelengths, as illustrated in Figure 9.6c and d. In Figure 9.6, the number of spatial samples in each direction is equal to the number of bands; hence, the ratio of the number of pixels on the detector to the number of spatial samples of

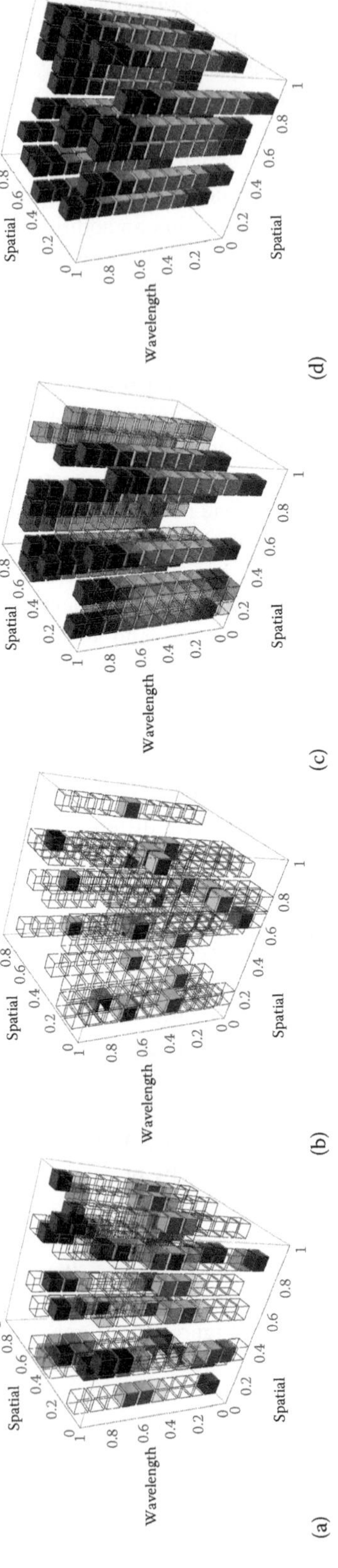

Figure 9.5 Examples of spectral–spatial masks of generalized coded apertures in the standard (x,y,λ) *3D* cube representation: (a) wide spectral band coded aperture, (b) color-coded aperture, (c) partially transparent patterned coded aperture, and (d) binary coded aperture.

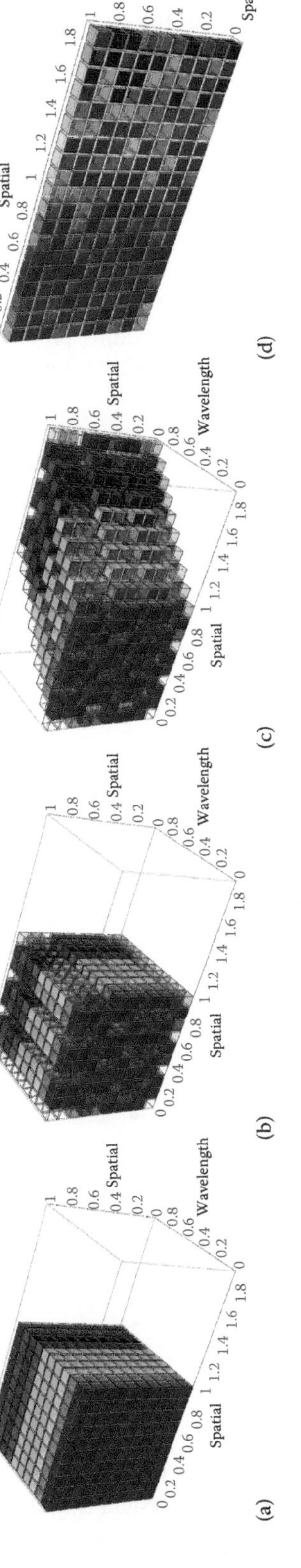

Figure 9.6 An illustration of the spectral optical flow in CASSI. (a) The SI cube. (b) Spatially encoding of the cube. (c) Lateral shearing transformation of the encoded cube. (d) The projected intensity over the sensor.

the cube is greater than 1. From Figure 9.6b and c, several characteristics resulting form the use of the dispersive prism can be seen. The first characteristic is that the number of samples associated with the detector (Figure 9.6d) in real cases is slightly larger than the amount of spatial samples in the spectroscopic cube. Another property resulting from the use of the prism is the independent encoding of rows. As seen from Figure 9.6c and d, the dispersion shearing only affects the spectral images in a single direction; hence, the system encoding is achieved in a single row. Taking these characteristics into consideration, one can accelerate the recovery process, for example, by using parallel computing sparse solvers.

Since its introduction, CASSI has attracted a lot of interest. In its basic version, CASSI is able to capture a reasonably good spectral image with a single exposure. Several variants of CASSI were developed and implemented by a multishot imaging process or with a dual-disperser architecture, producing even higher-quality results. The reader is referred to Arce et al. (2014) for an overview of CASSI.

Single Spectrometer Pixel Camera

Another technique for CS SI is generalizing the single-pixel camera (SPC; Figure 9.7a) to a single spectrometer pixel camera. A single-pixel spectrometer system, also called a single-pixel hyperspectral camera, can be built from a spatial encoding subsystem with the addition of a spectral scanning subsystem. In principle, in the SPC, all spectral components of the object undergo the same spatial encoding, as presented in Figure 9.5c or d. Hence, if we separate and capture all

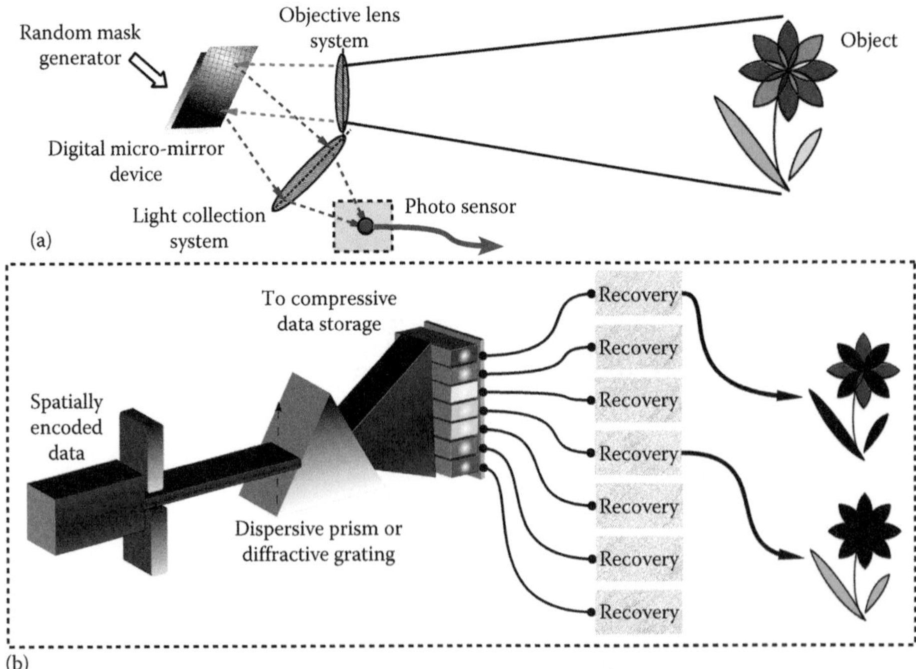

Figure 9.7 An illustration of the single spectrometer pixel camera. (a) Regular SPC based on DMD and (b) regular spectrometer. By combining both systems, that is, by replacing the photo seneor in (b) with the spectrometer (b), a compressive hyperspectral system is achieved. Each sensor element in the linear photosensor can be used in order to recover the image within its spectral band.

the spectral components that are already encoded spatially, enough information is acquired in order to provide complete reconstruction of the spectral datacube. The realization of spectral separation can be achieved in different ways; for example, we realize it by introducing a spinning color spectral filter wheel into the single-pixel (Duarte et al. 2008, 83) camera setup. In this case, sequences of CA mask patterns are generated by the DMD for each chosen spectral filter. By the end of a spatial masking sequence, a different spectral filter from the spectral filter wheel is chosen and a new sequence of spatial masks is applied and so on. This approach for implementation provides a high degree of spatial encoding per color filter but is somewhat inefficient, because only a narrow spectrum passes through the spatial encoder at each measurement and, thus, many measurements must be made per spectral filter.

A more efficient way for SI realization based on an SPC is by spectral separation of the spatially encoded data with a dispersive or refractive system (Sun and Kelly 2009, TuA5). In this case, the spectral separation can be implemented with a mechanical moving optical sensor (e.g., spectral scanning) or, alternatively, using a parallel sampling scheme by using linear photosensor arrays, as shown in Figure 9.7b. In addition to SI capabilities, a version with polarimetry capabilities can be designed (see Chapter 10).

Figure 9.7a shows a regular SPC, and the lower part of the figure illustrates a regular spectrometer system. This spectrometer uses a prism as a spectral splitter. Different wavelengths are captured at different positions on the line array sensor (i.e., a vector array). By replacing the photodiode component element in Figure 9.7a with the spectrometer system in Figure 9.7b, each elementary sensor in the linear photosensor array provides information that can be used to recover a spectral image $\mathbf{F}_k$ from the compressive measurements. One may write the mathematical model for the single spectrometer pixel camera in terms of the SPC, simply by applying it in parallel for all wavelengths k. Let us denote by $\mathbf{f}_k = vec(\mathbf{F}_k)$ the vectorized version of an image $\mathbf{F}_k$ at a specific wavelength k, where $\mathbf{g}_k$ is the compressive measurement vector and $\Phi_k \in \mathbb{R}^{N_x N_y \times M_{xy}}$ is the sensing matrix implemented by the DMD. The encoding of all spectral components k is parallel; therefore, the mathematical model of the sensing is as follows:

$$\begin{bmatrix} \mathbf{g}_1 \\ \mathbf{g}_2 \\ \vdots \\ \mathbf{g}_k \\ \vdots \\ \mathbf{g}_{N_\lambda} \end{bmatrix} = \begin{bmatrix} \Phi_k & \mathbf{0} & \cdots & \mathbf{0} \\ \mathbf{0} & \Phi_k & \cdots & \mathbf{0} \\ \vdots & \vdots & \ddots & \vdots \\ \mathbf{0} & \mathbf{0} & \cdots & \Phi_k \end{bmatrix} \begin{bmatrix} \mathbf{f}_1 \\ \mathbf{f}_2 \\ \vdots \\ \mathbf{f}_k \\ \vdots \\ \mathbf{f}_{N_\lambda} \end{bmatrix}. \tag{9.3}$$

This single spectrometer pixel camera is a relatively simple example of an extension of the SPC to a spectroscopic camera. However, with this architecture, the spatial information is encoded while the spectral information remains unencoded; therefore, it is suboptimal in the CS sense.

Separable Compressive Sensing Based on Coded Spectrometer

In this section, we present an expansion of the single spectrometer pixel camera system ("Compressive Spectral Imaging with Spectral Encoding in the Spatial Domain" section) to perform spatial–spectral compressing (August et al. 2013a, D46–D54, 2013b, 87170G-1–87170G-10). The first part of the section describes the physical implementation of such systems, and in the second part we provide useful details for technical computing.

To achieve compression in both the spatial and spectral domains, one can replace the regular spectrometer in the single spectrometer pixel camera system scheme presented in Figure 9.7b with a CA compressive spectrometer. CA spectrometers are spectrometers that use CAs, as shown in Figure 9.4c or d, to encode only the spectral information. This concept can be viewed as an extension of the Hadamard multiplex coded spectroscopy technique (Nelson and Fredman 1970, 1664–1669; Treado and Morris 1989, 723A–734A; Vickers et al. 1991, 42–49) but using general encoding patterns. The principle of the compressive coded spectrometer is based on spectral-to-spatial conversion followed by an operation of spatial encoding. Light that undergoes the coding operation converges to the sensor position, and the total power from all the wavelengths is measured.

Let us look at the scheme shown in Figure 9.8b. Light reaches the entrance pupil of the spectrometer and passes through a collimator. The collimated beam that reaches the diffractive element is split into its spectral components, yielding a 1D vector distribution representation of the spectrum. Next, the different spectral components propagate to a plane in which the coding operation is performed. Since the spectral splitting is performed along the vertical axis, a single-column CA can be used to encode the light, that is, the green marked area. Light that passes through the CA is collected, by means of anamorphic optics (August et al. 2013a, D46–D54), and

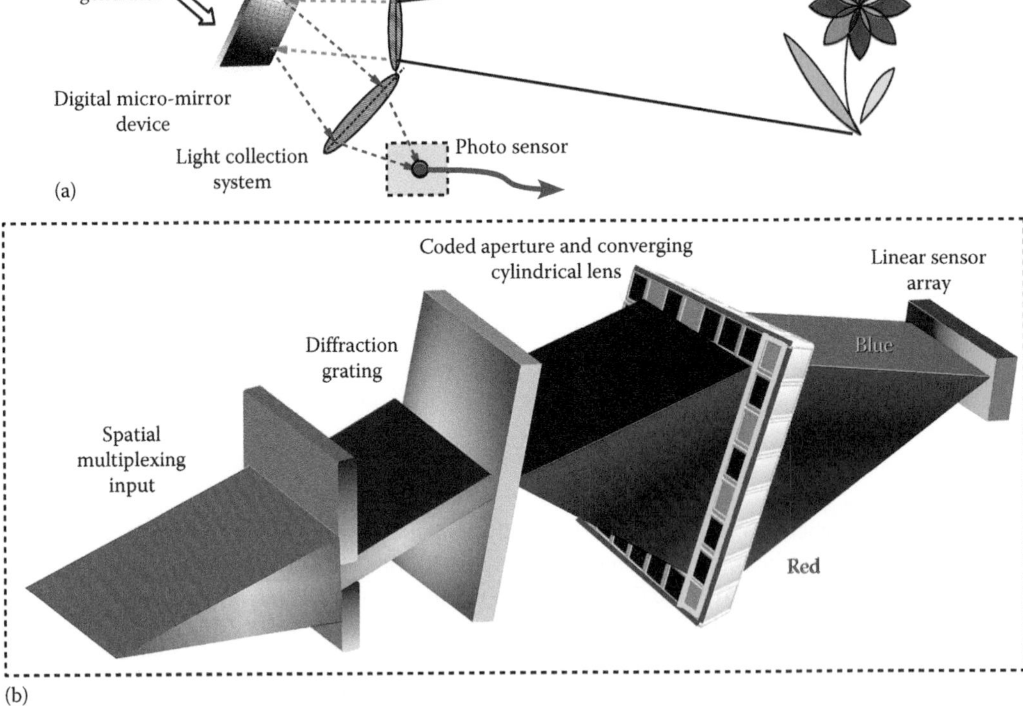

Figure 9.8 (a) The SPC and (b) the parallel CS coded spectrometer. Each column of apertures provides a pseudorandom encoding pattern and is used as a single spectral encoder. The set of all the columns and the corresponding measurements across the linear sensor array provides the CS data.

redirected to the point of measurement. The collected light is then measured by using a photodiode with broad spectral sensitivity. The spectral sensing process is repeated several times until enough information is captured for the signal reconstruction.

To enable different spectral encoding with the CA compressive spectrometer (Figure 9.8b), it is necessary to make use of SLMs. Since the required encoding is performed only in the vertical dimension, a single-column DMD is sufficient. This type of system requires a relatively long acquisition time due to the SLM switching time and the sensor integration time. Faster acquisition can be achieved by parallelization of the process. The light is fed through a number of parallel CA columns, and each encoded spectrum is then measured with a different point sensor. Figure 9.8b shows an illustration of a parallel coded spectrometer system. This system is built from a regular 2D CA (marked in red) that functions as a set of linear CAs. The system sensor is a linear (vector) sensor array with the same number of pixels as the number of columns in the CA. The use of a *2D* CA provides a parallel coded set of measurements and converts the coded spectrometer to a snapshot coded spectrometer.

The system described in Figure 9.8 performs spatial encoding and spectral encoding separately, and so it is natural to describe the mathematical model of the system with separable operators. The use of separable operators can simplify the description of the system and accelerates the reconstruction process. A mathematical model of the system in Figure 9.8a can be derived by extending the SPC model. In general, the SPC can be set to perform random multiplexing, which is described by a random matrix $\mathbf{\Phi}_{xy}$. Such encoding may work well for small images of size $N_x \times N_y$, which do not exceed the size of a few hundred by a few hundred pixels. However, for large encoding arrays, the size of $\mathbf{\Phi}_{xy}$ is huge, introducing technical limitations, as mentioned earlier in the "From Theoretical to Applied Compressive Spectral Imaging" section. To reduce the dimensionality problem, we can apply on the DMD patterned codes that are separable in the horizontal and the vertical dimensions. This reduces the computational and storage problem significantly (Rivenson and Stern 2009, 449–452; August et al. 2013a, D46–D54). The separable operation can be expressed by means of Kronecker products (or outer products; Rivenson and Stern 2009, 449–452); $\Phi_{xy} = \Phi_x^T \otimes \Phi_y$ where $\Phi_x^T, \Phi_y \in \mathbb{R}^{M\times N}$ are the encoding matrices and $\otimes$ is the Kronecker product operator. The process is applied for each spectral band $\mathbf{F}_k$. Technically, for the purpose of simulation of the system, and for reconstruction purposes, the separable operation is better described by

$$\mathbf{G}_k = \Phi_y \mathbf{F}_k \Phi_x, \tag{9.4}$$

where $\mathbf{F}_k$ is a matrix that represents the data at a single wavelength k so that the set $\{\mathbf{F}_k\}_{k=1}^{N_\lambda}$ builds the entire cube $\mathcal{F}$. $\mathbf{G}_k$ is the spatially encoded product of $\mathbf{F}_k$. Figure 9.9 illustrates the operation of Equation 9.4 for all the spectral bands $\mathbf{F}_k$ in the spectral cube. The sensing process model $\mathbf{F}_k \rightarrow \mathbf{G}_k$ is described in the same parallel way for all k wavelengths; due to the first encoding step (Figure 9.8a), all spectral bands undergo the same spectral encoding and, hence, all $\mathbf{\Phi}_k$ are the same. In order to describe the spectral sensing operation, the spatial encoded datacube $\Gamma \in \mathbb{R}^{M_x \times M_y \times N_\lambda}$ is first unfolded, that is, a matrix is created from a cube shape. As illustrated in Figure 9.10, each spectral slice $\mathbf{G}_k$ is unfolded to its lexicographic vector form $vec\{\mathbf{G}_k\} \rightarrow \mathbf{g}_k$ and the entire spatially encoded cube $\mathbf{\Gamma}$ is then written as a single matrix $\mathbf{G} \in \mathbb{R}^{M_x M_y \times N_\lambda}$. Figure 9.10 illustrates the operation of $\mathbf{\Phi}_\lambda$ on the spatially encoded data matrix $\mathbf{G}$. The product matrix can be used in its original form or unfolded form as an input for the recovery algorithm.

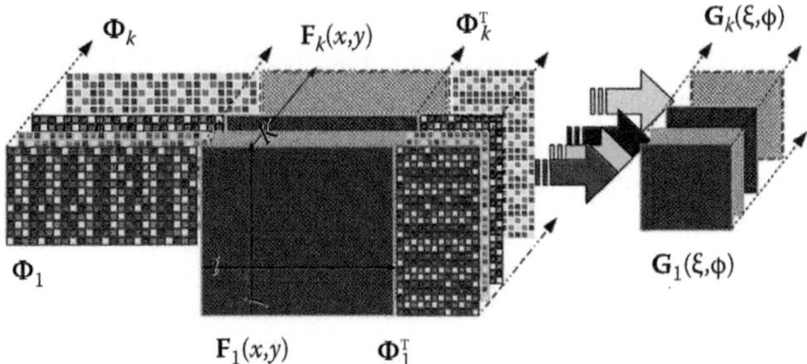

Figure 9.9 The spatial encoding model step with separable encoding in x and y directions. Parallel separable sampling is performed by using the same spatial encoding for all spectral bands.

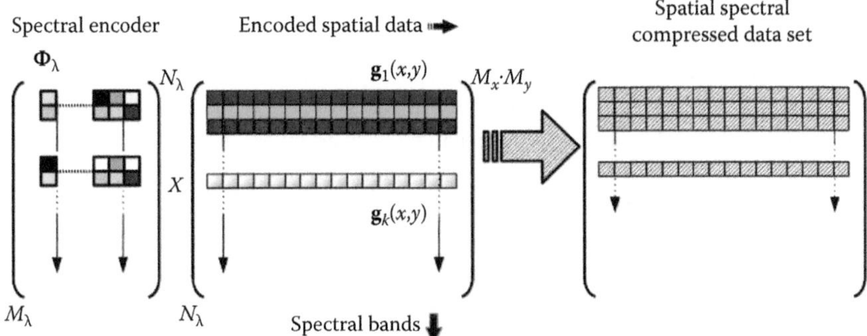

Figure 9.10 The spectral encoding model. The model for spectral encoding is based on multiple spectral encoding for each spatial encoding, that is, different mixing is done per column of **g**. This can be performed on the spatial captured data that are represented in its vector version.

Compressive Sensing System with Spectral Encoding in the Spectral Domain

By *spectral encoding in the spectral domain*, we wish to emphasize that this section presents systems that do not depend on *spectral-to-spatial* optical transforms. In addition, this section describes SI systems based on spectral encoding with continuous codes, in contrast to discrete codes, which were considered before. Such spectral multiplexing can be achieved with optical devices that act as generalized spectral filters. According to this approach, rather than using spectral filters to filter out only a single narrow spectral band at each scanning step (e.g., a color filter wheel or a tunable filter), generalized spectral filters transmit or reflect a combination of nonuniformly weighted components from the entire spectral range. Figure 9.11 illustrates this. The input spectral signal is illustrated in Figure 9.11a, and the system continuous spectral modulation is shown in Figure 9.11b. The transmitted or reflected light in Figure 9.11c is given as an inner product of the input signal and the system spectral transfer function, which is integrated by a photosensor to produce a single spectrally multiplexed measurement.

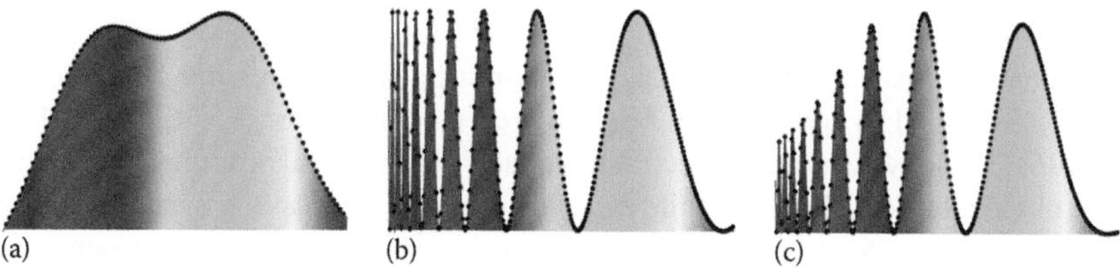

Figure 9.11 Illustration of sensing with generalized spectral filters: (a) is the input spectral signal, (b) is the system spectral response, and (c) is the spectral signal at the system output. The modulated spectrum is integrated, yielding a sample, $\mathbf{g}_i$. The process is repeated with different system spectral responses.

Repeated measurements using various generalized filters similar to those shown in Figure 9.11b may provide enough information for reliable recovery of the spectral signal. The multiplexed spectral measurement process can be mathematically described by the linear integral:

$$g(k) = \int_0^\infty \phi(\lambda, k) \cdot f(\lambda) d\lambda, \tag{9.5}$$

where

$f(\lambda)$ is the spectral input signal

$\phi(\lambda,k)$ is the generalized filter spectral response, which acts as a spectral encoder

The spectral encoding function $\phi(\lambda,k)$ depends on the parameter k in accordance with some physical mechanism. For each selection of k, the encoding function $\phi(\lambda,k)$ changes and a $g(k)$ measurement is recorded. If we look at Equation 9.5, we can see that, in contrast to systems that are based on spatial encoding (the "Compressive Spectral Imaging with Spectral Encoding in the Spatial Domain" section), the description here is based on continuous optical encoding through $\phi(\lambda,k)$. This feature may be used in order to achieve very high spectral resolution. In order to present the system in the form of Equation 9.5, some discretization of the sensing operation model is needed. By discretizing the spectrum, we can write Equation 9.6 in the form

$$g(k) \approx \sum_{\lambda=1}^{N_\lambda} \phi_\lambda(k) \cdot f_\lambda. \tag{9.6}$$

Another discretization procedure is associated with a number of filter configurations used throughout the imaging process. If we let k take discrete values $1, \ldots, k, \ldots, M_\lambda$, we can write Equation 9.6 in the form

$$g_k = \sum_{\lambda=1}^{N_\lambda} \phi_{\lambda k} \cdot f_\lambda \rightarrow {}_{M_\lambda}[\mathbf{g}] = {}_{M_\lambda}[\mathbf{\Phi}]^{N_\lambda}[\mathbf{f}]^{N_\lambda}. \tag{9.7}$$

This form is the standard for matrix–vector multiplexing, as presented earlier in Equation 9.1. In the following, different physical mechanisms implementing different $\phi(\lambda,k)$ are described.

Compressive Liquid Crystal Spectrometers for Imaging

The compressive LC spectrometer was introduced in August and Stern (2013, 4996–4999), and its application for SI was presented in August et al. (2014, FM3E. 4, 2016). These systems use a spectral modulator that is based on a single thick variable phase retarder cell. SI capability may be achieved by coupling the compressive LC spectrometers to an SPC. This can be done in a similar way to that presented in the "Compressive Spectral Imaging with Spectral Encoding in the Spatial Domain" and "Compressive Sensing System with Spectral Encoding in the Spectral Domain" sections, in which the spectral imaging capabilities can be achieved by replacing the photosensor with a compressive LC spectrometer. Other ways to transform the compressive LC spectrometer system into an SI system is by coupling the LC retarder to the optical detector or, alternatively, by mounting a large clear aperture of an LC retarder cell at the physical entrance pupil or at the aperture plane.

Both the compressive LC spectrometer and the common Lyot or Solc (Evans 1949, 229–237, 1958, 142–143; Aharon and Abdulhalim 2009, 11426–11433; Hamdi et al. 2012, pp. 1–4) spectral filters are based on birefringent spectral filters, but there is a fundamental difference between the ways they are used in CS. Lyot or Solc spectral filters use many layers of phase retarders in order to filter out a narrow spectral band. As the number of stages increases, the bandwidth narrows. This method may suffer from low optical efficiency and the need for N_λ measurements. In contrast, the compressive LC spectrometer is built from a single-stage, thick variable retarder, as illustrated in Figure 9.12. The phase retarder is made of an LC layer sandwiched between two flat glass plates and two linear cross polarizers, followed by an optical measurement unit. The glass plates are coated with indium tin oxide, comprising a transparent electrode, and a polymer alignment layer. The cavity is filled with a special blend of LC that has high birefringence. The LC birefringence is controlled by an electric field that is obtained by applying voltage on the electrodes. In the illustration, the alignment of the LC molecules is modified as a function of the applied voltage, leading to a change in the effective retardance in accordance.

The optical retardation is proportional to the induced birefringence, $\Delta n_i = n_{e,i} - n_o$, where n_e is the extraordinary refraction coefficient and n_o is the ordinary refraction coefficient. For a given cell with a gap d, the phase retardation (phase

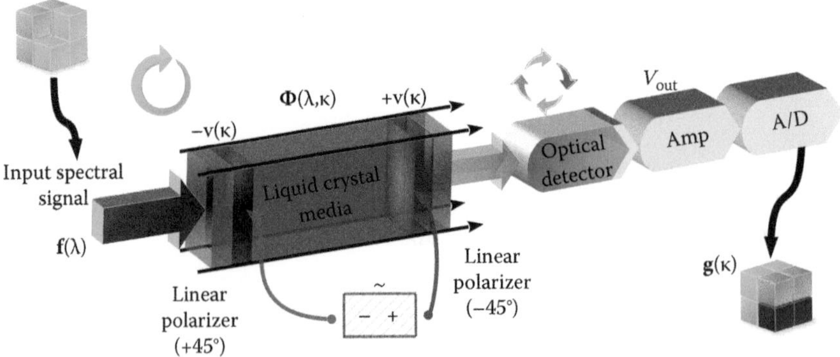

Figure 9.12 An LC variable retarder used as a spectral modulator and a photosensor that integrates the modulated spectrum. The spectral modulation at the output is a function of the birefringence, which can be controlled by the voltage applied over the cell.

difference) at wavelength λ is given by $\delta_k = (2 \cdot \pi \cdot \Delta n_k d)/\lambda$. The spectral modulation depends on the optical retardation and has the form (Yariv and Yeh 1984)

$$\phi_k(\lambda) = \frac{I_k(\lambda)}{I_0(\lambda)} \propto \sin^2\left(\frac{\delta_k(\lambda)}{2}\right), \tag{9.8}$$

where

$I_k(\lambda)$ denotes the kth modulated signal, obtained by tuning the cell to the kth state, corresponding to the birefringence Δn_k

$I_0(\lambda)$ is the input spectral power distribution

Figure 9.13 illustrates typical theoretical spectral responses as a function of birefringence Δn_i. The left graph presents the spectral transmission of a 10 μm gap cell, and the right graph presents the spectral transmission of a 50 μm cell gap. For both graphs, the induced birefringence Δn_i is in the range of zero to about 0.5. The number of modulation peaks increases as the cell gap and the birefringence increase. We see that using large gap cells with a large birefringence range, it is possible to modulate the spectral signal with more spectral modulation peaks than can be generated in cells with a small gap.

For the 1D spectral signal $\mathbf{f}$, the set of modulated measurements may be written as $\mathbf{g} = \mathbf{\Phi f}$, where, in our case, the sensing operator is

$$\mathbf{\Phi} = {}_{M_\lambda}\begin{pmatrix} \sin^2\left(\frac{1}{2}\delta_1\lambda_1\right) & \cdots & \sin^2\left(\frac{1}{2}\delta_1\lambda_{N_\lambda}\right) \\ \vdots & \sin^2\left(\frac{1}{2}\delta_i\lambda_j\right) & \vdots \\ \sin^2\left(\frac{1}{2}\delta_{M_\lambda}\lambda_{N_\lambda}\right) & \cdots & \sin^2\left(\frac{1}{2}\delta_{M_k}\lambda_{N_\lambda}\right) \end{pmatrix}^{N_\lambda}. \tag{9.9}$$

The sensing matrix in Equation 9.9 is obtained by sampling the continuous expression in Equation 9.8 on a grid of N_λ spectral bands by M_λ voltage points.

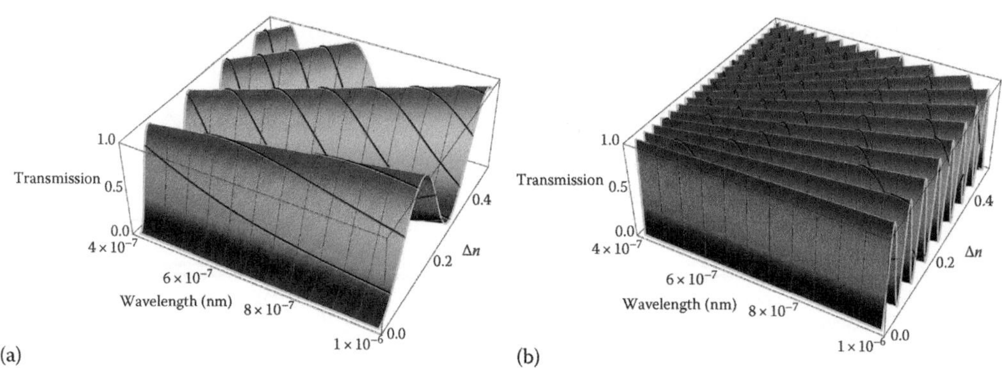

Figure 9.13 Two different spectral responses obtained by theoretical models of phase retarders with different retardation. The modulation of the phase retardation device with high retardation is denser. (a) Presents the spectral transmission of a 10 μm gap. (b) Presents the spectral transmission of a 50 μm gap.

Since the sampling process follows a continuous model, there is freedom in the choice of spectral points $\{\lambda_i\}_{i=1}^{N_\lambda}$ and voltage points $\{v_k\}_{k=1}^{M_\lambda}$. However, there is a constraint on the choice of steps $\Delta\lambda, \Delta k$ that is determined by uncertainties of the physical model and by the maximum number of coherence bands (Chapter 3). Uncertainties in the model arise from the fact that it only describes perpendicular incidence of the light and an ideal birefringence material. Therefore, in practice, one has to "calibrate" the system, that is, to measure its response. The calibration process is performed on a finite resolution spectral grid, so that, technically, the sensing matrix spectral grid is also limited by the calibration spectrometer grid and its precision.

Figure 9.14 compares the spectral signal reconstructed with compressive LC spectrometers to that measured directly with a conventional spectrometer. The compressive LC spectrometer was numerically simulated for a cell gap of 13 μm.

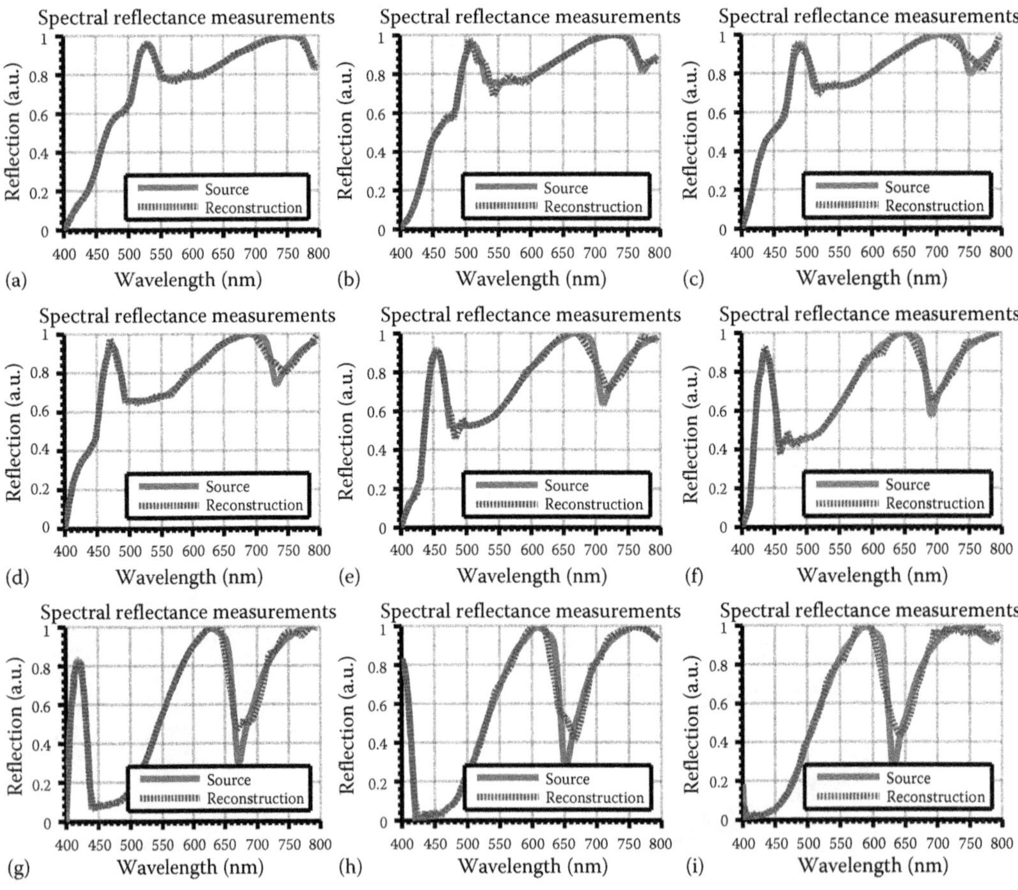

Figure 9.14 Comparison between measured and reconstructed spectra of directly measured (continuous line) and reconstructed (dashed line). The nine different biogenic crust-like spectra data have $N = 1024$ samples, and the CS spectrum is recovered from $M = 108$ measurements; Daubechies (db3) wavelet transform was used as the sparsifying operator. Reconstruction PSNR values are (a)–(i) 40.3, 36.4, 34.8, 33.2, 30.9, 31.1, 27.6, 25.5, and 25.7 (dB).

Compressive Modified Fabry–Pérot Spectrometers for Imaging

Classically, the Fabry–Pérot interferometer (FPI) has been used as a scanning spectrometer that has very high resolving power and high throughput. The transmitted wavelengths of an FPI are determined by the cavity resonance and can be tuned by varying the thickness, *d*, of the cavity or the index of refraction, *n*. The FPI spectral encoding function is given in terms of *d*, *n*, and the mirror reflection *R* (Hernández 1988):

$$\phi_k(\lambda) = \frac{I_k(\lambda)}{I_0(\lambda)} \propto \frac{1}{1+(4R(\lambda)/(1-R(\lambda))^2)\sin(2\pi p_k/\lambda)}, \tag{9.10}$$

where

$I_k(\lambda)$ denotes the *k*th modulated signal, obtained by tuning the FPI to the *k*th state, which corresponds to a specific optical path, $p_k = n(\lambda)d$

$I_0(\lambda)$ is the input spectral power distribution

Figure 9.15 shows typical theoretical spectral responses of FPIs. The right graph presents the spectral transmission as a function of the refractive index of the material in between the fixed distance mirrors. The illustrated case is for an FPI response where the gap between the mirrors is of width 1 μm. The left graph shows the spectral responses of the FPI with an air-filled cavity, where the distances between the mirrors are between 1 and 3 μm.

Figure 9.16 illustrates two ways in which an FPI can be used to form a scanning spectrometer. In its simplest description, an FPI consists of two highly reflective mirrors with a high degree of flatness and strictly parallel alignment. Figure 9.16a can represent an FPI spectrometer that is filled with an LC solution. The electric field applied between the electrodes changes the extraordinary refractive index n_e and, hence, the transmitted wavelength. As only the extraordinary refractive index n_e changes with this configuration, this method can only be used with polarized light. The second illustration (Figure 9.15b) is for an FPI device that can be tuned by changing the distance *d* between the flat mirrors. We realize this by placing a set of piezoelectric devices, and so the gap can be controlled precisely by using the applied voltage or by electrothermal devices (Wang et al. 2015, 1418–1421).

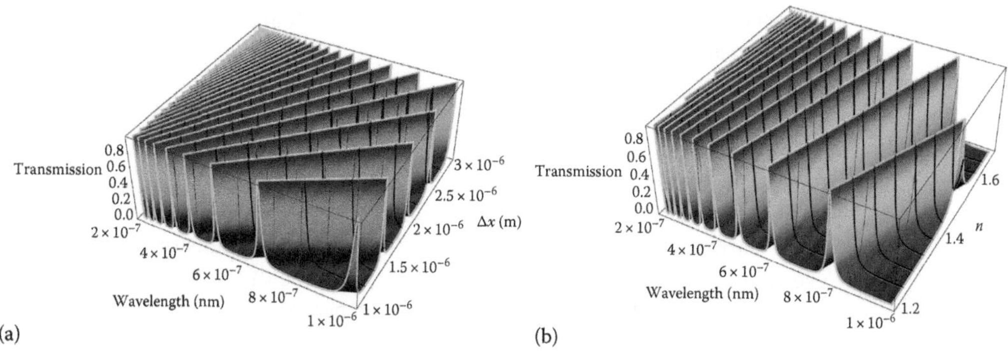

Figure 9.15 Spectral responses of FPI (a) as a function of the distance between the mirrors and (b) as a function of the refraction index of the cavity between the mirrors.

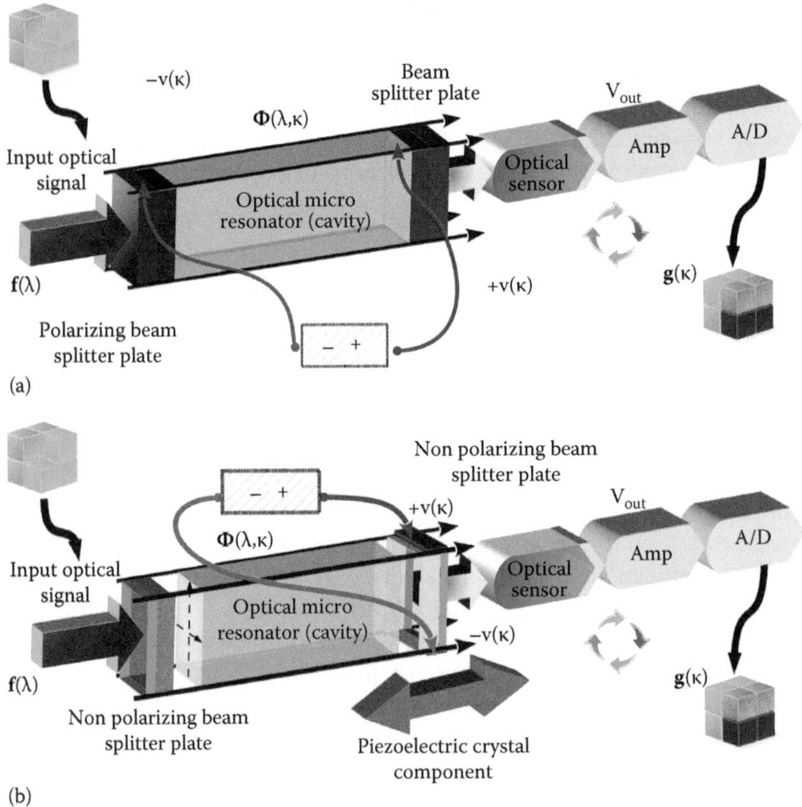

Figure 9.16 Schematic description of an FPI with LC (a) and with piezoelectric crystal (b).

The main difference between a classical FPI and a compressive modified FPI is the way of operation. In the classical case, the FPI works as a narrow spectral scanning filter, while with the compressive FPI, wideband multiplexed components pass through and are integrated by the detector.

Figure 9.17 shows the spectral signal reconstructed with a simulated compressive FPI spectrometer and with a conventional spectrometer signal. The compressive modified FPI spectrometer was numerically simulated for a cell gap varying from 4 to about 5 mm.

Other Compressive Sensing Spectrometers

Compressive Fourier Transform Spectrometers

Fourier transform spectroscopy (or Fourier transform infrared spectroscopy; Griffiths and De Haseth 2007) is a common spectroscopic method mainly used in the infrared spectrum. One variation of the Fourier transform spectrometer uses the Michelson interferometer. In this case, one of the reflecting mirrors is able to move at a steady speed, or in a sequence of small steps of known size. The change in the optical path due to the reflector movement provides a phase difference.

In classical Fourier transform spectroscopy, an N-point multiplex spectrum is captured through the sensing process and, by using numerical Fourier

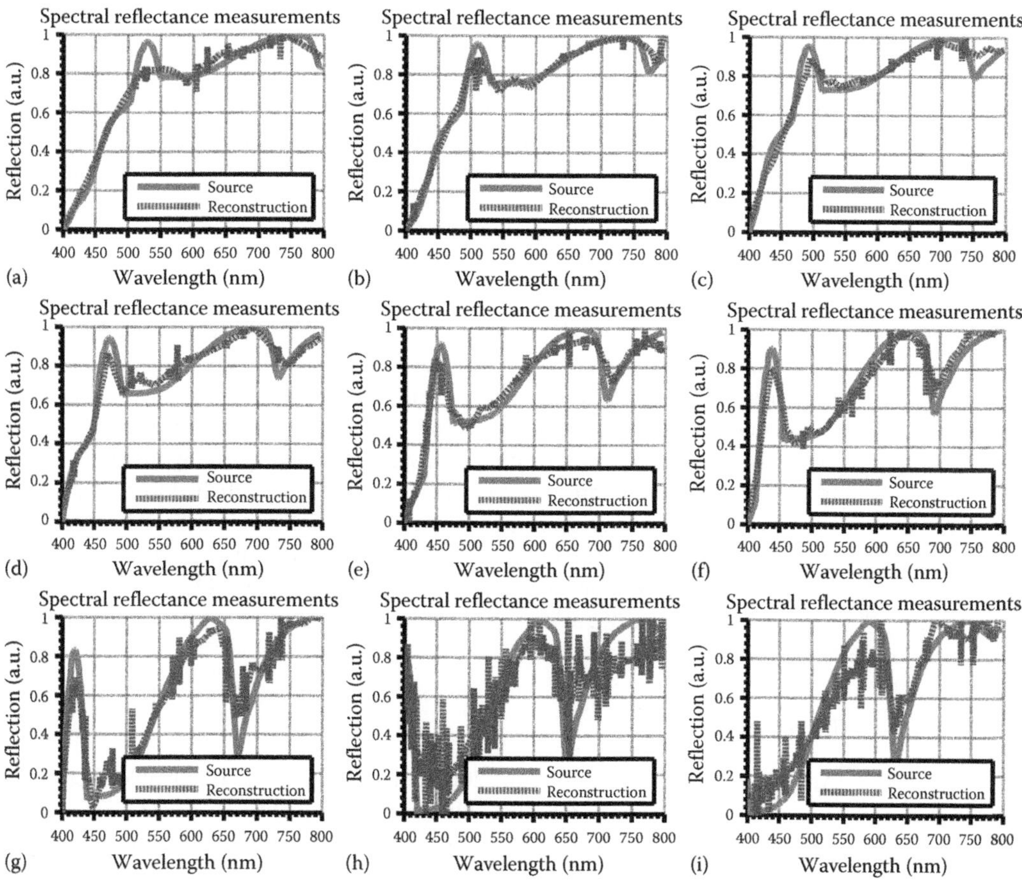

Figure 9.17 Comparison between measured and reconstructed power spectral density signals. Directly measured (continuous line) and reconstructed (dashed line). The nine different biogenic crust-like spectra data have $N = 1024$ samples, and the CS spectrum is recovered from $M = 153$ measurements; Daubechies (db3) wavelet transform was used as the sparsifying operator. Reconstruction PSNR values are (a)–(i) 32.7, 30.3, 32.4, 31.1, 32.8, 29.3, 23.6, 25.5, and 29.1 (dB).

transforming, a set of N equally spaced spectral bands is recovered. Generally, a CS system based on Fourier transform spectroscopy can be implemented with a set of only $M < N$ measurements taken from random positions (Katz et al. 2010; Sanders et al. 2012, 2697–2702). In this case, the system takes measurements in nonequally spaced positions and a sparse recovery program is conducted.

Structured Spectral Illumination

An alternative method for spectral imaging can be implemented by structured illumination of the scene using a spectrally structured light field without any additional spectral separation on the imager side. In this case, the illumination source must have the ability to illuminate with a broad encoded spectrum and to provide uniform spatial illumination. In principle, this approach can be applied with any of the mentioned spectral encoding principles. The illumination source can be attached to the spectral illuminator, and a synchronization link between the spectral encoder and the camera shutter is necessary. Structured illumination may bypass some of the limitations of direct spectroscopic techniques. For instance, it removes the incidence angle limitation of the system described in the "Compressive Sensing System with Spectral Encoding in the Spectral Domain" section. This opens up the possibility of using such spectral imaging techniques in applications that require high numerical apertures, such as in microscopy. For intermediate distance imaging, one can envisage a scheme that uses spatial modulation in the illumination part, in addition to the spectral encoding (Zhang et al. 2015).

Summary

Compressive sensing provides a way to reduce the required effort in the acquisition stage, which is normally high in spectroscopy and SI. CS SI achieves similar performance to the classical methods, but with significant reduction in acquisition time, system size, complexity, and measurement noise.

Recently, there has been increasing interest in the implementation of CS theory in the fields of spectroscopy and spectral imaging. Many methods for CS spectroscopy and compressive SI systems have been published in recent years. In this chapter, we have described the main principles of compressive spectroscopy and compressive imaging spectroscopy. We have provided the taxonomy of CS spectroscopy and outlined SI design approaches. The sections of this chapter have been categorized in accordance with the way that the encoding is realized.

References

Aharon, O. and I. Abdulhalim. 2009. Liquid crystal Lyot tunable filter with extended free spectral range. *Optics Express* 17(14): 11426–11433.

Akbari, H., L. V. Halig, D. M. Schuster, A. Osunkoya, V. Master, P. T. Nieh, G. Z. Chen, and B. Fei. 2012. Hyperspectral imaging and quantitative analysis for prostate cancer detection. *Journal of Biomedical Optics* 17(7): 076005-1–076005-10.

Arce, G. R., D. J. Brady, L. Carin, H. Arguello, and D. S. Kittle. 2014. Compressive coded aperture spectral imaging: An introduction. *IEEE Signal Processing Magazine* 31(1): 105–115.

Arguello, H. and G. R. Arce. 2011. Code aperture optimization for spectrally agile compressive imaging. *Journal of the Optical Society of America A* 28(11): 2400–2413.

Arguello, H. and G. R. Arce. 2014. Colored coded aperture design by concentration of measure in compressive spectral imaging. *IEEE Transactions on Image Processing* 23(4): 1896–1908.

August, I., Y. Oiknine, M. AbuLeil, I. Abdulhalim, and A. Stern. 2016. Miniature compressive ultra-spectral imaging system utilizing a single liquid crystal phase retarder. *Scientific Reports*, 6, 23524.

August, Y., Y. Oiknine, A. Stern, and D. G. Blumberg. 2014. Hyperspectral compressive imaging based on spectral modulation in the spectral domain. In *Frontiers in Optics. Optical Society of America*, p. FM3E. 4.

August, Y., Y. Oiknine, A. Stern, and D. G. Blumberg. 2014a. Hyperspectral compressive imaging based on spectral modulation in the spectral domain. *Optical Society of America.*

August, Y., A. Stern, and D. G. Blumberg. 2014. Single-pixel spectroscopy via compressive sampling. In *Computational Optical Sensing and Imaging. Optical Society of America*, p. CTu2C. 2.

August, Y. and A. Stern. 2013. Compressive sensing spectrometry based on liquid crystal devices. *Optics Letters* 38(23): 4996–4999.

August, Y., C. Vachman, Y. Rivenson, and A. Stern. 2013a. Compressive hyperspectral imaging by random separable projections in both the spatial and the spectral domains. *Applied Optics* 52(10): D46–D54.

August, Y., C. Vachman, and A. Stern. 2013b. Spatial versus spectral compression ratio in compressive sensing of hyperspectral imaging. In *SPIE Defense, Security, and Sensing*, pp. 87170G–87170G. International Society for Optics and Photonics.

Born, M. and E. Wolf. 1999. *Principles of Optics: Electromagnetic Theory of Propagation, Interference and Diffraction of Light.* Cambridge University Press, Melbourne, Australia.

Delaney, J. K., J. G. Zeibel, M. Thoury, R. Littleton, M. Palmer, K. M. Morales, E. R. de La Rie, and A. Hoenigswald. 2010. Visible and infrared imaging spectroscopy of Picasso's Harlequin Musician: Mapping and identification of artist materials in situ. *Applied Spectroscopy* 64(6): 584–594.

Duarte, M. F., M. A. Davenport, D. Takhar, J. N. Laska, T. Sun, K. E. Kelly, and R. G. Baraniuk. 2008. Single-pixel imaging via compressive sampling. *IEEE Signal Processing Magazine* 25(2): 83.

Dupuis, G., M. Elias, and L. Simonot. 2002. Pigment identification by fiber-optics diffuse reflectance spectroscopy. *Applied Spectroscopy* 56(10): 1329–1336.

Eismann, M. T. 2012. *Hyperspectral Remote Sensing.* SPIE Press Book, Bellingham, WA.

Evans, J. W. 1949. The birefringent filter. *JOSA* 39(3): 229–237.

Evans, J. W. 1958. Solc birefringent filter. *JOSA* 48(3): 142–143.

Fellgett, P. 1967. Conclusions on multiplex methods. *Le Journal De Physique Colloques* 28(C2): C2-165–C2-171.

Golub, M., M. Nathan, A. Averbuch, A. Kagan, V. Zheludev, and R. Malinsky. 2015. Snapshot spectral imaging based on digital cameras. U.S. Patent No 9,013,691.

Griffiths, P. R. and J. A. De Haseth. 2007. *Fourier Transform Infrared Spectrometry*, Vol. 171. John Wiley & Sons, West Sussex, UK.

Hamdi, R., R. M. Farha, S. Redadaa, B.-E. Benkelfat, D. Abed, A. Halassi, and Y. Boumakh. 2012. Optical bandpass Lyot filter with tunable bandwidth. In *19th International Conference on Telecommunications (ICT)*, Jounieh, Lebanon.

Hernández, G. 1988. *Fabry-Perot Interferometers*, Vol. 3. John Wiley & Sons Ltd, West Sussex, UK.

Hirschfeld, T. 1976. Fellgett's advantage in UV-VIS multiplex spectroscopy. *Applied Spectroscopy* 30(1): 68–69.

Iordache, M.-D., J. Bioucas-Dias, and A. Plaza. 2009. Unmixing sparse hyperspectral mixtures.

Iordache, M. D., J. M. Bioucas Dias, and A. Plaza. 2011. Sparse unmixing of hyperspectral data. *IEEE JGRS* 49(6): 2014–2039.

Katz, O., J. M. Levitt, and Y. Silberberg. 2010. Compressive Fourier transform spectroscopy. In *Frontiers in Optics. Optical Society of America*, p. FTuE3.

Keshava, N. and J. F. Mustard. 2002. Spectral unmixing. *IEEE Signal Processing Magazine* 19(1): 44–57.

Kiers, H. A. L. 2000. Towards a standardized notation and terminology in multiway analysis. *Journal of Chemometrics* 14(3): 105–122.

Li, Q., X. He, Y. Wang, H. Liu, D. Xu, and F. Guo. 2013. Review of spectral imaging technology in biomedical engineering: Achievements and challenges. *Journal of Biomedical Optics* 18(10): 100901.

Lim, S., K. H. Sohn, and C. Lee. 2001a. Principal component analysis for compression of hyperspectral images. In *International Geoscience and Remote Sensing Symposium. IGARSS '01.* IEEE. University of New South Wales, Sydney, Australia.

Lim, S., K. H. Sohn, and C. Lee. 2001b. Compression for hyperspectral images using three dimensional wavelet transform. In *International Geoscience and Remote Sensing Symposium. IGARSS '01.* IEEE. University of New South Wales, Sydney, Australia.

Lin, X., Y. Liu, J. Wu, and Q. Dai. 2014. Spatial-spectral encoded compressive hyperspectral imaging. *ACM Transactions on Graphics* (*TOG*) 33(6): 233.

Lukin, V., N. Ponomarenko, M. Zriakhov, and A. Kaarna. 2008. Two aspects in lossy compression of hyperspectral aviris images. In *12th International Conference on Mathematical Methods in Electromagnetic Theory.* IEEE, Odesa, Ukraine, pp. 375–377.

Mallat, S. 2009. *A Wavelet Tour of Signal Processing: The Sparse Way*, 3rd edn. Elsevier, Amsterdam, The Netherlands.

Manolakis, D., D. Marden, and G. A. Shaw. 2003. Hyperspectral image processing for automatic target detection applications. *Lincoln Laboratory Journal* 14(1): 79–116.

Manolakis, D. and G. Shaw. 2002. Detection algorithms for hyperspectral imaging applications. *IEEE Signal Processing Magazine* 19(1): 29–43.

Martin, M., M. Wabuyele, K. Chen, P. Kasili, M. Panjehpour, M. Phan, B. Overholt et al. 2006. Development of an advanced hyperspectral imaging (HSI) system with applications for cancer detection. *Annals of Biomedical Engineering* 34(6): 1061–1068.

Martinez, K., J. Cupitt, D. Saunders, and R. Pillay. 2002. Ten years of art imaging research. *Proceedings of the IEEE* 90(1): 28–41.

Meggers, W. F., C. H. Corliss, and B. F. Scribner. 1975. Tables of spectral line intensities. In *Arranged by Wavelengths*, Vol. v.2. U.S. National Bureau of Standards, Washington, DC, N.B.S. Monograph, 145.

Nelson, E. D. and M. L. Fredman. 1970. Hadamard spectroscopy. *JOSA* 60(12): 1664–1669.

Qian, S.-E., A. B. Hollinger, M. Dutkiewicz, H. Zwick, H. Tsang, and J. R. Freemantle. 2001. Effect of lossy vector quantization hyperspectral data compression on retrieval of red-edge indices. *IEEE JGRS* 39(7): 1459–1470.

Ribés, A. 2013. Image spectrometers, Color high fidelity, and fine-art paintings. In *Advanced Color Image Processing and Analysis*, C. Fernandez-Maloigne, ed. Springer, New York, pp. 449–483.

Rivenson, Y. and A. Stern. 2009. Compressed imaging with a separable sensing operator. *IEEE Signal Processing Letters* 16(6): 449–452.

Rueda, H., H. Arguello, and G. R. Arce. 2014. Compressive spectral imaging based on colored coded apertures. In *IEEE International Conference on Acoustic, Speech and Signal Processing* (*ICASSP*), Florence, Italy.

Ryan, M. J. and J. F. Arnold. 1997a. The lossless compression of AVIRIS images by vector quantization. *IEEE JGRS* 35(3): 546–550.

Ryan, M. J. and J. F. Arnold. 1997b. Lossy compression of hyperspectral data using vector quantization. *Remote Sensing of Environment* 61(3): 419–436.

Sanders, J. N., S. K. Saikin, S. Mostame, X. Andrade, J. R. Widom, A. H. Marcus, and A. Aspuru-Guzik. 2012. Compressed sensing for multidimensional spectroscopy experiments. *The Journal of Physical Chemistry Letters* 3(18): 2697–2702.

Sansonetti, C. J., M. L. Salit, and J. Reader. 1996. Wavelengths of spectral lines in mercury pencil lamps. *Applied Optics* 35(1): 74–77.

Shaw, G. A. and K. B. Hsiao-hua. 2003. Spectral imaging for remote sensing. *Lincoln Laboratory Journal* 14(1): 1–28.

Sloane, N. J. 1979. Multiplexing methods in spectroscopy. *Mathematics Magazine* 52(2): 71–80.

Smith, D. M. 1952. *Visual Lines for Spectroscopic Analysis.* Hilger & Watts, London, UK.

Sun, T. and K. Kelly. 2009. Compressive sensing hyperspectral imager. In *Computational Optical Sensing and Imaging. Optical Society of America,* p. CTuA5.

Thenkabail, P. S., J. G. Lyon, and A. Huete. 2012. *Hyperspectral Remote Sensing of Vegetation.* CRC Press, Boca Raton, FL.

Treado, P. J. and M. D. Morris. 1989. A thousand points of light: The Hadamard transform in chemical analysis and instrumentation. *Analytical Chemistry* 61(11): 723A–734A.

Vetterli, M. and J. Kovaseviss. 1995. *Wavelets and Subband Coding.* Prentice-Hall, Englewood Cliffs, NJ. Reprinted in 2007.

Vickers, T. J., C. K. Mann, and J. Zhu. 1991. Hadamard multiplex multichannel spectroscopy to achieve a spectroscopic power distribution advantage. *Applied Spectroscopy* 45(1): 42–49.

Wang, W., S. R. Samuelson, J. Chen, and H. Xie. 2015. Miniaturizing Fourier transform spectrometer with an electrothermal micromirror. *IEEE Photonics Technology Letters* 27(13): 1418–1421.

Warner, T. A., M. Duane Nellis, and G. M. Foody. 2009. *The Sage Handbook of Remote Sensing.* Sage, Thousand Oaks, CA.

Wolfe, W. L. 1997. *Introduction to Imaging Spectrometers.* SPIE-International Society for Optical Engineering, Bellingham WA.

Wu, D. and D.-W. Sun. 2013. Advanced applications of hyperspectral imaging technology for food quality and safety analysis and assessment: A review—Part I: Fundamentals. *Innovative Food Science & Emerging Technologies* 19(0): 1–14.

Wu, Y., I. O. Mirza, G. R. Arce, and D. W. Prather. 2011. Development of a digital-micromirror-device-based multishot snapshot spectral imaging system. *Optics Letters* 36(14): 2692–2694.

Yariv, A. and P. Yeh. 1984. *Optical Waves in Crystals*, Vol. 5. Wiley, New York.

Zhang, Z., X. Ma, and J. Zhong. 2015. Single-pixel imaging by means of Fourier spectrum acquisition. *Nature Communications* 6, article no. 24752.

10 Compressive Polarimetric Sensing

Fernando Soldevila, Vicente Durán,
Pere Clemente, Esther Irles,
Mercedes Fernández-Alonso,
Enrique Tajahuerce, and Jesús Lancis

Contents

Introduction

Polarimetric imaging (PI) has been established as a key tool in several scientific areas such as remote sensing, material sciences, and biomedicine. In remote sensing, PI techniques enhance the quality of the recovered images and improve the detection of targets allowing an exhaustive analysis of the scene (Tyo et al. 2006). Polarization provides information about spatial features of a surface such as shape and texture. For that reason, PI techniques have been used in material sciences in order to segment rough surfaces (Terrier et al. 2008). In the field of biomedical imaging, PI has been applied to improve the visualization of samples at different penetration depths (Demos and Alfano 1997). Furthermore, PI has been used in in vivo techniques to detect and diagnose cancerous tumors in living tissues (Baba et al. 2002; Laude-Boulesteix et al. 2004). In addition, the flexibility of PI systems has allowed building optical coherence tomography systems (Nadkarni et al. 2007), adaptive ophthalmic optics (Song et al. 2008), and hyperspectral systems (Glenar et al. 1994; Oka and Kato 1999) with better specifications than nonpolarimetric approaches.

In general, the measurement of multidimensional images, such as polarimetric images, involves the acquisition of a large amount of information, which causes both storage and transmission difficulties. In addition, techniques such as polarimetric or multispectral imaging require a sequential acquisition of images leading to an increase of the measurement time. A recent approach to polarimetric and multispectral imaging is based on the use of miniaturized polarimetric or spectral filters (Geelen et al. 2013; Zhao et al. 2009a) that are incorporated to each pixel of the sensor, which allows acquiring multidimensional images in one shot. However, the development of such systems implies the use of high-end micro-optical components.

In this chapter, we describe several single-pixel PI systems based on compressive sensing (CS). First, in the "Polarimetric Cameras" section, we discuss the concept of polarimetric images and we review different methods to obtain them. Second, in the "Single-Pixel Imaging and Compressive Sensing" section, we describe our approach for imaging by single-pixel detection with sequential pattern illumination. We show also that this technique is very well adapted to apply CS techniques. Next, in the main section of the chapter, the "Single-Pixel Polarimetric Imaging" section, we explain our technique to record polarimetric images based on single-pixel detection and CS. In the "Single-Pixel Spectropolarimetric Imaging" section, we show how to extend this technique to measure both polarimetric and spectral information simultaneously. In one case, we measure only the spatial distribution of orthogonal polarization states together with the spectral information. In the second case, we include full-Stokes PI. Finally, in the "Conclusions" section, we review the main conclusions of this work.

Polarimetric Cameras

In general, objects in a scene modify the polarization state of the reflected, transmitted, or scattered light with respect that of the incident light. This change in the polarization state provides information about the scene. PI has the aim of measuring spatially resolved polarization properties of light fields, objects, or optical systems (Solomon 1981). The physical parameters chosen to describe the polarimetric information define two different types of polarimetric systems. Passive imaging polarimeters perform Stokes PI. They provide the Stokes parameters of light in each pixel of the image. Active imaging polarimeters achieve Mueller PI. In this case, each pixel of the image provides information about the full Mueller matrix characterizing the scene. Stokes polarimetric images usually show the Stokes parameters, the degree of polarization, and the degree of depolarization. When working with Mueller polarimetric images, the parameters represented are usually the retardation, the diattenuation, and the depolarization. Those parameters are simply deduced from the polar decomposition of the Mueller matrix (Lu and Chipman 1996).

The simplest system that deals with polarization combined with conventional imaging contains a linear polarizer before the camera sensor. Adjusting the orientation of the polarizer, the contrast between the object and the background is maximized. This technique is of common use in commercial portraits. Similar techniques have been used in underwater photography to reduce scattering (Duntley 1974). Another option is differential polarimetry, which can provide some of the Stokes parameters of light to be characterized (Demos and Alfano 1997; Tyo et al. 1996). By taking multiple images with different polarizer orientations, the Stokes parameters S_0 and S_1 can be calculated. If multiple polarizers are used, the full linear polarization state can be determined, which is adequate for natural scenes with negligible circular polarized light. The next step consists of combining retarder plates and linear polarizers. By doing this, the full-Stokes vector can be acquired (Solomon 1981).

In Mueller PI, the Mueller matrix of the sample is measured by controlling the polarization state of the incident light by using active polarimeters. A set of known polarization states is generated by the illumination system. Once light is transmitted, scattered, or reflected by the object, the resultant polarization state is determined. Considering an adequate number of polarization states, the full Mueller matrix of the sample can be mathematically recovered (Pezzaniti 1995). This idea has been used in liquid-crystal (LC) screen characterization (Clemente et al. 2008).

Single-Pixel Imaging and Compressive Sensing

Single-pixel imaging is a remarkable alternative to conventional imaging techniques with pixelated detectors. It enables to obtain spatial information of an object, such as the reflectance distribution or other optical properties, by sampling the scene with a set of microstructured light patterns (Duarte et al. 2008). A light detector such as a photodiode records the signal associated to each pattern, and the final image is reconstructed by mathematical algorithms. Single-pixel imaging has been applied in many different fields, for example, color imaging (Welsh et al. 2013), time-of-flight imaging (Kirmani et al. 2011, 2014), or holography (Clemente et al. 2013). With this technique, it is possible to add new degrees of freedom to the sensing process, allowing one to use very sensitive light sensors (Howland et al. 2013), to explore unusual spectral bands for imaging (Chan et al. 2008; Watts et al. 2014), or to use exotic photodetectors such as spectropolarimeters (Soldevila et al. 2014), as shown in this chapter.

One of the main properties of imaging techniques using single-pixel detectors is that they are very well adapted to apply the theory of compressive sampling (CS) (Candes and Wakin 2008; Donoho 2006; Duarte et al. 2008; Howland et al. 2013; Studer et al. 2012). As shown in the previous chapters of this book, this theory exploits the fact that natural images tend to be sparse, that is, only a small set of expansion coefficients is nonzero when a suitable basis of function to express the image is chosen. In this way, images can be retrieved with a number of measurements lower than that established by the Shannon–Nyquist limit.

The basic procedure for single-pixel imaging by CS can be briefly described as follows. A sequence of N microstructured light patterns with light irradiance distribution $\Psi_\ell[m,n]$, m and n being discrete spatial coordinates and $\ell = 1, \ldots, N$, is projected onto the input object. This sampling operation determines the spatial resolution of the technique. Therefore, the object is characterized by a 2D distribution $f[m,n]$ with the same number of pixels, N. Light reflected, transmitted, or scattered by the object is collected by a single photodiode. If we consider the set of patterns $\Psi = \{\Psi_\ell\}$ as a basis of functions, then the irradiance corresponding to the inner product between the patterns and the object provides the coefficients of the image expansion in the new basis. In mathematical terms

$$\mathbf{f} = \Psi \cdot \mathbf{g}, \tag{10.1}$$

where

$\boldsymbol{\Psi}$ is a $N \times N$ matrix that has vectors $\{\Psi_\ell\}$ as columns

$\mathbf{g}$ is the $N \times 1$ vector that contains the expansion coefficients of $f[n,m]$ in the chosen basis

The introduction of CS improves the performance of these single-pixel architectures exploiting sparsity (Candes and Wakin 2008). In this way, images can be retrieved without measuring all the projections of the object on the chosen base. The object under study, $f[m,n]$, can be reconstructed from just a random subset of projection functions. To this end, we randomly choose M different functions of the basis ($M<N$) and measure the projections of the object. This process can be expressed in matrix form as

$$\mathbf{J} = \mathbf{\Phi} \cdot \mathbf{f} = \mathbf{\Phi}(\Psi \cdot \mathbf{g}) = \Theta \cdot \mathbf{g}, \tag{10.2}$$

where

$\mathbf{J}$ is a $M \times 1$ vector that contains the measured projections
$\mathbf{\Phi}$ is a $M \times N$ matrix called sensing matrix

Each row of $\mathbf{\Phi}$ is a function of $\mathbf{\Psi}$ chosen randomly, and the product of $\mathbf{\Phi}$ and $\mathbf{\Psi}$ gives the matrix $\mathbf{\Theta}$ acting on $\mathbf{g}$. If the chosen basis is orthonormal, every row of $\mathbf{\Theta}$ randomly selects a unique element of $\mathbf{g}$. As $M<N$, the underdetermined matrix relation obtained after the measurement process is resolved through an off-line algorithm such as those based on convex optimization or greedy algorithms. In our approach, we use a convex optimization algorithm based on the minimization of the ℓ_1-norm of $\mathbf{g}$ subjected to the constraint given by Equation 10.2. That is, the solution given by the algorithm has to be compatible with the performed measurements. In this case, the proposed reconstruction $\mathbf{f}^*$ is given by the following optimization algorithm:

$$\min_{f^*} \left\| \Psi^{-1} f^* \right\|_{l_1} \quad \text{subject to } \mathbf{\Phi} \cdot \mathbf{f}^* = \mathbf{J} \tag{10.3}$$

In the approaches described in this chapter, the basis $\mathbf{\Psi}$ is a family of binary intensity patterns derived from the Walsh–Hadamard basis. This basis was first proposed in image coding and transmission techniques (Pratt et al. 1969). This choice provides several advantages. First, these patterns are members of an orthonormal basis. Second, natural images tend to be sparse in the Hadamard basis, making these functions very useful also for CS purposes. Third, these patterns are binary functions with values +1 and −1. Therefore, it is very easy to codify them with fast binary amplitude modulators. A Walsh–Hadamard matrix of order N, $\mathbf{H}_N$, is an $N \times N$ matrix with ± 1 entries that satisfies $\mathbf{H}_N^T \cdot \mathbf{H}_N = N \cdot \mathbf{I}_N$, where $\mathbf{I}_N$ is the identity matrix and $\mathbf{H}_N^T$ denotes the transposed matrix. By shifting and rescaling the values of the different columns of $\mathbf{H}_N$, it is possible to generate binary 2D patterns taking value 0 or 1 that can be simply encoded onto the spatial light modulator (SLM) as an intensity modulation.

The key elements to implement a single-pixel camera, according to this approach, are an SLM, to sample the input object with microstructured light patterns, and a light detector. A suitable SLM for a single-pixel camera is a display formed by voltage-controlled LC cells, as those found in video projection systems (Magalhães et al. 2011). This SLM is used to design the single-pixel polarimetric camera described in the "Single-Pixel Polarimetric Imaging" section. Another option to project microstructured light patterns is to use a digital micromirror device (DMD), which is composed of an array of micromirrors that can rotate between two positions. In this way, only selected portions of the incoming light beam are reflected in a given direction (Duarte et al. 2008).

This kind of SLM is used to implement the spectropolarimetric camera described in the "Single-Pixel Spectropolarimetric Imaging" section. Regarding detection, in general, a photodiode is used as a single-pixel camera, which measures the irradiance of the light coming from an object for each pattern generated by the SLM. In the optical systems described in this chapter, we use more complex single-pixel detectors than just a photodiode, such as a beam polarimeter or a fiber spectrometer.

Single-Pixel Polarimetric Imaging

In this section, we describe a passive polarimetric camera by applying the concept of single-pixel imaging with CS (Durán et al. 2012). This camera provides the full spatial distribution of the Stokes parameters of a scene by just using a commercial beam polarimeter. This commercial beam polarimeter is designed for free-space and fiber-based measurements and provides the state of polarization (SOP) of an optical beam as a whole, that is, without spatial resolution. The PI system exhibits high dynamic range (up to 70 dB), broad wavelength range, and high accuracy on the Poincaré sphere thanks to the use of the beam polarimeter. This fact simplifies the development of polarimetric cameras with respect to those based on pixelated image sensors. The key to obtain polarization images is to multiplex the spatial polarization distribution in the time domain by using a programmable SLM.

The operation of the beam polarimeter used as a single-pixel sensor is based on measuring the irradiance of a beam under modulated states of polarization. This is performed with a polarization state analyzer (PSA). In the commercial polarimeter used here, which is sketched in Figure 10.1, the PSA is formed by two voltage-controlled LC variable retarders (LCVRs) and a polarizing beam splitter (PBS). Two photodiodes are respectively located at the output ports of the PBS. By using the Mueller formalism, the input SOP, $\vec{S}$, and the SOP $\vec{S}_{PD1}$ at a photodiode placed in one of the beam splitter arms can be related by the following equation:

$$\vec{S}_{PD1} = \mathrm{M} \cdot \vec{S} \Rightarrow \begin{pmatrix} I_{PD1} \\ S_{1,PD1} \\ S_{2,PD1} \\ S_{3,PD1} \end{pmatrix} = \begin{pmatrix} m_{00} & m_{01} & m_{02} & m_{03} \\ m_{10} & m_{21} & m_{12} & m_{13} \\ m_{20} & m_{31} & m_{22} & m_{23} \\ m_{30} & m_{31} & m_{32} & m_{33} \end{pmatrix} \cdot \begin{pmatrix} I_0 \\ S_1 \\ S_2 \\ S_3 \end{pmatrix}, \quad (10.4)$$

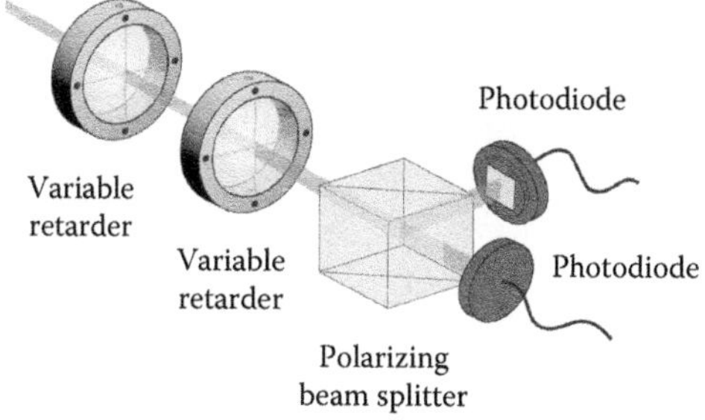

Figure 10.1 Single-pixel polarimetric detector based on a beam Stokes polarimeter.

where **M** is the Mueller matrix of the PSA. The irradiance of the input beam, I_0, is obtained by adding the signals provided by the two photodiodes located at the output ports of the PBS.

The first equation of the matrix system in Equation 10.4 can be written in the following way:

$$I_{PD1}(\delta_1,\delta_2) = m_{00}(\delta_1,\delta_2)I_0 + \sum_{q=1}^{3} m_{0q}(\delta_1,\delta_2)S_q, \tag{10.5}$$

which relates the input beam polarization state to the measured photocurrent. By applying different retardation values to the LCVR, the Mueller matrix of the PSA can be modified. By doing this, the measured photocurrent also changes and Equation 10.5 allows creating an overdetermined equation system. By using a least-squares technique, this system can be solved, obtaining the Stokes parameters of the incident light beam.

The outline of the polarimetric single-pixel camera is shown in Figure 10.2. It has three key elements: an SLM, a beam reducer, and a commercial beam polarimeter as that described earlier. The SLM is used to generate the binary masks necessary for single-pixel reconstruction. Those masks are projected onto the object under study, and the transmitted light is collected onto the beam polarimeter with the aid of the beam reducer. The beam polarimeter measures the Stokes vector for each of the illumination masks, and the different Stokes images are recovered with single-pixel imaging techniques based on CS.

The light source used in our experiment is a He-Ne laser emitting at 632.8 nm. The SLM is a SONY LCX016AL configured as a binary intensity-transmitting modulator by using the proper polarization elements. The masks are projected onto the object by using the appropriate optical system. The commercial beam polarimeter used is PolarVIEW 3000, by Meadowlark Optics. The binary masks used in the experiment are based on Walsh–Hadamard functions. The masks projected onto the object have a resolution of 64×64 pixels, the pixel pitch being 64 μm. In our experiment, 1225 of those masks are projected onto the object (~33% of the Nyquist criterion). Custom software written in LabVIEW is used to synchronize both the illumination and the detection by the beam polarimeter.

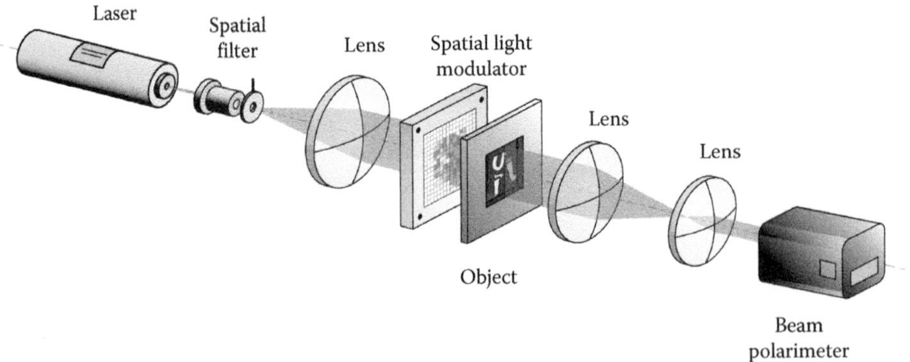

Figure 10.2 Experimental setup of the polarimetric single-pixel camera. Linear polarizers before and after the SLM are not shown.

For each realization, the values of the normalized Stokes parameters, $\sigma_i = S_i/I_0$ ($i = 1,\ldots,3$), as well as the signals of both photodiodes, which are added to obtain the intensity I_0, are measured. The maximum measurement rate of the polarimeter (10 Stokes vectors per second) was the speed limiting factor, since the refreshing frequency of the SLM was 60 Hz. Images representing the spatial distribution of the Stokes parameters are recovered with the CS algorithm l1eq-pd in MATLAB® (Candes 2015).

In order to test the camera, we use a binary amplitude object—a black and white slide of our university logo, which is shown in Figure 10.3a. To generate a polarimetric spatial distribution, one of the letters is covered by a plastic film operating as a half-wave retarder plate. As the light emerging from the SLM is linearly polarized, two different polarization states are created when it is transmitted through the object, one for the U and I letters and another for the J letter. The polarimetric images acquired by the camera are shown in Figure 10.3b through d. The normalized Stokes parameters recovered for each letter are in good agreement with those measured previously by scanning with a beam polarimeter, which validates the system for PI.

The aforementioned results demonstrate the possibility of performing spatially resolved Stokes polarimetry by using CS techniques. In particular, the optical system described here converts a commercial beam Stokes polarimeter into a PI system. However, the operation of the camera is restricted by the acquisition time necessary to project the set of microstructured light patterns.

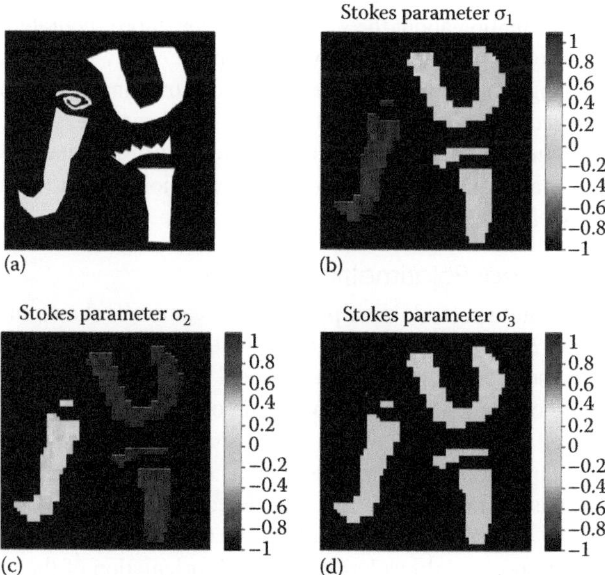

Figure 10.3 (a) High-resolution image of the object under study, an amplitude mask representing the logo of the university UJI, with a cellophane film over the letter in yellow. (b), (c), and (d) show pseudocolor pictures of the spatial distribution of the normalized Stokes parameters. (Reproduced from Durán, V., Clemente, P., Fernández-Alonso, M., Tajahuerce, E., and Lancis, J., Single-pixel polarimetric imaging, *Opt. Lett.*, 37(5), 824–826, 2012. With permission of Optical Society of America.)

One way to improve our setup is to speed up the acquisition process. By substituting the LC SLM with a fast SLM such as a DMD and the detectors with photometers operating at higher frequencies, the total acquisition time could be reduced enough to acquire images with exposure times under a second, better suited for real-time imaging.

Single-Pixel Spectropolarimetric Imaging

Multispectral imaging is a useful optical technique providing 2D images of an object for a set of specific wavelengths within a selected spectral range (Brady 2009). Multispectral imaging provides both spatial and spectral information and represents a powerful analysis tool in different scientific fields as medicine (Stamatas et al. 2006), pharmaceutics (Hamilton and Lodder 2002), astronomy (Scholl et al. 2008), and agriculture (Dale et al. 2013). In industry, new techniques have emerged that use imaging in the visible and near infrared light spectrum to make quality and safety control, for example, in detection of surface properties on fruits (Mehl et al. 2004). Acquiring both spectral and polarimetric information in the same device brings up new imaging possibilities. The increase in dimension of the recovered images provides more information about the sample. For example, in the field of biomedical optics, the measure of the angular distribution of the intensity, the diffuse spectral reflectance, or the degree of polarization of both reflected and transmitted light can be used to distinguish abnormal zones of a tissue. With these ideas in mind, multispectral polarimetric imaging (MPI) has been used to characterize human colon cancer (Pierangelo et al. 2011) and pathological analysis of skin (Zhao et al. 2009b). As spectropolarimetric images provide more information about the scene, they allow more accurate analysis and better identification of pathologies in living samples.

In the following section, we describe a hyperspectral linear polarimetric camera with integrated detection based on the single-pixel camera scheme (Soldevila et al. 2013). Moreover, we also show an improvement of the linear polarimetric camera, which is able to provide full-Stokes information about a scene in several chromatic channels.

Multispectral Linear Polarimetric Camera

The single-pixel multispectral linear polarimetric camera is shown in Figure 10.4 (Soldevila et al. 2013). The object is illuminated by a white light source and imaged onto the surface of a DMD by an optical system. Binary masks are codified sequentially on the DMD for sampling the object. The light reflected by the DMD is collected by an optical system, and a single-pixel detector measures the integrated irradiance transmitted by the masks. In order to acquire both polarimetric and multispectral information, the detector consists of a linear polarizer mounted on a rotating holder, working as an analyzer, and a commercial spectrometer with no spatial resolution. For each orientation of the analyzer, the modulator sequentially generates a set of binary intensity patterns that sample the image of the object under consideration enabling the application of CS techniques. For each intensity pattern, the spectrometer provides a full spectrum with the required spectral bandwidth and resolution.

In the experiment described here, the scene under study consists of two small square capacitors with different colors. In order to add polarimetric information, the light projected to each one has a different linear SOP. The images acquired

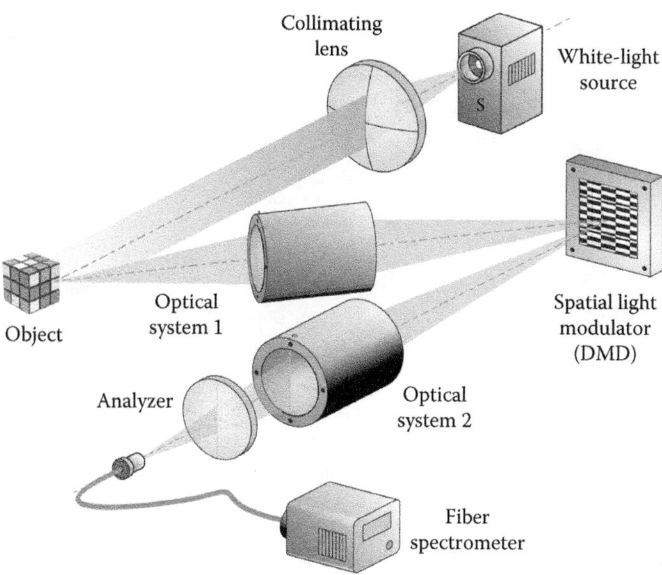

Figure 10.4 Optical system for MPI using a single-pixel detector and CS. (Reproduced with permission from Springer Science+Business Media: *Appl. Phys. B*, Single-pixel polarimetric imaging spectrometer by compressive sensing, 113(4), 2013, 551–558, Soldevila, F., Irles, E., Durán, V., Clemente, P., Fernández-Alonso, M., Tajahuerce, E., and Lancis, J.)

have a resolution of 128 × 128 pixels, the pixel pitch being 43.2 μm. The integration time of the spectrometer is set to 500 ms to provide a good SNR in the measurement stage. The total number of projected masks is roughly 20% of the number required by the Nyquist criterion, so the images are recovered with the aid of CS techniques. The analyzer is rotated to four different orientations. For each orientation of the analyzer, different spectra are measured, and multiple images are recovered in different spectral bands. In Figure 10.5, we show the reconstruction for eight different spectral bands with 20 nm width. Each column corresponds to a spectral band, and each row is linked to an orientation of the transmission axis of the analyzer.

This MPI system is able to provide the spatial distribution of the Stokes parameters, S_0, S_1, and S_2, of the light reflected by the sample in several spectral bands. In order to acquire a full polarization analysis of the scene, one must add a circular or elliptic polarizer to the analyzer and gather enough data to compute S_3. By replacing the linear analyzer with a PSA based on LCVR plates, it is possible to acquire the spatial distribution of the full-Stokes parameter of the scene with no moving elements. We show this improvement of the system in the following section.

Multispectral Full-Stokes Imaging Polarimeter

In the previous optical system in Figure 10.4, only the spatial distribution of the Stokes vectors S_0, S_1, and S_2 can be derived as just a linear polarizer is used as an analyzer. However, a full-Stokes polarimeter able to measure S_3 requires adding at least a linear retarder. The scheme of such a full-Stokes polarimeter is shown in

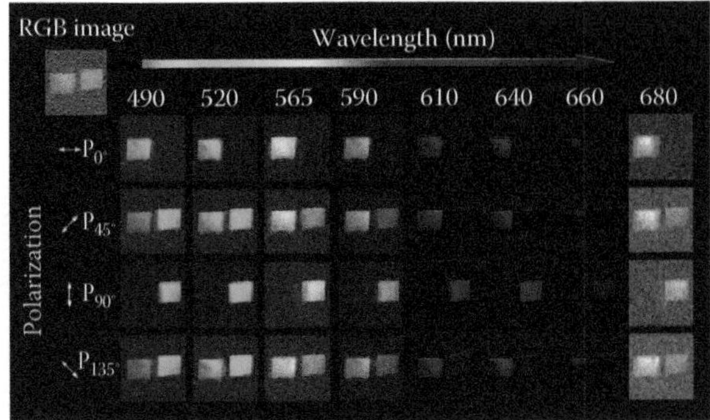

Figure 10.5 Multispectral image cube reconstructed by CS algorithm for four different configurations of the polarization analyzer in Figure 10.4. The RGB image of the object is also included. In the visible spectrum, all channels are represented by pseudocolor images and a gray-scale representation is used for the wavelength closer to the NIR spectrum. (Reproduced with permission from Springer Science+Business Media: *Appl. Phys. B*, Single-pixel polarimetric imaging spectrometer by compressive sensing, 113(4), 2013, 551–558, Soldevila, F., Irles, E., Durán, V., Clemente, P., Fernández-Alonso, M., Tajahuerce, E., and Lancis, J.)

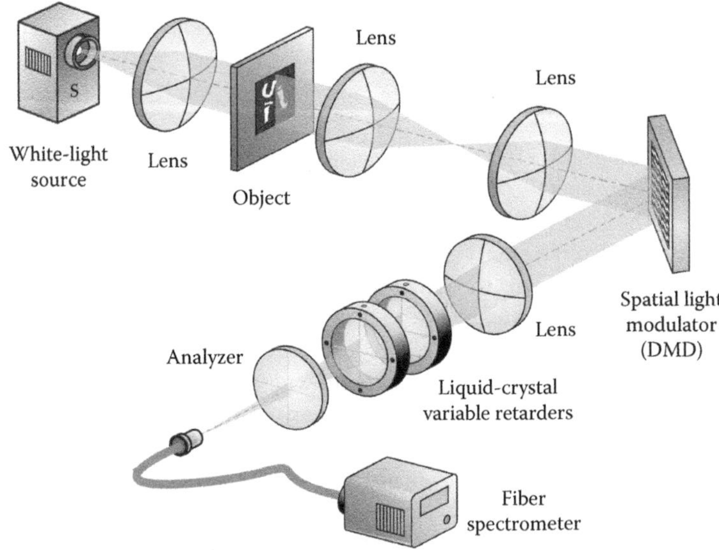

Figure 10.6 Layout of the multispectral full-Stokes imaging polarimeter. (From Soldevila, F., Irles, E., Durán, V., Clemente, P., Fernández-Alonso, M., Tajahuerce, E., and Lancis, J. / Javidi, B., Tajahuerce, E., and Andrés, P.: *Multidimensional Imaging*. 2014. Copyright Wiley-VCH Verlag GmbH & Co. KGaA. Reproduced with permission.)

Figure 10.6 (Soldevila et al. 2014). A white light beam generated by a Xenon lamp is collimated by a lens and illuminates the input object, whose image is formed on a DMD by a pair of lenses. The light emerging from the DMD is focused onto a single-pixel detector with the aid of another lens. In order to achieve both polarimetric and spectral information, the single-pixel detector consists of two

LCVRs with their slow axis oriented at 45° and 0°, followed by a linear analyzer with its transmission axis oriented at 45° and a commercial fiber spectrometer. Each LCVR is precalibrated to introduce controlled retardances in each chromatic channel of interest. The commercial fiber spectrometer is the same as used in the preceding section.

To compute the Stokes parameters at each pixel of the scene, four images are acquired for different retardances of LCVRs. These images are generated by applying the CS algorithm described in the "Polarimetric Cameras" section, providing in this way an intensity map that corresponds to the spatial distribution of the Stokes parameter S_0'. By using Stokes–Mueller calculus, it is possible to relate the value of the Stokes vector to the measured irradiance S_0' for each pixel. From the Mueller matrix expressions of a retarder wave plate and a linear polarizer, it is possible to show that the relationship between the recovered irradiance and the original Stokes parameters is

$$
\begin{aligned}
S_0'(2\delta_1, 2\delta_2) = \frac{1}{2}S_0 + \frac{1}{2}\sin(2\delta_1)\sin(2\delta_2)S_1 \\
+\frac{1}{2}\cos(2\delta_2)S_2 - \frac{1}{2}\cos(2\delta_1)\sin(2\delta_2)S_3,
\end{aligned}
\tag{10.6}
$$

where $2\delta_1$ and $2\delta_2$ are the phase retardances introduced, respectively, by the two LCVRs. Equation 10.6 establishes an undetermined system with four unknown quantities, the Stokes parameters of the incident light. To solve this system, a minimum of four pairs of phase retardances must be applied to the LCVR pair. After the off-line reconstructions, the Stokes vector in each point of the scene is given by $\mathbf{S}_0' = \mathbf{M} \cdot \mathbf{H}_N$, where

$$
\mathbf{M} = \frac{1}{2}\begin{pmatrix}
1 & \sin\left(2\delta_1^{(1)}\right)\sin\left(2\delta_2^{(1)}\right) & \cos\left(2\delta_2^{(1)}\right) & -\cos\left(2\delta_1^{(1)}\right)\sin\left(2\delta_2^{(1)}\right) \\
1 & \sin\left(2\delta_1^{(2)}\right)\sin\left(2\delta_2^{(2)}\right) & \cos\left(2\delta_2^{(2)}\right) & -\cos\left(2\delta_1^{(2)}\right)\sin\left(2\delta_2^{(2)}\right) \\
1 & \sin\left(2\delta_1^{(3)}\right)\sin\left(2\delta_2^{(3)}\right) & \cos\left(2\delta_2^{(3)}\right) & -\cos\left(2\delta_1^{(3)}\right)\sin\left(2\delta_2^{(3)}\right) \\
1 & \sin\left(2\delta_1^{(4)}\right)\sin\left(2\delta_2^{(4)}\right) & \cos\left(2\delta_2^{(4)}\right) & -\cos\left(2\delta_1^{(4)}\right)\sin\left(2\delta_2^{(4)}\right)
\end{pmatrix}.
\tag{10.7}
$$

In Equation 10.7, the subscripts in the elements of **M** denote the LCVR and the superscripts denote each one of the four acquisitions. The solution of this linear system provides the spatial distribution of the Stokes parameters.

We carried out a photoelasticity measurement on a piece of polystyrene as a direct application of the single-pixel spectral Stokes polarimeter. Due to the fabrication process of the polystyrene piece, the material presents stresses that cause a spatial distribution of birefringence. This distribution can be seen directly when the piece is placed between crossed linear polarizers and illuminated with white light, as is shown in Figure 10.7. In the experiment to obtain spectropolarimetric information of this object, we used the optical system in Figure 10.6. The Walsh–Hadamard patterns addressed to the DMD had a resolution $N = 128 \times 128$ pixels. The number of measurements was chosen to be $M = 3249$, which corresponds to approximately 20% of N. The integration time of each spectrometer measurement was set to 20 ms. The experimental results are shown in Figure 10.8. The color

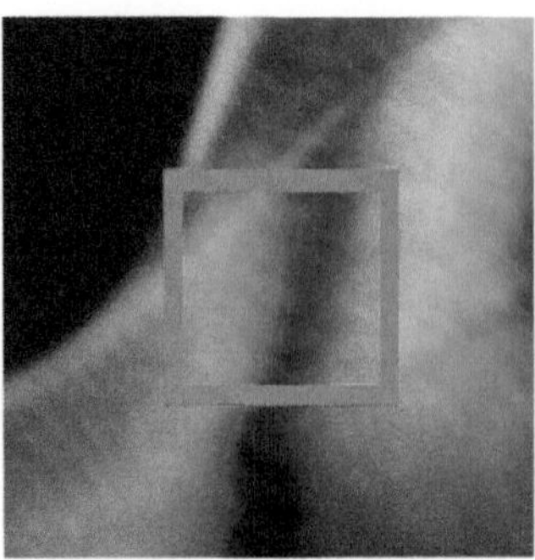

Figure 10.7 Color picture of the polystyrene piece used for the experiment of spectropolarimetry. It is placed between two crossed linear polarizers and illuminated with white light. Color fringes are a consequence of the different states of polarization produced by the stress in the piece. The square indicates the region of interest imaged by the spectropolarimetric single-pixel camera. (From Soldevila, F., Irles, E., Durán, V., Clemente, P., Fernández-Alonso, M., Tajahuerce, E., and Lancis, J. / Javidi, B., Tajahuerce, E., and Andrés, P.: *Multidimensional Imaging*. 2014. Copyright Wiley-VCH Verlag GmbH & Co. KGaA. Reproduced with permission.)

pictures of the figure show the normalized Stokes parameters for eight chromatic channels, each one with 20 nm width. To simplify data display, image reconstructions are arranged in a table. Each column corresponds to a spectral channel and each row shows the spatial distribution of a normalized Stokes parameter. By comparing the results shown in Figures 10.7 and 10.8, it is possible to see that the expected fringe distribution of the Stokes parameters is recovered. We note that for wavelengths near the IR, reconstructions are a bit noisy. This is caused by the low amount of light the source emits in this zone of the spectrum. However, this problem can be solved by increasing the integration time of the spectrometer or by using a light source with a flat spectrum.

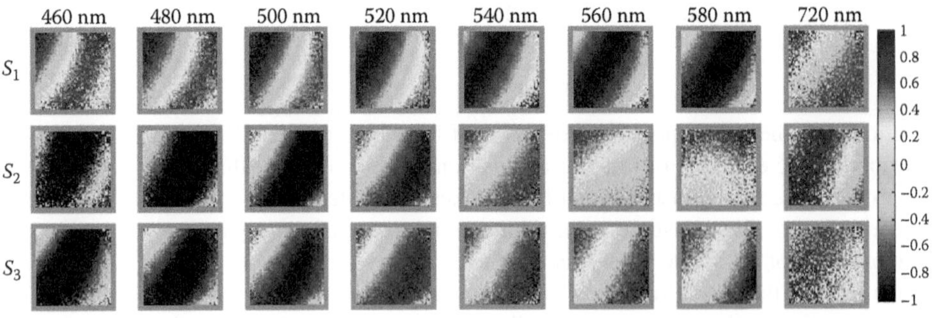

Figure 10.8 Spatial distribution of the Stokes parameters of the polystyrene piece for several spectral bands. Each spatial distribution is represented by a pseudocolor picture with 128 × 128 pixels. The values range from −1 (blue) to 1 (red). (From Soldevila, F., Irles, E., Durán, V., Clemente, P., Fernández-Alonso, M., Tajahuerce, E., and Lancis, J./Javidi, B., Tajahuerce, E., and Andrés, P.: *Multidimensional Imaging*. 2014. Copyright Wiley-VCH Verlag GmbH & Co. KGaA. Reproduced with permission.)

Conclusions

We have described several polarimetric cameras based on single-pixel imaging techniques and CS algorithms. They are able to provide the spatial distribution of the Stokes parameters of an input scene. In all cases, the key element of the optical system is an SLM that sequentially generates a set of binary masks for sampling the input scene. In this way, it is possible to apply the theory of CS to data acquired with a single-pixel sensor. In particular, we have described a single-pixel imaging polarimeter that is able to provide spatially resolved measurements of the Stokes parameters with a bucket detector. In this case, the SLM is an LC display, and the sensor is a commercial beam polarimeter. Moreover, we have shown a single-pixel hyperspectral imaging polarimeter. This setup provides spatially resolved measurements of the Stokes parameters for different spectral channels. The SLM used in this case has been a DMD, and the sensor is composed of polarizing elements followed by a commercial fiber spectrometer. Experimental results for color objects with an inhomogeneous polarization distribution show the ability of the method to measure the spatial distribution of polarization properties for multiple spectral components.

Acknowledgments

This work has been partly funded by project FIS2010-15746 from the Spanish Ministry of Education and projects ISIC/2012/013 and PROMETEO/2012/021 from Generalitat Valenciana.

References

Baba, J.S., J.-R. Chung, A.H. DeLaughter, B.D. Cameron, and G.L. Coté. 2002. Development and calibration of an automated Mueller matrix polarization imaging system. *Journal of Biomedical Optics* 7(3): 341.

Brady, D.J. 2009. *Optical Imaging and Spectroscopy*. Hoboken, NJ: John Wiley & Sons.

Candes, E.J. 2005. l1-Magic. Available at: http://statweb.stanford.edu/~candes/l1magic/, accessed June 17, 2013.

Candes, E.J. and M.B. Wakin. 2008. An introduction to compressive sampling. *IEEE Signal Processing Magazine* 25(2): 21–30.

Chan, W.L., K. Charan, D. Takhar, K.F. Kelly, R.G. Baraniuk, and D.M. Mittleman. 2008. A single-pixel terahertz imaging system based on compressed sensing. *Applied Physics Letters* 93(12): 121105.

Clemente, P., V. Durán, L. Martínez-León, V. Climent, E. Tajahuerce, and J. Lancis. 2008. Use of polar decomposition of Mueller matrices for optimizing the phase response of a liquid-crystal-on-silicon display. *Optics Express* 16(3): 1965–1974.

Clemente, P., V. Durán, E. Tajahuerce, P. Andrés, V. Climent, and J. Lancis. 2013. Compressive holography with a single-pixel detector. *Optics Letters* 38(14): 2524.

Dale, L.M., A. Thewis, C. Boudry, I. Rotar, P. Dardenne, V. Baeten, and J.A. Fernández Pierna. 2013. Hyperspectral imaging applications in agriculture and agro-food product quality and safety control: A review. *Applied Spectroscopy Reviews* 48(2): 142–159.

Demos, S.G. and R.R. Alfano. 1997. Optical polarization imaging. *Applied Optics* 36(1): 150–155.

Donoho, D.L. 2006. Compressed sensing. *IEEE Transactions on Information Theory* 52(4): 1289–1306.

Duarte, M.F., M.A. Davenport, D. Takhar, J.N. Laska, K.F. Kelly, and R.G. Baraniuk. 2008. Single-pixel imaging via compressive sampling. *IEEE Signal Processing Magazine* 25(2): 83–91.

Duntley, S.Q. 1974. Underwater visibility and photography. In *Optical Aspects of Oceanography*, eds. N.G. Jerlov and E.S. Nielsen. London and New York: Academic Press, Inc.

Durán, V., P. Clemente, M. Fernández-Alonso, E. Tajahuerce, and J. Lancis. 2012. Single-pixel polarimetric imaging. *Optics Letters* 37(5): 824–826.

Geelen, B., N. Tack, and A. Lambrechts. 2013. A snapshot multispectral imager with integrated tiled filters and optical duplication. In, eds. G. von Freymann, W.V. Schoenfeld, and R.C. Rumpf, *Advanced Fabrication Technologies for Micro/Nano Optics and Photonics VI, Proc. SPIE 8613*, Bellingham: SPIE.

Glenar, D.A., J.J. Hillman, B. Saif, and J. Bergstralh. 1994. Acousto-optic imaging spectropolarimetry for remote sensing. *Applied Optics* 33(31): 7412.

Hamilton, S.J. and R.A. Lodder. 2002. Hyperspectral imaging technology for pharmaceutical analysis. In, eds. D.J. Bornhop, D.A. Dunn, R.P. Mariella, Jr., C.J. Murphy, D.V. Nicolau, S. Nie, M. Palmer, and R. Raghavachari, *Biomedical Nanotechnology Architectures and Applications, Proc. SPIE 4626*, Bellingham: SPIE, pp. 136–147.

Howland, G.A., D.J. Lum, M.R. Ware, and J.C. Howell. 2013. Photon counting compressive depth mapping. *Optics Express* 21(20): 23822–23837.

Kirmani, A., A. Colaço, F.N.C. Wong, and V.K. Goyal. 2011. Exploiting sparsity in time-of-flight range acquisition using a single time-resolved sensor. *Optics Express* 19(22): 21485.

Kirmani, A., D. Venkatraman, D. Shin, A. Colaco, F.N.C. Wong, J.H. Shapiro, and V.K. Goyal. 2014. First-photon imaging. *Science* 343(6166): 58–61.

Laude-Boulesteix, B., A. De Martino, B. Drévillon, and L. Schwartz. 2004. Mueller polarimetric imaging system with liquid crystals. *Applied Optics* 43(14): 2824–2832.

Lu, S.-Y. and R.A. Chipman. 1996. Interpretation of Mueller matrices based on polar decomposition. *Journal of the Optical Society of America A* 13(5): 1106.

Magalhães, F., F.M. Araújo, M.V. Correia, M. Abolbashari, and F. Farahi. 2011. Active illumination single-pixel camera based on compressive sensing. *Applied Optics* 50(4): 405–414.

Mehl, P.M., Y.-R. Chen, M.S. Kim, and D.E. Chan. 2004. Development of hyperspectral imaging technique for the detection of apple surface defects and contaminations. *Journal of Food Engineering* 61(1): 67–81.

Nadkarni, S.K., M.C. Pierce, B. Hyle Park, J.F. de Boer, P. Whittaker, B.E. Bouma, J.E. Bressner, E. Halpern, S.L. Houser, and G.J. Tearney. 2007. Measurement of collagen and smooth muscle cell content in atherosclerotic plaques using polarization-sensitive optical coherence tomography. *Journal of the American College of Cardiology* 49(13): 1474–1481.

Oka, K. and T. Kato. 1999. Spectroscopic polarimetry with a channeled spectrum. *Optics Letters* 24(21): 1475–1477.

Pezzaniti, J.L. 1995. Mueller matrix imaging polarimetry. *Optical Engineering* 34(6): 1558.

Pierangelo, A., A. Benali, M.-R. Antonelli, T. Novikova, P. Validire, B. Gayet, and A. De Martino. 2011. Ex-vivo characterization of human colon cancer by Mueller polarimetric imaging. *Optics Express* 19(2): 1582–1593.

Pratt, W., J. Kane, and H.C. Andrews. 1969. Hadamard transform image coding. *Proceedings of the IEEE* 57(1): 58–68.

Scholl, J.F., E. Keith Hege, M. Hart, D. O'Connell, and E.L. Dereniak. 2008. Flash hyperspectral imaging of non-stellar astronomical objects. In, eds. M.S. Schmalz, G.X. Ritter, J. Barrera, and J.T. Astola, *Mathematics of Data/Image Pattern Recognition, Compression, and Encryption with Applications XI, Proc. SPIE 7075*, Bellingham: SPIE.

Soldevila, F., E. Irles, V. Durán, P. Clemente, M. Fernández-Alonso, E. Tajahuerce, and J. Lancis. 2013. Single-pixel polarimetric imaging spectrometer by compressive sensing. *Applied Physics B* 113(4): 551–558.

Soldevila, F., E. Irles, V. Duran, P. Clemente, M. Fernandez-Alonso, E. Tajahuerce, and J. Lancis. 2014. Spectro-polarimetric imaging techniques with compressive sensing. In, eds. B. Javidi, E. Tajahuerce, and P. Andrés, *Multidimensional Imaging*, Chichester, U.K.: John Wiley & Sons.

Solomon, J.E. 1981. Polarization imaging. *Applied Optics* 20(9): 1537–1544.

Song, H., Y. Zhao, X. Qi, Y.T. Chui, and S.A. Burns. 2008. Stokes vector analysis of adaptive optics images of the retina. *Optics Letters* 33(2): 137–139.

Stamatas, G.N., M. Southall, and N. Kollias. 2006. In vivo monitoring of cutaneous edema using spectral imaging in the visible and near infrared. *The Journal of Investigative Dermatology* 126: 1753–1760.

Studer, V., J. Bobin, M. Chahid, H.S. Mousavi, E. Candes, and M. Dahan. 2012. Compressive fluorescence microscopy for biological and hyperspectral imaging. *Proceedings of the National Academy of Sciences of the United States of America* 109(26): E1679–E1687.

Terrier, P., V. Devlaminck, and J.M. Charbois. 2008. Segmentation of rough surfaces using a polarization imaging system. *Journal of the Optical Society of America A, Optics, Image Science, and Vision* 25(2): 423–430.

Tyo, J.S., D.L. Goldstein, D.B. Chenault, and J.A. Shaw. 2006. Review of passive imaging polarimetry for remote sensing applications. *Applied Optics* 45(22): 5453–5469.

Tyo, J.S., M.P. Rowe, E.N. Pugh, and N. Engheta. 1996. Target detection in optically scattering media by polarization-difference imaging. *Applied Optics* 35(11): 1855–1870.

Watts, C.M., D. Shrekenhamer, J. Montoya, G. Lipworth, J. Hunt, T. Sleasman, S. Krishna, D.R. Smith, and W.J. Padilla. 2014. Terahertz compressive imaging with metamaterial spatial light modulators. *Nature Photonics* 8(8): 605–609.

Welsh, S.S., M.P. Edgar, R. Bowman, P. Jonathan, B. Sun, and M.J. Padgett. 2013. Fast full-color computational imaging with single-pixel detectors. *Optics Express* 21(20): 23068–23074.

Zhao, X., F. Boussaid, A. Bermak, and V.G. Chigrinov. 2009a. Thin photo-patterned micropolarizer array for CMOS image sensors. *IEEE Photonics Technology Letters* 21(12): 805–807.

Zhao, Y., L. Zhang, and Q. Pan. 2009b. Spectropolarimetric imaging for pathological analysis of skin. *Applied Optics* 48(10): D236–D246.

IV
Compressive Sensing Microscopy

11 STORM Using Compressed Sensing

Lei Zhu and Bo Huang

Contents

Introduction

Since its invention more than 400 years ago, light microscopy has become an indispensable tool for biological research (Kasper and Huang 2011). Among various light microscopy modalities, fluorescence microscopy (FM) is probably the most versatile and the most widely used (Lichtman and Conchello 2005). By labeling target molecules with fluorophores—molecules that can be excited by light in one wavelength and then emit light in another wavelength—FM can provide 3D images of a wide range of biomolecules in the context of cells up to whole living animals.

Despite the ubiquitous application in biological and biomedical research, classic FM has a fundamental limit in its spatial resolution due to light diffraction, inconveniently at the subcellular level. Not surprisingly, the newly emerged superresolution microscopy techniques have thus generated enormous interests

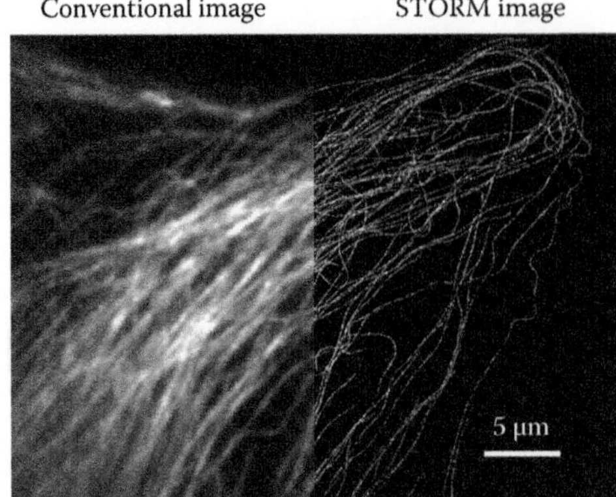

Figure 11.1 FM and STORM images of a mammalian cell, showing the microtubule cytoskeleton.

for achieving spatial resolutions far surpassing this diffraction limit (Hell 2009, Huang et al. 2010) and were awarded the Nobel Prize in Chemistry in 2014. Among the two categories of superresolution microscopy techniques, the single-molecule-switching-based approach, commonly known as stochastic optical reconstruction microscopy (STORM) or photoactivated localization microscopy (PALM) (Betzig et al. 2006, Hess et al. 2006, Rust et al. 2006) (referred to as STORM hereafter), has received more attention because of its instrument simplicity. Figure 11.1 compares the FM image and the STORM image of the same cellular structure, showing the superior performance of STORM on revealing structural details.

Although all superresolution microscopy methods are compatible with live imaging (Shroff et al. 2008, Westphal et al. 2008, Shao et al. 2011), in exchange for higher spatial resolution, they all have compromised temporal resolution compared to their conventional FM counterparts. STORM is no exception. For example, earlier live STORM reports were generally limited to a temporal resolution of 20–30 s to achieve a spatial resolution of approximately 60 nm (Shroff). This resolution is able to capture slow processes such as focal adhesion rearrangement during cell migration (Shroff et al. 2008), but it is too slow for many other cellular processes that occur at the seconds to subseconds time scale (Huang et al. 2013).

This chapter is focused on a faster STORM imaging technique using an improved data processing algorithm derived from the compressed sensing (CS) theory. The method improves the temporal resolution of STORM over the classic fitting approach by a factor of more than 10. We first review the imaging mechanism of STORM and reveal the limitation of classic STORM data processing algorithm. We then introduce the method formulation of STORM using CS and present implementation examples. Finally, we discuss several practical issues in further improvement of the CS-based algorithm.

Imaging Mechanism of STORM

Physical Principles of Wide-Field FM

STORM imaging is developed on the same instrumentation of the basic FM, also known as wide-field FM. In wide-field FM, as shown in Figure 11.2, the objective lens projects the excitation light to the sample, illuminating an area with relatively

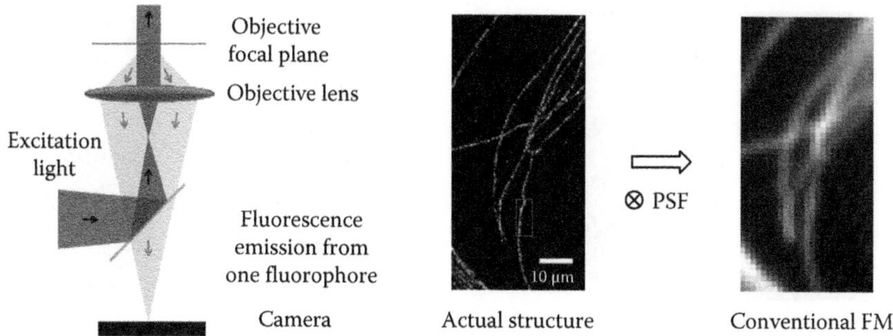

Figure 11.2 Schematic diagram of wide-field FM.

uniform intensity. The fluorescence signal from the sample is collected by the same objective and imaged to a camera. The resolution of wide-field microscopy is limited by two facts: in the transverse plane (*x*–*y*), photons from one point in the sample spreads into a blurry spot over multiple pixels due to light diffraction; in the sample-lens direction (*z*), the camera cannot distinguish photons from different *z* positions but along the same light path in the sample because it is a 2D detector. The point spread function (PSF) is used to describe these two effects, which is defined as the 3D image of a point object and is intrinsically determined by the imaging physics. Under the condition that the PSF is linear and shift-invariant, the fluorescence image on the camera is the convolution of the original object and the PSF. In the highest-end wide-field fluorescence microscope, the PSF width is approximately half the wavelength in the *x*–*y* plane (~250 nm) and one wavelength in the *z* direction (~600 nm).

Data Acquisition and Classic Data Processing for STORM

The superresolution microscopy method of STORM (or PALM) improves the spatial resolution by using a different imaging mechanism and image postprocessing technique (Betzig et al. 2006, Hess et al. 2006, Rust et al. 2006, Bates et al. 2007, Huang et al. 2008b). It has two key elements: a sensitive wide-field microscope that can image single fluorophores and fluorophores that can be switched from a nonfluorescent chemical state to a fluorescent state. For the latter, photoswitching is the most commonly used mechanism, in which certain organic dyes and fluorescent proteins can convert to the fluorescent state after absorbing a specific wavelength of light (activation light). In many cases, the excitation light can also cause fluorophore activation, resulting in spontaneous "blinking" of the fluorophore.

During acquisition, STORM records a sequence of camera images (see Figure 11.3). In each image, fluorophores labeling the sample structure are switched to the fluorescent state by the activation light, whose intensity is adjusted so that only a small, random faction of fluorophores are activated. These activated fluorophores absorb excitation light and emit fluorescence signals that are captured as sparse, isolated spots on the camera. In the image processing, single-fluorophore spots are identified and fit to the PSF to determine the fluorophore positions. A final superresolution image is reconstructed from these positions. The precision of determining the position is approximately proportional to the width of the PSF and inversely proportional to the square root of the number of photons detected from one fluorophore (Thompson et al. 2002), typically between 200 and 5000. Other factors affecting the precision include pixel size and background noise. By imaging a small portion of fluorophores in a sample in each frame,

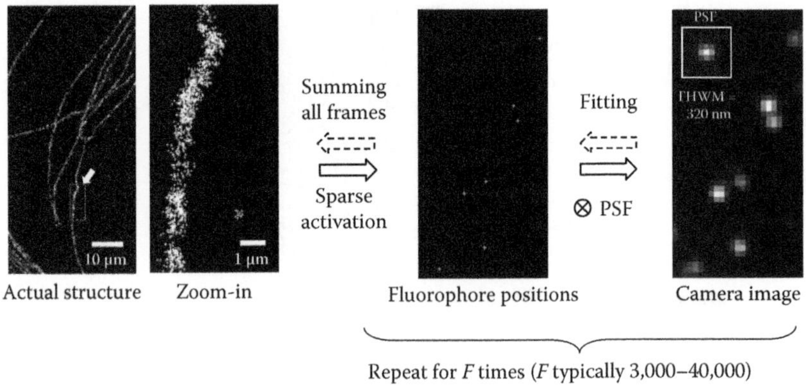

Figure 11.3 Illustration of STORM imaging procedure.

the superresolution methods overcome the diffraction barrier of light microscopy and achieve a resolution up to about 10 nm (Shtengel et al. 2009), at the price of degraded temporal resolution (Shroff et al. 2008, Jones et al. 2011). Researchers have used superresolution microscopy to successfully image a wide variety of cellular structures not observable before, including cytoskeleton, organelles, protein complexes, and even synapse in brain sections (Bates et al. 2007, Huang et al. 2008a,b, Dani et al. 2010, Wu et al. 2010, Beaudoin et al. 2012).

Formulation of STORM Using CS

Limitation of Fitting-Based STORM Algorithms

STORM images consist of localized fluorophore points, whose density determines the signal-to-noise ratio and thus the effective spatial resolution. The generation of high-quality STORM images thus requires a large number of camera frames (typically 3,000–40,000) to accumulate sufficient localization points, limiting the temporal resolution, especially when the frame rate of the camera cannot be increased due to intrinsic constraints by the photophysical properties of the fluorophores. In this case, to achieve high temporal resolution, more fluorophores should be activated and sampled in each frame of raw image. However, oversampling inevitably causes overlap of fluorophore spots, which creates difficulties in molecule detection using the single-molecule fitting algorithm (Huang et al. 2008a,b, 2010). Consequently, information carried by photons from these fluorophores is lost.

Advanced molecule localization algorithms may allow lower sparsity without losing detection efficiency. One example is DAOSTORM that handles overlapped photon clusters (Holden et al. 2011). The algorithm starts with an estimated number of molecules N and then perform fitting of N PSFs on the camera image. N is iteratively updated based on the resulting fitting errors. At the cost of increased computation, DAOSTORM is able to identify some molecules with overlapping signal distributions on the camera image, which are rejected from signal processing in the conventional single-molecule fitting algorithm.

Brief Introduction of the CS Theory

In search of advanced algorithms that are able to efficiently detect densely activated fluorophores, the recently developed CS technique becomes a viable option. If the original signals are known sparse (i.e., mostly zeroes) or compressible via a known transformation, CS is able to obtain an accurate recovery using very few measurements from a linear system, without knowing in advance the support of the signal (Donoho 2006, Tsaig and Donoho 2006, Candès 2008). An exhaustive review of the CS theory is beyond the scope of this chapter. Here, we provide a brief introduction of the CS optimization framework toward a basic understanding of the CS-based STORM algorithm in the next section:

$$\text{Minimize } \| \vec{x} \|_1, \text{ s.t.} \| A\vec{x} - \vec{b} \|_2 \leq \varepsilon \tag{11.1}$$

Equation 11.1 shows the general optimization framework of CS for recovery of sparse signals $\vec{x}$, which is the input of a linear system. Vector $\vec{b}$ is the system output with noise. A is the system matrix modeling the relationship between the input and output signals as $A\vec{x} = \vec{b}$ in the absence of noise. $\| \vec{t} \|_p$ calculates the L-p norm of a vector $\vec{t}$, defined as

$$\| \vec{t} \|_p = \left(\sum_i |t(i)|^p \right)^{1/p} \tag{11.2}$$

Therefore, the objective of Equation 11.1, $\| \vec{x} \|_1$, is simply the summation of the absolute values of $\vec{x}$ elements. $\| A\vec{x} - \vec{b} \|_2$, often referred to as the data fidelity error, is the root of mean square error of the estimated system output $A\vec{x}$, compared with the measured output $\vec{b}$. ε is the error tolerance on the data fidelity, a user-defined algorithm parameter.

Different from a least-square approach that minimizes $\| A\vec{x} - \vec{b} \|_2$, the CS algorithm finds an estimated input signal with the minimum $\| \vec{x} \|_1$ from a large pool of feasible solutions defined by the data fidelity constraint $\| A\vec{x} - \vec{b} \|_2 \leq \varepsilon$. It is important to note that the optimization objective is the L-1 norm of $\vec{x}$. It has been shown that the scheme of L-1 norm minimization enforces data sparsity on the estimated solution and thus significantly reduces the measurement data required for exact data recovery. If the input signal is sparse, the CS algorithm finds a solution more accurately than the least-square methods.

Figure 11.4 shows a 1D simulation example illustrating the performances of signal recovery using different algorithms. The input signals of the linear system, that is, $\vec{x}$, are sparse, with only a few nonzero data points (see Figure 11.4a). The system PSF is shown in Figure 11.4b and the Figure 11.4c is the measured system output, that is, $\vec{b}$, which is calculated by convolving the true input signals with the system PSF and then adding Gaussian noise. Figure 11.4d shows the recovered input signals by minimizing the least-square errors, that is, $\| A\vec{x} - \vec{b} \|_2$, where the system matrix A is constructed from the system PSF as described in detail in the next section. It is seen that the least-square method fails to generate a reliable and accurate estimate of the input signals. Figure 11.4e and f is the results of estimated input signals by minimizing L-2 and L-1 norm of $\vec{x}$, respectively, with the constraint of $\| A\vec{x} - \vec{b} \|_2 \leq \varepsilon$, where ε is the estimated system noise variance. The comparison clearly reveals the advantage of the L-1 norm minimization method, that is, CS, on the recovery of sparse signals.

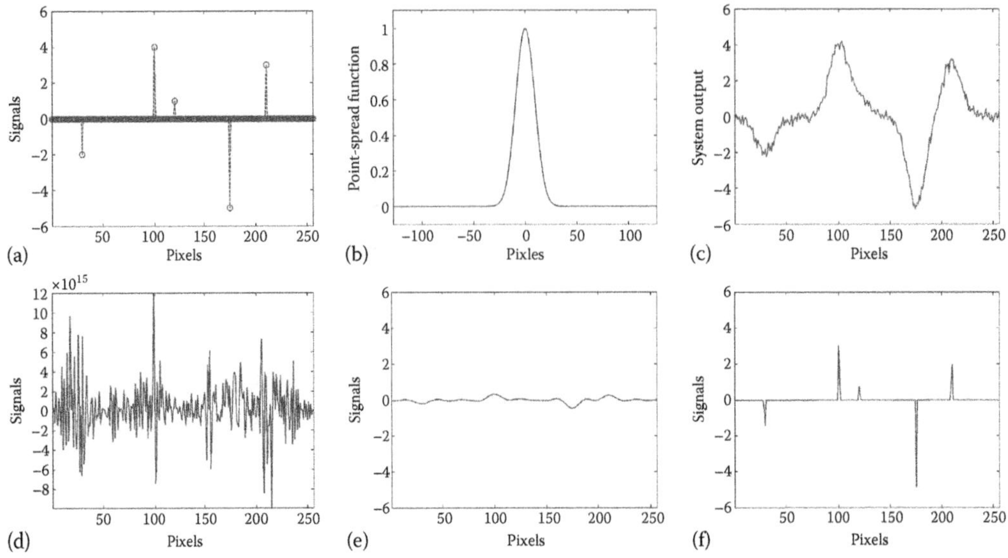

Figure 11.4 A simulation example of signal recovery using different algorithms. (a) The true input signals, with nonzero data points at sparse locations; (b) the system PSF; (c) the measured system output signals, which is modeled as the true signals convolved with the system PSP plus Gaussian noise; (d) the recovered input signals using a least-square method; (e) the recovered input signals using optimization similar to Equation 11.1 but with a minimization objective of L-2 norm; and (f) the recovered input signals using CS (i.e., Equation 11.1).

CS-Based STORM Algorithm

The CS optimization framework potentially improves over the fitting-based STORM algorithm since it accurately recovers sparse input signals. Nonetheless, a direct implementation of CS for STORM processing is not straightforward. The CS optimization framework is based on a linear relationship in the absence of noise between the signal to be estimated and the measured data, that is, $A\vec{x} = \vec{b}$. In the conventional molecule localization for each frame of camera image, as in the fitting-based signal processing, the target output is the spatial location of each identified molecule. The measured raw image, if Poisson noise is not considered, is the summation of PSFs centered at the molecule locations, weighted by the individual number of total emitted photons. Such a nonlinear model undermines the mathematical foundation of CS implementation in STORM imaging.

To make the procedure of molecule localization of STORM compatible with the framework of CS, we first construct a linear model to describe the relationship between the signal to be estimated and the camera raw image. Instead of a list of molecule positions, we use a different representation of the superresolution image. As shown in Figure 11.5, we place the molecules on to a discrete grid, which is finer than the camera pixel with a zoom-in factor of much larger than 1 (e.g., 8 × 8 subdivision). The superresolution image is then described by the fluorescence intensity of molecules at the grid points, which is considered as the signal to be estimated in the CS optimization. Note that, because of the random subsampling in each camera frame, most of the grid points will have no molecules fluorescing and thus will have an intensity of zero. The final superresolution image of the structure can then be generated as the summation of individual superresolution images from each camera frame.

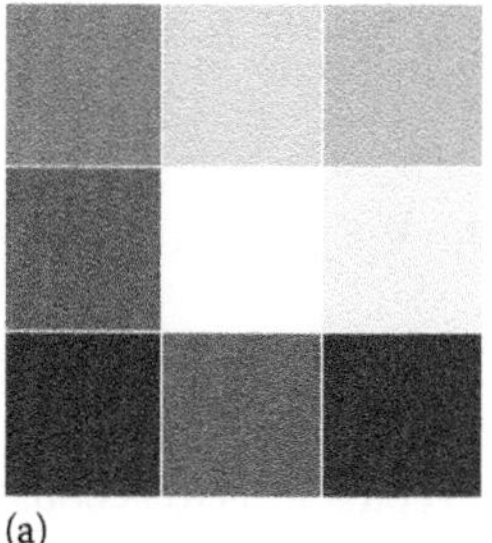
(a)

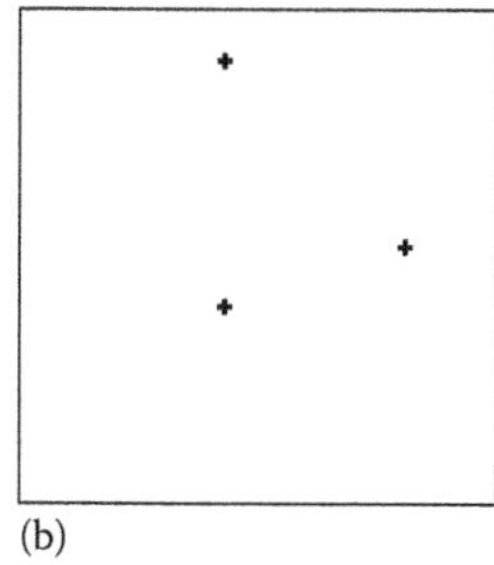
(b)

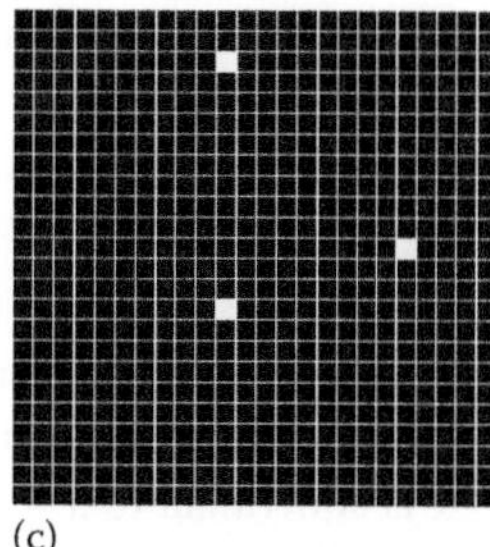
(c)

Figure 11.5 STORM image presentation in the CS-based algorithm. (a) Raw image pixels; (b) fluorophore position (STORM representation); (c) oversampled grid representation (zoom-in factor = 8).

In the conventional STORM image processing, the obtained locations and the fluorescence levels of the molecules are eventually converted into a discrete image for display, that is, signals on a discrete grid. As such, the proposed image presentation for a CS-based algorithm does not necessarily lose biological information of a STORM image if the zoom-in factor is sufficiently large. However, it should be noted that the CS method implicitly assumes that the possible molecule locations are only on the predetermined uniform grid points. This grid approximation becomes less accurate when the grid pitch size is large. In STORM imaging, fortunately, the zoom-in factor of the superresolution image compared with the raw camera image is typically larger than 8, leading to a negligible error from the grid approximation as supported by results shown in the next section.

Now, in the context of STORM imaging, we are able to construct the mathematical variables of CS used in the "Brief Introduction of the CS Theory" section based on the physical principles of FM. $\vec{b}$ is one frame of the measured camera image in the unit of photon counts, and $\vec{x}$ is the superresolution image (except for the last element, see more explanations later). If photon noise is not considered, $\vec{b}$ has a linear relationship with $\vec{x}$:

$$\vec{b} = A\vec{x} \tag{11.3}$$

where the 1D vectors, $\vec{b}$ and $\vec{x}$, consist of row-wise concatenations of the camera and the superresolution images, respectively. The matrix A is determined by the PSF of the imaging system. The ith column of A (except for the last column, see more explanations later) corresponds to the row-wise concatenations of the acquired raw image if only one molecule emits fluoroscopic photons at the position index i of $\vec{x}$.

The goal of the superresolution molecular imaging is to obtain $\vec{x}$ from the measured $\vec{b}$, provided that the system matrix A is exactly known and $\vec{x}$ is sparse (i.e., mostly zeroes). Several practical issues should be noted when implementing the CS algorithm, that is, Equation 11.1, for STORM imaging. First, nonnegativity constraint should be included on the superresolution image. Second, the error tolerance on the data fidelity term is determined by the Poisson statistics of the measured camera image. Lastly, the image background obeys a model different from convolution with camera PSF and must be removed during the molecule localization.

With these considerations, we implement the CS algorithm via the formulation shown in the following:

$$\text{Minimize}: \vec{c}\cdot\vec{x}, \text{ s.t. } \| A\vec{x}-\vec{b} \|_2 \leq \varepsilon\cdot(\Sigma b_j)^{1/2}, \quad x_i \geq 0 \tag{11.4}$$

Compared with Equation 11.1, the objective function in Equation 11.4 changes to $\vec{c}\cdot\vec{x}$ due to the nonnegativity constraint on $\vec{x}$. As each element of $\vec{x}$ is nonnegative, $\vec{c}\cdot\vec{x}$ is equivalent to a weighted L-1 norm of $\vec{x}$. The weight vector, $\vec{c}$, is to account for the difference of the total contribution on the measured raw image from one fluoroscopic molecule at different locations. The value of the *i*th element of $\vec{c}$ equals the summation of the *i*th column of *A*. If $\vec{c}$ is set to be an all-one vector, the algorithm may generate strong artifacts around the edges of the camera image window.

Because $(\| A\vec{x}-\vec{b} \|_2)^2$ is the sum of variances of the estimated camera image based on the optimization result, $A\vec{x}$, compared with the raw camera image, the constraint on $\| A\vec{x}-\vec{b} \|_2$ set an upper limit of χ^2 on how well the optimization result match the original image. The Poisson statistics of photon counting implies that the variance of photon counts on pixel *j* equals to the expectation of the photon counts on that pixel. We use the measured photon number b_j to approximate the variance of photon counts of that pixel. Assuming random signals are independent on different camera pixels, we obtain the sum of variances of camera signals as Σb_j, which is the theoretical upper limit of $(\| A\vec{x}-\vec{b} \|_2)^2$. To account for the approximation error on the signal variances, we include a user-defined parameter, ε, in Equation 11.4. ε^2 sets the maximum ratio between the sum of variances of estimation errors and the sum of variances of camera signals, equivalent to the reduced χ^2 without weighting each pixel with their variances individually.

Equation 11.3 does not consider the background signals of a camera image. To ensure the success of the CS-based algorithm on real data, we introduce one additional element in $\vec{x}$. The corresponding element in *c* is set to zero and all elements in the corresponding column of *A* are set to 1. In this case, the value of this extra *x* element represents a uniform image background, and there is no sparsity constraint imposed on this element.

Implementation Details and Examples

An analytical solution to the CS-based optimization problem, Equation 11.4, is not available; thus, iterative algorithms of linear programming with quadratic constraints have to be employed to find the global optimum. The implementation of the CS method for STORM, however, is hindered by two issues: intensive computation and high memory usage. Each raw camera image requires at least hundreds of iterations, with the computation complexity of each iteration close to that of the conventional fitting method. The issue of heavy computation load is exacerbated by the large size of the matrix *A*. The number of columns of *A* is determined by the size of $\vec{x}$. To formulate the linear equation (11.3), we apply a discrete grid on the superresolution image to describe the positions of the molecular distribution (Figure 11.5). The grid spacing is much smaller than the camera pixel size to ensure sufficient accuracy, which results in a fat matrix *A*. For example, if the grid spacing is one-eighth of the camera pixel size, *A* has a size of *N*-by-64*N*, where *N* is the size of the camera image, which often contains $>10^5$ pixels. A large *A* significantly increases the computation complexity as well as the memory load.

As shown in Zhu et al. (2012), we can reduce the size of the optimization problem such that it is small enough to be solved on a standard PC. Note that the PSF of the imaging system is typically very narrow. We can therefore carry out the optimization separately on small patches of the original image without compromising the overall performance. Specifically, the following algorithm is designed to substantially improve the computational efficiency:

```
Given: camera image B, super-resolution image x, patch size u-by-v
pixels, oversampling factor R, reduced chi-squared target ε, the
imaging matrix A and the vector c for a small u-by-v image.
Dividing the camera image, B, into a set of small u-by-v image
patches, b, with 2-pixel overlap in between.

For each small image, b, do:

1. Create an oversample grid, x, with a size of R× (u+4) by R×
(v+4). The margin of extra 2 pixels on each edge of the patch
accounts for the contribution from molecules outside.
2. Minimize cᵀx'
     subject to xᵢ ≥ 0
          and ||Ax' - b'||₂ ≤ ε × sqrt(sum(b))

For each patch of the optimization result, x, set the outmost
3-pixel boundary to zero to avoid overlap (2 pixel extra margin + 1
pixel overlap between patches).

Add all x together to form the result of the full image.
```

Existing optimization software can be used to implement this algorithm. In the examples we demonstrate later, we use the CVX software (Grant and Boyd 2013) in MATLAB® to generate all the results. The MATLAB code and one set of sample data can be found online at Zhu et al. (2012).

We first use simulations to evaluate the CS algorithm for single-molecule detection. Different numbers of molecules are simulated in a small region with a PSF and photon statistics derived from real experiments. Figure 11.6 shows one extremely high-density example of single-molecule detection with 75 molecules simultaneously activated in one camera image (~6/μm²). The conventional fitting method detect 4 molecules, respectively, while the CS-based method is able to identify

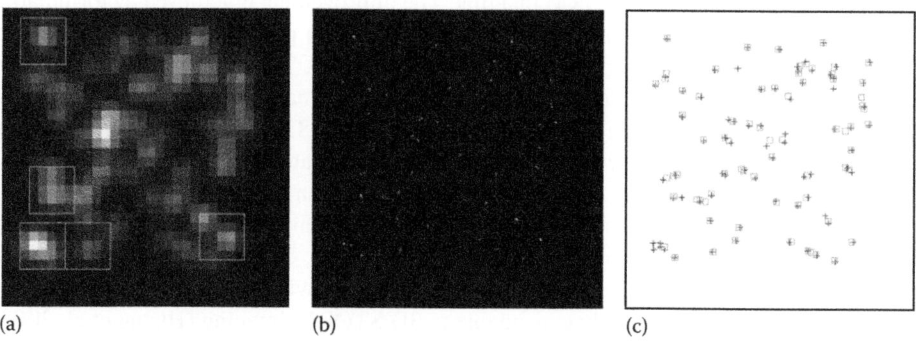

Figure 11.6 One example of single-molecule detection on simulated data. (a) Simulated STORM raw image and molecule detection by fitting (□); (b) CS result (8 × 8 oversampled grid); (c) comparison of CS results (□) and ground truth (+).

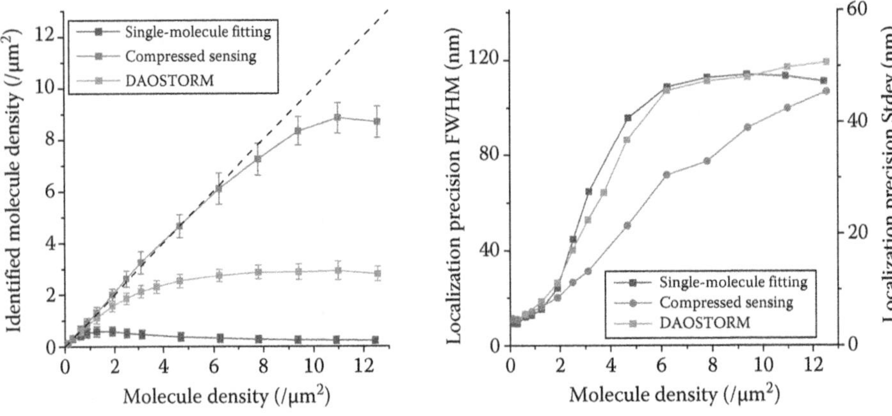

Figure 11.7 Identified molecule densities and localization precision using different algorithms on simulated data. (Courtesy of Zhu, L. et al., *Nat. Methods*, 9(7), 721, 2012.)

73 molecules, out of which 64 correctly matched the simulation ground truth. Figure 11.7 summarizes the simulation results with different numbers of excited molecules in the field. It is seen that CS identifies up to 15-fold as many molecules as by the single-molecule fitting method that rejects all overlapping spots (8.8/μm^2 compared to 0.58/μm^2). CS also outperforms the DAOSTORM method. The localization error of all three methods follows similar increasing trend when molecule density increases, with CS being substantially better at high densities.

Figure 11.8 shows our implementation of CS on STORM imaging of immunostained microtubules in *Drosophila* S2 cells. The nearby microtubules can be resolved by CS using as few as 100 camera frames, whereas they are not discernible by the single-molecule fitting method. The significantly enhanced image quality enables live STORM on microtubules in S2 cells (Figure 11.7b) with 3-second resolution, much faster than previously reported (Shroff et al. 2008).

Summary and Discussion

Using a data processing framework distinct from that in conventional algorithms, the CS-based method significantly enhances the performance of single-molecule localization in STORM imaging. The improvement is achieved by formulating the STORM signal processing as an optimization problem, and therefore, the resultant images are optimal in a statistical sense. The CS approach maximizes the number of detected molecules obtained from each camera raw image, which not only results in higher temporal resolution of STORM but also greatly improves the photon efficiency. These features are very attractive in live cell imaging tasks, where dynamics of molecular structures often requires fast imaging and photo bleaching effects limit the total number of available photons. In this chapter, we review the physical principles of STORM, discuss the formulation of the CS algorithm, and demonstrate its initial success with implementation examples. The algorithm is readily expandable to 3D STORM imaging (Huang et al. 2008b) by constructing a different system matrix based on the 3D system PSF.

We focus on demonstrations of the improved STORM image quality provided by the CS-based method. An existing software package of convex

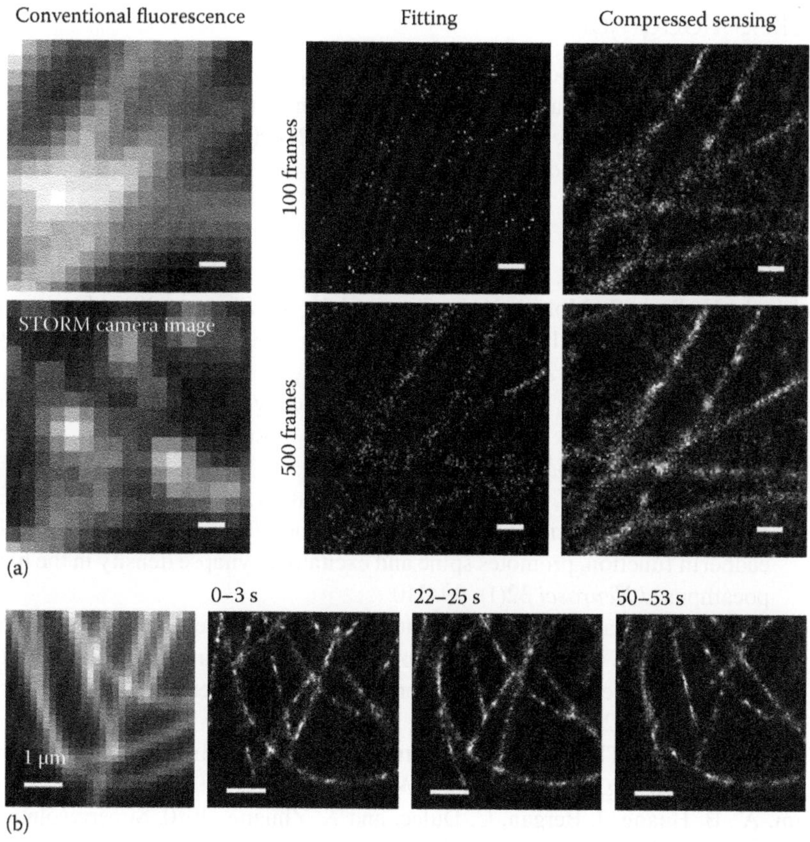

Figure 11.8 STORM imaging using CS. (a) STORM of immunostained microtubules in fixed S2 cells. Left: conventional FM and one frame of the raw STORM images. *Middle*: result of single-molecule fitting. *Right*: result by CS using the same raw images. *Scale bars*: 300 nm. (b) STORM of mEos2-tubulin in a living S2 cell. The conventional FM is shown on the leftmost. Three snapshots from the STORM movie are displayed, each with 3-second integration time. (Courtesy of Zhu, L. et al., *Nat. Methods*, 9(7), 721, 2012.)

optimization is used in our implementations. The method performance can be further enhanced via more sophisticated algorithms. For example, L1-homotopy is able to accelerate the computation of Equation 11.4 (Babcock et al. 2013). A multiresolution scheme will significantly reduce the memory consumption (Donelli et al. 2005) by first using a large grid spacing of the superresolution image and then gradually increasing the spatial resolution of the molecular distribution by a factor of >1 until the finest resolution is reached. Furthermore, the current CS-based method independently generates snapshots of STORM images in a time sequence by using raw camera images frame by frame. As most imaged objects have slowly changing structures, a modified CS algorithm of processing the entire STORM image sequence together by exploiting the redundant structural information among different frames will increase the molecule detection efficiency. These improvements will be of high interest in future research on CS-based STORM algorithms.

Acknowledgments

Lei Zhu receives support from the U.S. National Institutes of Health (1R21EB012700). Bo Huang receives support from the U.S. National Institute of Health Director's New Innovator Award (DP2OD008479) and the Packard Fellowship for Science and Engineering.

References

Babcock, H. P., J. R. Moffitt, Y. L. Cao, and X. W. Zhuang. 2013. Fast compressed sensing analysis for super-resolution imaging using L1-homotopy. *Opt Express* **21**(23): 28583–28596.

Bates, M., B. Huang, G. T. Dempsey, and X. Zhuang. 2007. Multicolor super-resolution imaging with photo-switchable fluorescent probes. *Science* **317**(5845): 1749–1753.

Beaudoin, G. M., 3rd, C. M. Schofield, T. Nuwal, K. Zang, E. M. Ullian, B. Huang, and L. F. Reichardt. 2012. Afadin, a Ras/Rap effector that controls cadherin function, promotes spine and excitatory synapse density in the hippocampus. *J Neurosci* **32**(1): 99–110.

Betzig, E., G. H. Patterson, R. Sougrat, O. W. Lindwasser, S. Olenych, J. S. Bonifacino, M. W. Davidson, J. Lippincott-Schwartz, and H. F. Hess. 2006. Imaging intracellular fluorescent proteins at nanometer resolution. *Science* **313**(5793): 1642–1645.

Candès, E. J. 2008. The restricted isometry property and its implications for compressed sensing. *Comptes Rendus Mathematique* **346**(9): 589–592.

Dani, A., B. Huang, J. Bergan, C. Dulac, and X. Zhuang. 2010. Superresolution imaging of chemical synapses in the brain. *Neuron* **68**(5): 843–856.

Donelli, M., D. Franceschini, A. Massa, M. Pastorino, and A. Zanetti. 2005. Multi-resolution iterative inversion of real inhomogeneous targets. *Inverse Probl* **21**(6): S51–S63.

Donoho, D. L. 2006. Compressed sensing. *IEEE Trans Inf Theory* **52**(4): 1289–1306.

Grant, M. and S. Boyd. 2013. CVX: Matlab software for disciplined convex programming, version 2.0 beta. http://cvxr.com/cvx, accessed September, 2013.

Hell, S. W. 2009. Microscopy and its focal switch. *Nat Methods* **6**(1): 24–32.

Hess, S. T., T. P. Girirajan, and M. D. Mason. 2006. Ultra-high resolution imaging by fluorescence photoactivation localization microscopy. *Biophys J* **91**(11): 4258–4272.

Holden, S. J., S. Uphoff, and A. N. Kapanidis. 2011. DAOSTORM: An algorithm for high-density super-resolution microscopy. *Nat Methods* **8**(4): 279–280.

Huang, B., H. Babcock, and X. Zhuang. 2010. Breaking the diffraction barrier: Super-resolution imaging of cells. *Cell* **143**(7): 1047–1058.

Huang, B., S. A. Jones, B. Brandenburg, and X. Zhuang. 2008a. Whole-cell 3D STORM reveals interactions between cellular structures with nanometer-scale resolution. *Nat Methods* **5**(12): 1047–1052.

Huang, B., W. Wang, M. Bates, and X. Zhuang. 2008b. Three-dimensional super-resolution imaging by stochastic optical reconstruction microscopy. *Science* **319**(5864): 810–813.

Huang, F., T. M. Hartwich, F. E. Rivera-Molina, Y. Lin, W. C. Duim, J. J. Long, P. D. Uchil et al. 2013. Video-rate nanoscopy using sCMOS camera-specific single-molecule localization algorithms. *Nat Methods* **10**(7): 653–658.

Jones, S. A., S. H. Shim, J. He, and X. Zhuang. 2011. Fast, three-dimensional super-resolution imaging of live cells. *Nat Methods* **8**(6): 499–508.

Kasper, R. and B. Huang. 2011. SnapShot: Light microscopy. *Cell* **147**(5): 1198.el.

Lichtman, J. W. and J. A. Conchello. 2005. Fluorescence microscopy. *Nat Methods* **2**(12): 910–919.

Rust, M. J., M. Bates, and X. Zhuang. 2006. Sub-diffraction-limit imaging by stochastic optical reconstruction microscopy (STORM). *Nat Methods* **3**(10): 793–795.

Shao, L., P. Kner, E. H. Rego, and M. G. Gustafsson 2011. Super-resolution 3D microscopy of live whole cells using structured illumination. *Nat Methods* **8**(12): 1044–1046.

Shroff, H., C. G. Galbraith, J. A. Galbraith, and E. Betzig 2008. Live-cell photoactivated localization microscopy of nanoscale adhesion dynamics. *Nat Methods* **5**(5): 417–423.

Shtengel, G., J. A. Galbraith, C. G. Galbraith, J. Lippincott-Schwartz, J. M. Gillette, S. Manley, R. Sougrat et al. 2009. Interferometric fluorescent super-resolution microscopy resolves 3D cellular ultrastructure. *Proc Natl Acad Sci USA* **106**(9): 3125–3130.

Thompson, R. E., D. R. Larson, and W. W. Webb. 2002. Precise nanometer localization analysis for individual fluorescent probes. *Biophys J* **82**(5): 2775–2783.

Tsaig, Y. and D. L. Donoho. 2006. Extensions of compressed sensing. *Signal Process* **86**(3): 549–571.

Westphal, V., S. O. Rizzoli, M. A. Lauterbach, D. Kamin, R. Jahn, and S. W. Hell. 2008. Video-rate far-field optical nanoscopy dissects synaptic vesicle movement. *Science* **320**(5873): 246–249.

Wu, M., B. Huang, M. Graham, A. Raimondi, J. E. Heuser, X. Zhuang, and P. De Camilli. 2010. Coupling between clathrin-dependent endocytic budding and F-BAR-dependent tubulation in a cell-free system. *Nat Cell Biol* **12**(9): 902–908.

Zhu, L., W. Zhang, D. Elnatan, and B. Huang. 2012. Faster STORM using compressed sensing. *Nat Methods* **9**(7): 721–723. http://www.nature.com/nmeth/journal/v9/n7/full/nmeth.1978.html.

Compressive Sampling-Based Decoding Methods for Lensfree On-Chip Microscopy and Sensing

Ikbal Sencan and Aydogan Ozcan

Contents

Introduction

Recent advances in light microscopy (Betzig et al. 2006; Gustafsson 2005; Hell 2003; Hess et al. 2006; Rust et al. 2006) transformed our toolset in biomedical research and clinical problems. Nowadays, optical microscopes are routinely used to nondestructively observe live biological samples with high spatial and temporal resolution. However, the overall space-bandwidth product of these imaging platforms is usually limited (Neifeld 1998), and users need to often trade off between spatial resolution and field of view, depending on the application. Moreover, most advanced microscopes are still relatively complex, expensive, and bulky. All these factors hinder easy access to advanced light microscopes at resource-limited settings or developing countries. Computational imaging, particularly lensfree on-chip microscopy, aims to democratize access to advanced optical microscopy tools by offering imaging platforms with high space-bandwidth products, low operational complexity and costs (Ozcan 2014). These unique advantages result from the simplicity of the on-chip imaging design of these lensfree platforms and the lack of expensive optical components. One should also note that the structural simplicity of lensfree on-chip microscopy also brings challenges that can be addressed computationally (Greenbaum et al. 2012). Lensfree microscopy platforms provide imaging solutions, which are portable for field use, compatible with lab-on-a-chip devices, and provide extremely wide field of view to address various challenges and needs of global health, telemedicine, and microfluidics/lab-on-a-chip applications, among others (Greenbaum et al. 2013; Mudanyali et al. 2010; Seo et al. 2010; Su et al. 2010; Zhu et al. 2013).

This chapter provides a summary of lensfree on-chip microscopy techniques that uniquely benefit from compressive decoding framework. It is organized as follows: the "Overview of Different Lensfree On-Chip Imaging Systems" section is an introduction to the basics of different lensfree on-chip microscopy geometries and their working principles. The "Application Examples on the Use of Compressive Decoding in Lensfree On-Chip Imaging" section includes several application examples, which we present under two categories: (1) fluorescent and incoherent imaging and (2) partially coherent holographic imaging. Finally, we conclude with a brief discussion on potential future applications of compressive sensing/sampling techniques in lensfree microscopy.

Overview of Different Lensfree On-Chip Imaging Systems

On-Chip Fluorescence and Incoherent Microscopy

Fluorescent microscopy is widely used to visualize biological structures and study their functions/dynamics by selectively labeling target biomolecules or analytes. Fluorescent labels are excited with high-intensity, single-color light, and they emit light with a different color (red shifted in spectrum, lower-energy photons). Consequently, there are two main components in a fluorescence microscope: excitation and emission interfaces. Since fluorescent emission is in all directions, and orders of magnitude weaker in power than the excitation light, rejection of the excitation beam should be effective. Otherwise, weak fluorescent signal of interest will be buried under the strong background due to the excitation light. Since most conventional microscopes use infinity-corrected objective lenses, they have plenty of space to place beam splitters, color filter wheels, etc., to achieve effective excitation rejection. However, these essential tasks of

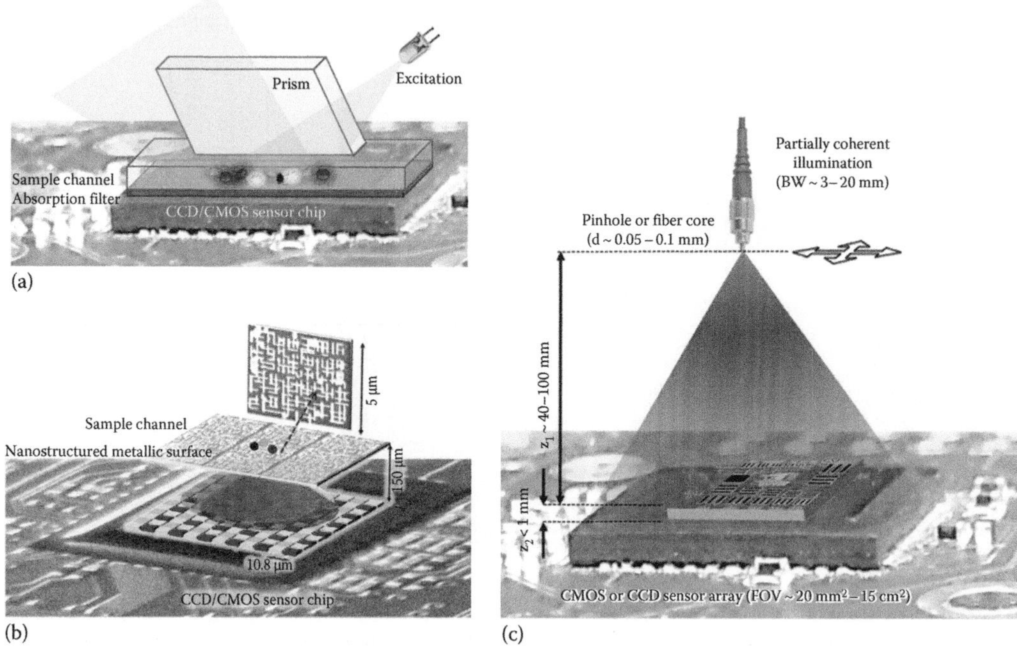

Figure 12.1 Examples of lensfree on-chip imaging platforms: (a) FLUOCHIP. (b) MONA setup. (c) Partially coherent, on-chip holographic microscopy setup. ([a–c] Courtesy of Sencan, I., Lensfree computational microscopy tools and their biomedical applications, Dissertation, University of California, Los Angeles, CA, 2013, http://gradworks.umi.com/35/94/3594279.html, accessed July 12, 2015; [c] Courtesy of Sencan, I. et al., *Sci. Rep.*, 4, article no. 3760, 1, 2014. doi:10.1038/srep03760.)

fluorescence imaging are quite challenging when it comes to an on-chip imaging geometry due to space restrictions between the sample and sensor planes. To overcome these design challenges, we created a new platform termed Fluorescent, Lensfree, Ultra-wide field-of-view, On-Chip, High-throughput Imaging Platform (FLUOCHIP) as illustrated in Figure 12.1a, where a large sample volume is illuminated through the side of a prism, with an illumination angle that is larger than the critical angle so that most of this excitation beam is rejected by total internal reflection (TIR) on the glass–air interface at the bottom of the sample channel. We should note that this imaging geometry should not be confused with TIR fluorescence (TIRF) microscopy as in our case the excitation occurs through travelling waves as opposed to evanescent waves in a TIRF microscope. The scattered portion of the illumination, which no longer satisfies the TIR angle requirement, is filtered out by a thin (30–100 μm) absorption filter placed under the sample channel. The fluorescent signal, however, passes through this thin absorption filter and reaches to the detector array to be sampled on a chip.

In lensfree on-chip microscopy, imaging performance is substantially affected by the close-up structure of the sensor geometry: pixel fill factor, pixel size, pixel circuitry, microlens arrays, color filters, and protective coatings on the chip all contribute to the resulting point spread function (PSF, i.e., the response of the imaging system to a point emitter). For FLUOCHIP geometry, it is fair to assume that PSF is spatially invariant in lateral dimensions. We experimentally characterized this unique, geometry- and chip-dependent, PSF for each experimental setup

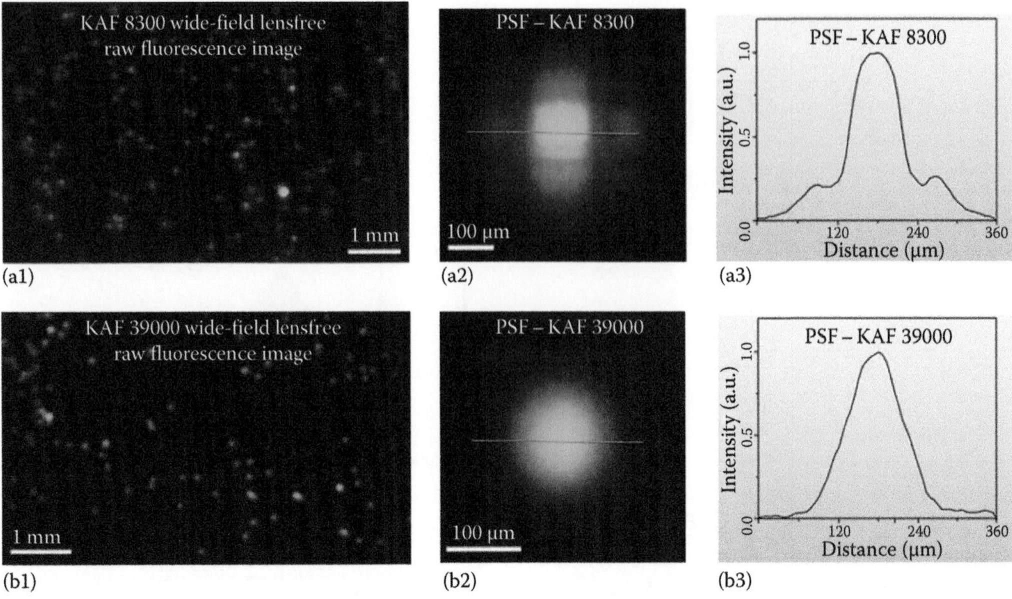

Figure 12.2 (a1 and b1) Wide-field lensfree raw fluorescence images corresponding to 4 μm fluorescent beads recorded using KAF-8300 and KAF-39000 charge-coupled device (CCD) sensor chips. (a2 and b2) Experimentally characterized point spread functions (PSF) of each system (a3, b3). Cross-sectional profiles for each PSF, which are marked with blue lines on (a2 and b2). (Courtesy of Coskun, A.F. et al., *PLoS ONE*, 6(1), e15955, 2011b.)

(see Figure 12.2) and measured it by center-aligning and averaging the signals from multiple small fluorescent microspheres sparsely seeded over a wide field of view at a constant vertical distance (typically 50–200 μm) from the chip active area. Then, this measured PSF, together with other system parameters such as pixel sampling function and the arrangement of color filters (if a red–green–blue [RGB] chip is employed), is used to define the forward model of the lensfree on-chip imaging system. This forward model, which depends on the choice of the charge-coupled device (CCD) or complementary metal–oxide–semiconductor (CMOS) imager chip, forms our measurement basis for the compressive decoding step. Equation 12.1 summarizes this general decoding problem for lensfree on-chip incoherent microscopy:

$$\tilde{f} = \arg\min\left(\| A(f) - g \|_2^2 + \beta \| f \|_1 \right) \tag{12.1}$$

Here

"A(f)" represents our forward model for the imaging system
g is the raw measurement
β is a regularization parameter
$f = [f_1, f_2, \ldots, f_N]$ is the object distribution

The lensfree raw measurements (g) initially look blurred due to the rapidly diverging fluorescence emission. In this cost function defined in Equation 12.1, we minimize the mismatch between the measured and estimated values, which is regularized by the l_1-norm of the object. This regularization term is assuring the convergence to a sparse solution. Accordingly, microscopic distributions of fluorescent emitters can be computed from these raw lensfree measurements by using a compressive solver such as ℓ_1 type (Kim et al. 2007) or TwIST (Bioucas-Dias and Figueiredo 2007), which are used to quickly solve Equation 12.1, even for large-scale lensfree measurement data.

Compressive decoding of these raw lensfree images using the measured PSF of a given setup permits significant increase in the resolving power through digital reconstruction of the fluorescent distribution at the object plane(s). For example, by using FLUOCHIP (Figure 12.1a), we achieved a spatial resolution that is on the order of the pixel size of the detector used despite the unit magnification (1×) resulting from our "on-chip" design. To reach subpixel resolution under the same unit magnification, however, we needed to break the symmetry of the measurement basis by using a design with spatially varying PSF. For this end, we introduced a new technique (illustrated in Figure 12.1b) that is referred to as Microscopy with On-chip Nano-Apertures (MONA). In this approach, the light emitted from incoherent objects of interest, for example, fluorescently labeled cells, which are directly positioned on a nanostructured thin-metallic film is first modulated by the presence of these nanostructures. After diffracting over a short distance (<1 mm), this spatially modulated light is sampled by a detector array (e.g., a CMOS imager) without the use of any lenses. Since the structured metallic film provides spatial modulation at the subpixel level, it effectively creates a spatially varying PSF over the lensfree imaging field of view as shown in Figure 12.3a. As a result of this, each chip needs to be calibrated, "only once," by recording the unique diffraction patterns resulting from a point source (e.g., a tightly focused spot/beam) as it moves over the surface of the nanostructured mask with a deeply subpixel step size. These calibration measurements only need to be performed once for a given mask design and are also used to iteratively optimize the nanopatterns for breaking the correlations among the diffraction patterns of closely spaced points (see Figure 12.3).

After the calibration of a given nanostructured chip/mask is completed, high-resolution object distributions are decoded from the lensfree raw images solving the same optimization problem described in Equation 12.1, except that "A" is now a high-rank measurement matrix formed by these calibration images.

So far, we have introduced two techniques for incoherent lensfree on-chip imaging: FLUOCHIP and MONA. Before we report some compressive decoding results using these two platforms for various applications as detailed in the "Application Examples on the Use of Compressive Decoding in Lensfree On-Chip Imaging" section, next we will introduce another lensfree on-chip imaging technique that is based on partially coherent holographic microscopy.

Partially Coherent On-Chip Holographic Microscopy

On-chip imaging of biological objects based on partially coherent holography has great potential to complement conventional optical microscopy tools with its compact, cost-effective, and high-throughput imaging geometry. In this imaging scheme, as illustrated in Figure 12.1c, objects of interest are placed directly on the top of an optoelectronic sensor array (CCD or CMOS), where the typical distance between the objects and the sensor active area (z_2) is <1–2 mm. A partially coherent light source, for example, a light-emitting diode (LED), which is placed $z_1 = 40–100$ mm above the object plane is first filtered with a large pinhole, with a diameter of, for example, $d = 0.05–0.1$ mm before it uniformly illuminates the object plane. While a small portion of the illumination is scattered by the objects,

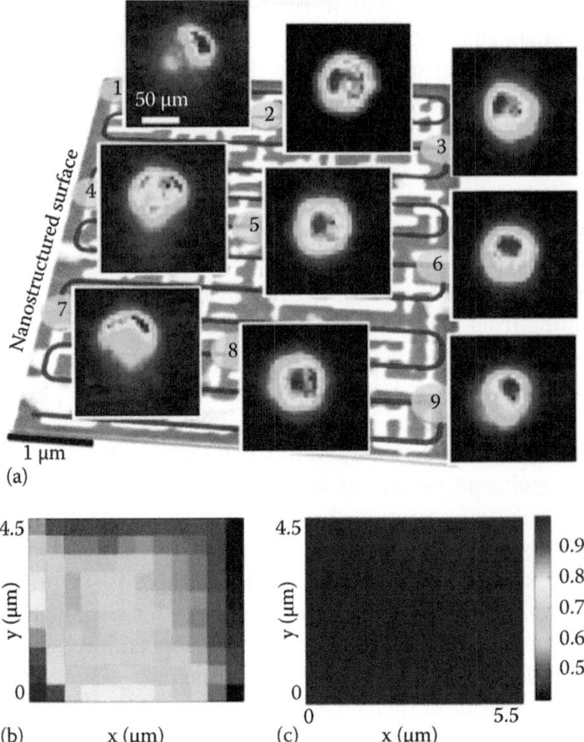

Figure 12.3 (a) To calibrate the nanostructured chip/mask, light from a fiber-coupled light-emitting diode (LED) (~30 nm bandwidth) is tightly focused onto the nanostructured surface, and the resulting far-field transmission images are recorded as the spot is spatially scanned (in x–y). (b) 2D map of cross-correlation coefficients between the first and the rest of these calibration images. (c) Same cross-correlation coefficient map as in (b), when the nanostructured chip/mask is replaced with a bare glass substrate. (Reproduced with permission from Khademhosseinieh, B., Sencan, I., Biener, G., Su, T.-W., Coskun, A.F., Tseng, D., and Ozcan, A., Lensfree on-chip imaging using nanostructured surfaces, *Appl. Phys. Lett.*, 96, 171106, 2010b. Copyright 2010, American Institute of Physics.)

s(x,y), the remaining light forms the unperturbed reference wave r(x,y). Based on this simple formalism, the intensity to be sampled by the optoelectronic sensor array (I_{det}) can be written as

$$\begin{aligned} I_{det} &= \left| r(x,y,z_0) + s(x,y,z_0) \right|^2 \\ &= \left| s(x,y,z_0) \right|^2 + \left| r(x,y,z_0) \right|^2 + r(x,y,z_0) \cdot s^*(x,y,z_0) + s(x,y,z_0) \cdot r^*(x,y,z_0) \end{aligned} \tag{12.2}$$

The scattered light intensity in itself is usually too weak and is therefore buried under the strong uniform reference wave intensity. However, the interference term allows the heterodyne detection of the object's spatial frequencies. Assuming that the illumination is quasimonochromatic, the complex field right after the object plane can be recovered by using a phase retrieval algorithm, which iteratively improves the field estimation by enforcing additional constraints such as a rough estimate of the object boundaries, phase or amplitude only assumption on the object, and/or intensity measurements at multiple object-to-sensor distances. This technique has been successfully implemented for various mobile imaging or cytometry applications (Bishara et al. 2012; Greenbaum et al. 2013; Isikman et al. 2010, 2012; Mudanyali et al. 2010; Seo et al. 2010; Su et al. 2010, 2013; Tseng et al. 2010) and has been improved to achieve diffraction-limited spatial resolution over wide imaging fields of view (FOV) (Greenbaum et al. 2012; McLeod et al. 2013; Mudanyali et al. 2013). However, within the context of this chapter, we will focus on one example, where we used compressive decoding for spectral demultiplexing, which will be detailed next.

When the bandwidth of the illumination source is wider than 10–15 nm, quasimonochromatic assumption is no longer valid, and the recorded lensfree hologram is in fact an incoherent superposition of weighted monochromatic hologram intensities, that is,

$$I_{\lambda_1} = |R_{\lambda_1} + S_{\lambda_1}|^2 = |R_{\lambda_1}|^2 + |S_{\lambda_1}|^2 + 2\mathrm{Re}\{R_{\lambda_1} S^*_{\lambda_1}\} \quad (12.3)$$

$$I_{total} = c_1 I_{\lambda_1} + c_2 I_{\lambda_2} + c_3 I_{\lambda_3} + \cdots + c_i I_{\lambda_i} \quad (12.4)$$

In this notation, the spectrum of the illumination source and sensor spectral response curve defines the contribution (c_i) of each wavelength. This spectral overlap causes a smearing effect in the measured hologram and decreases the fringe visibility. As an example, Figure 12.4 illustrates the degradation in the reconstruction quality due to this effect.

We demonstrated that we can partially release this bandwidth-related artifact using compressive decoding to spectrally demultiplex the holograms recorded with broadband light sources to quasimonochromatic holograms, that is,

$$\tilde{I} = \arg\min(\|cI - I_{total}\|_2^2 + \beta \cdot TV(I)) \quad (12.5)$$

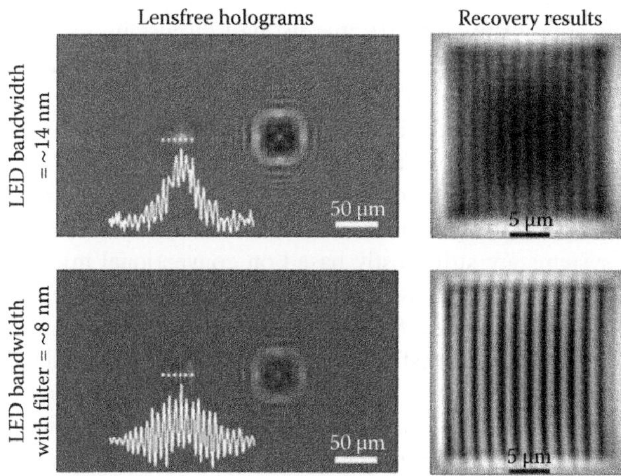

Figure 12.4 Broad bandwidth of the illumination source degrades the quality of lensfree on-chip microscopy reconstructions. (Courtesy of Sencan, I. et al., *Sci. Rep.*, 4, article no. 3760, 1, 2014. doi:10.1038/srep03760.)

$$I = [I_{\lambda_1}\ I_{\lambda_2}\ I_{\lambda_3} \ldots I_{\lambda_i}] \quad (12.6)$$

where
"c" contains the combined spectra of illumination and sensor pixel sensitivity
"TV" represents total variation

Assuming that multispectral spatial image data are naturally redundant, we can minimize the squared l_2 norm of the mismatch between the measured total hologram (I_{total}) and our estimation for narrower band holograms (I), regularized with the total variation of the multispectral hologram stack (Sencan et al. 2014). After this step, the demultiplexed holograms can be individually processed as before, by, for example, using an iterative monochromatic phase retrieval approach (Fienup 1982). Experimental results of this approach under direct sunlight illumination will be reported at the end of the "Application Examples on the Use of Compressive Decoding in Lensfree On-Chip Imaging" section.

Application Examples on the Use of Compressive Decoding in Lensfree On-Chip Imaging

On-Chip Incoherent Microscopy Applications Using FLUOCHIP and MONA

Incoherent and fluorescent lensfree imaging techniques are quite suitable to benefit from compressive sensing/sampling approaches because incoherent imaging platforms are linear in intensity and fluorescent objects are mostly sparse in spatial domain. Also, lensfree imaging platforms have the potential to optimally serve a large range of applications, each of which has different set of requirements in terms of throughput, spatial resolution, and structural simplicity. Starting with the next subsection, we report different applications of two main incoherent computational microscopy techniques, namely, FLUOCHIP and MONA.

High-Throughput Dual-Mode Imaging of Transgenic Caenorhabditis elegans *Using FLUOCHIP*

The first application of FLUOCHIP that we will discuss in this chapter is lensfree on-chip fluorescent imaging of transgenic *Caenorhabditis elegans* (Coskun et al. 2011b). Since *C. elegans* is a widely used model organism in various fields (Lehner et al. 2006; Mellem et al. 2008; Pinkston-Gosse and Kenyon 2007), several microfluidic platforms (Chokshi et al. 2010; Rohde et al. 2007; Strange 2006) have been developed to sort and manipulate these worms at high throughputs. On the other hand, optical imaging methods used for screening of *C. elegans* in these microfluidic systems are still mostly based on conventional microscopy, which are limited in their space-bandwidth products and are not entirely compatible with microfluidics due to their relatively bulky designs. Based on FLUOCHIP design (see Figure 12.5a), a lensfree on-chip platform can be used to image fluorescent *C. elegans* samples over large FOVs. The sparse distribution of fluorescence emission sites of these transgenic *C. elegans* samples allows compressive decoding of the emitter locations without representing them in any sparsifying basis and achieving a spatial resolution of ~10 μm even under unit magnification. Moreover, the unique design of FLUOCHIP (see Figure 12.5a) is also suitable for sequential on-chip holographic imaging of the same region of interest. In this dual-mode microscopy scheme, after on-chip imaging of fluorescence, the same

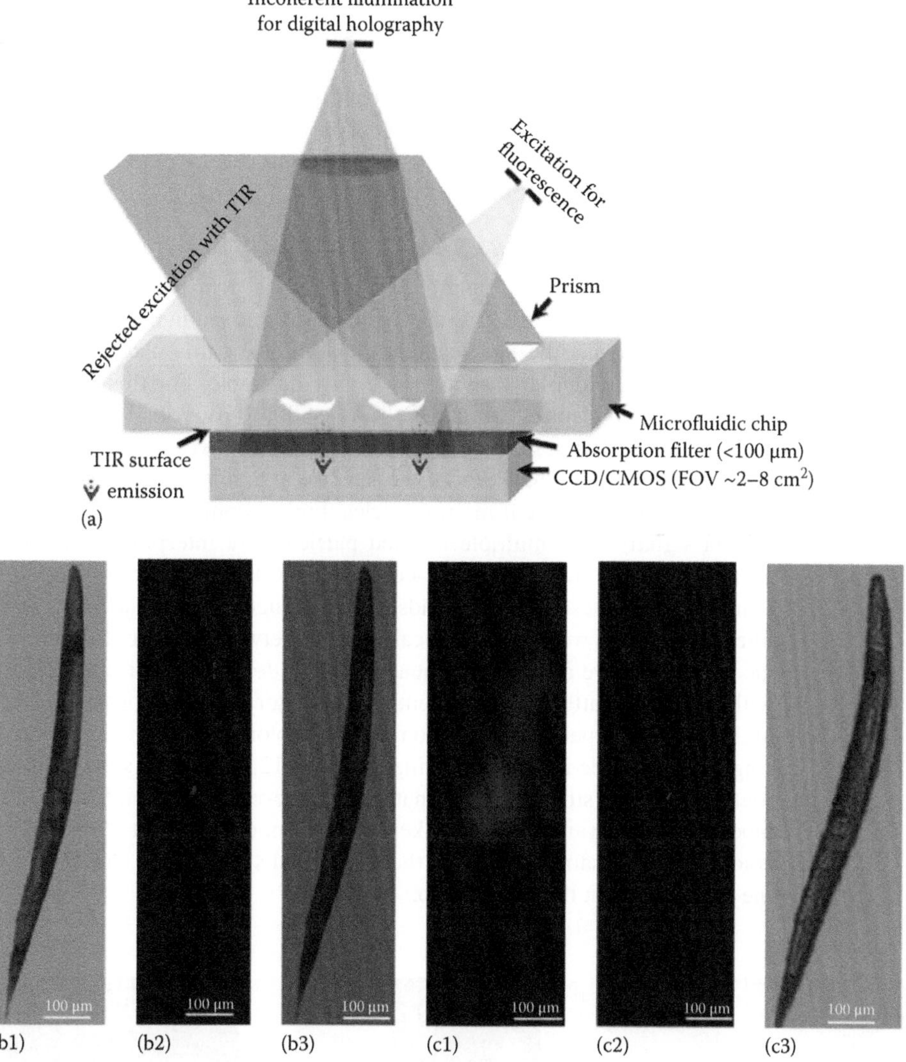

Figure 12.5 (a) Dual-mode lensfree fluorescent and holographic transmission imaging platform. (b) Conventional microscopy images of an individual *C. elegans* worm using a 10× objective lens with (b1) bright-field, (b2) fluorescence, and (b3) overlaid/superimposed. (c) Lensfree imaging results for the same sample. (c1) Corresponding lensfree raw fluorescent image, which is cropped from a wide field of view image; (c2) decoded lensfree fluorescence image; and (c3) lensfree on-chip holographic transmission image superimposed with the decoded fluorescence image. (Courtesy of Coskun, A.F. et al., *PLoS ONE*, 6(1), e15955, 2011b.)

samples are then illuminated using a partially coherent light source located above the top surface of the rhomboid prism to create a transmission holographic image of the specimen. Dual-mode (fluorescent and bright-field) imaging over such wide FOVs (e.g., >2–8 cm^2) can particularly be beneficial for biomedical applications requiring high-throughput screening of samples.

We should emphasize that raw lensfree signatures of the worms, as shown in Figure 12.5c1, are highly blurred due to rapid divergence of the fluorescence emission. These raw lensfree images together with the experimentally characterized PSF are used to compute, based on Equation 12.1, higher-resolution fluorescence images of the transgenic *C. elegans* samples, as illustrated in Figure 12.5c2. These decoded fluorescent distributions can then be superimposed onto the holographic transmission image of the same objects as demonstrated in Figure 12.5c3, the results of which are also confirmed by conventional microscopy images (Figure 12.5b).

Color Imaging Using FLUOCHIP

Performing fluorescence detection without lenses by using an on-chip imaging geometry can improve imaging throughput as discussed in the previous sections. To further increase the multiplexing capabilities of this platform, performing simultaneous multicolor detection is highly desirable. For this end, we used a raw-format color sensor chip (CCD, KAF-8300, RGB, with 5.4 μm pixel pitch) in our FLUOCHIP design (see Figure 12.6a) and experimentally characterized the red, green, and blue PSFs of the lensfree imaging system using the smallest observable fluorescent microparticles. For this purpose, the measured fluorescent signals from multiple isolated particles are interpolated, aligned, and averaged, as was done for single-color lensfree PSF measurements. The functional form of these PSFs depends on the structure of the sensor chip, color filter response, as well as the vertical distance between the particles/emitters and the detector active area. These measured PSFs for each color channel, together with the Bayer pattern arrangement of the sensor chip, are then used to decode high-resolution spatial distribution of multicolor fluorescent emitters from a single raw lensfree image by solving Equation 12.1 using the same optimization routine used for single-color signatures. However, in this multicolor case, we modified the forward model to take into account (1) different PSFs for different colors, (2) undersampling due to the large pixel size, and (3) the distribution of the color filters on the sensor chip.

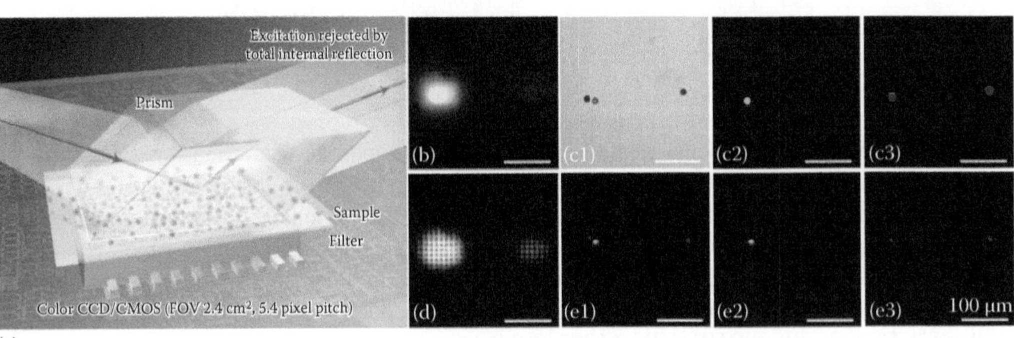

Figure 12.6 (a) Lensfree, on-chip multicolor fluorescence imaging platform. (b) RGB debayered lensfree image of green and red fluorescent microbeads with 10 μm diameter. (c) Conventional microscope images are included for verification in (c1) bright field, (c2) green fluorescent channel, and (c3) red fluorescent channel. (d) Raw, lensfree fluorescence signatures of microbeads, cropped from a wide field of view of ~2.42 cm^2. The checkerboard-like texture is due to the RGB color filters on the sensor chip. (e) Lensfree signatures are decoded to achieve multicolor imaging. (e1) Superimposed decoding results for (e2) red and (e3) green fluorescent channels. (Courtesy of Sencan, I. et al., *Sci. Rep.*, 4, article no. 3760, 1, 2014. doi:10.1038/srep03760.)

The performance of this simultaneous on-chip color imaging approach is tested with 10 μm red and green fluorescent beads, as shown in Figure 12.6. The raw lensfree signatures were modulated with the Bayer pattern and severely overlapped in space (see Figure 12.6d) due to the lensfree operation of the fluorescence imaging platform. The spatial distribution of these multicolor emitters was successfully recovered (see Figure 12.6e) and verified with fluorescent and bright-field microscope images of the same samples (see Figure 12.6c). This lensfree multicolor imaging platform improves the throughput and multiplexing capabilities of FLUOCHIP and can be beneficial for applications in, for example, imaging cytometry.

FLUOCHIP with Fiber-Optic Faceplates and Tapers

The next FLUOCHIP implementation is designed to further improve the resolution of this lensfree imaging system, where a fiber-optic faceplate is added between the bottom end of the object chamber and the detector array as shown in Figure 12.7a. A fiber-optic faceplate is a passive optical element, composed of a dense bundle of fibers, which act as a relay in our on-chip imaging system, also providing thermal and mechanical isolation of the objects from the sensor active area. Furthermore, it narrows down the lensfree imaging PSF and increases the signal-to-noise ratio (SNR) of our on-chip imaging platform. As a result of this improved SNR, ~10 μm spatial resolution can be achieved (see Figure 12.7d) despite a unit magnification-based on-chip imaging geometry with a large CCD pixel pitch of 9 μm.

It should also be noted that the individual fibers that make up of the faceplate are dense and aligned regular enough for these fluorescent images to assume a shift-invariant PSF over the object plane, although they distort the holographic images due to the deterministic but unknown phase mixing that occurs within the fiber-optic array.

Moreover, the presented lensfree approach is also able to reconstruct fluorescent micro-objects distributed at multiple microchannels that are stacked vertically (Figure 12.8). This feature increases the throughput of FLUOCHIP multiple folds by decoding the 2D lensfree fluorescent image captured at the detector array into a 3D image stack through compressive sensing framework.

To further improve the resolution of this on-chip imaging platform, a tapered fiber-optic faceplate can be added to the FLUOCHIP setup such that the period of the fiber cores on the bottom facet is significantly larger compared to the fiber period on the top facet as illustrated in Figure 12.9a, insets. As shown in Figure 12.9b and c, the spatial resolution of this taper-based system is improved to ~3–4 μm by taking advantage of the 2.4-fold magnification factor that is due to the fiber-optic taper that is used in this on-chip imaging experiment.

Incoherent Microscopy with On-Chip Nano-Apertures

To improve the resolution to subpixel levels in an on-chip imaging system under unit magnification, there is a need to break the coherence between the sparsifying and measurement bases. For this end, the shift-invariant on-chip imaging setup of FLUOCHIP can be turned into a position-dependent system by introducing a nanostructured metallic thin-film or mask to spatially modulate the PSF at each point on the chip (see Figure 12.10a). This technique is referred to as MONA (Khademhosseinieh et al. 2010b).

To emphasize the compressive nature of this MONA method, the experimental performance of the decoding algorithm is reported in Figure 12.10b and c,

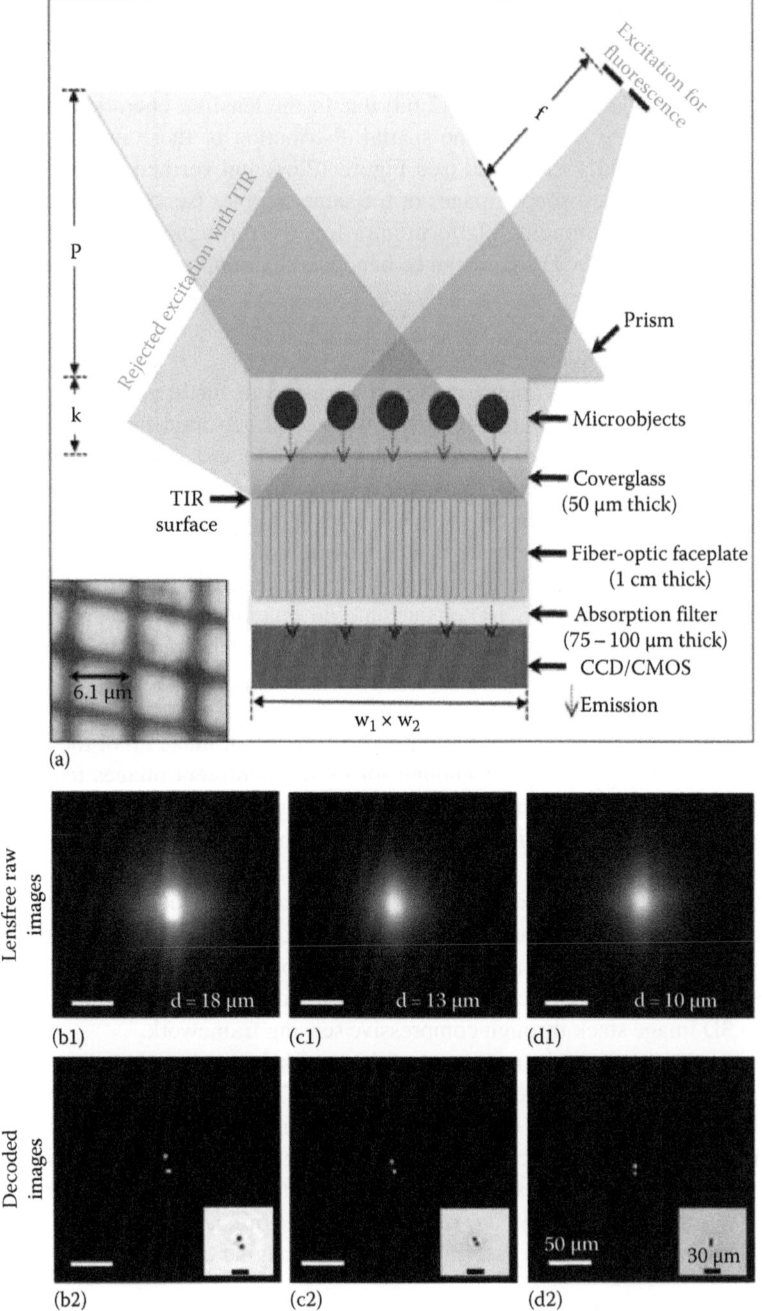

Figure 12.7 (a) FLUOCHIP with a fiber-optic faceplate. $w_1 \times w_2 = 25$ mm × 35 mm; p = 1.7 cm; k = 10–100 μm; f = 1–2 cm. (a, inset) Microscope image of the fiber-optic faceplate with a numerical aperture of ~0.3. (b1, c1, and d1) Lensfree raw fluorescent signatures of closely spaced (d = 18 μm, 13 μm, and 10 μm) micro-beads, which are 10 μm in diameter (b2, c2, and d2). Compressive decoding results for each bead pair are provided. (b2, c2, and d2 insets) are regular bright-field transmission microscope images provided for comparison purposes. (From Coskun, A.F. et al., *Opt. Express*, 18(10), 10510, 2010. With permission.)

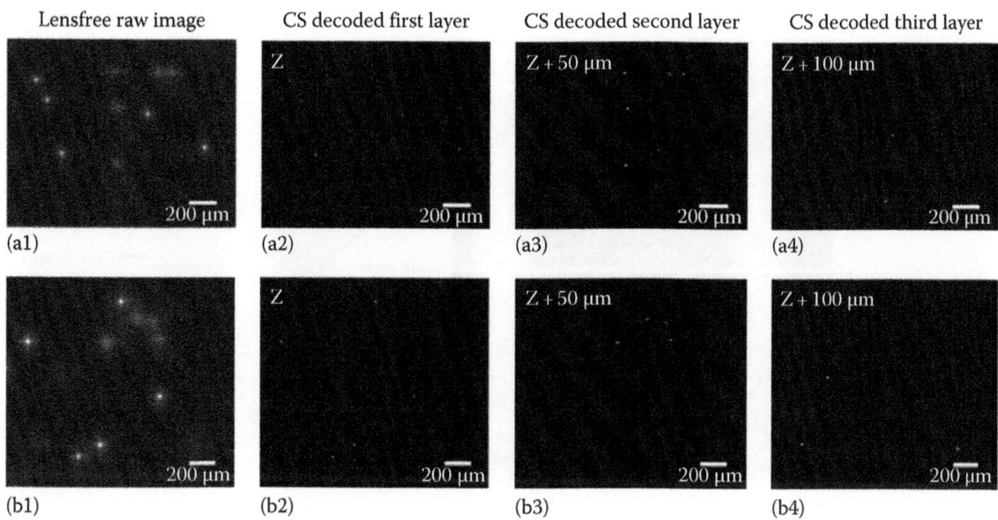

Figure 12.8 (a1 and b1) Raw lensfree fluorescence images from two different regions of interest cropped from a large FOV of >8 cm². Lensfree signatures of the emitters located within three vertically stacked sample channels are simultaneously recorded. (a2–a4 and b2–b4) Spatial distributions of the emitters within three channels vertically separated by 50 μm are successfully decoded. (From Coskun, A.F. et al., *Opt. Express*, 18(10), 10510, 2010. With permission.)

when the number of pixels (i.e., number of measurements, M) is far less than the recovered number of points (N > M). The raw lensfree diffraction signatures that resulted from two tightly focused spots are shown in Figure 12.10b (for M = 120, 64, 36, and 25). These undersampled lensfree images are synthesized by binning multiple pixels of the experimentally measured lensfree image in order to show the effects of larger pixel size and/or shorter distance between the nanostructured surface and the sensor array plane. Notice that none of these lensfree images show any resemblance to the real object distribution, and yet the compressive decoding results agree well with the microscope comparisons (Figure 12.10c, insets), even when the number of measurements (M) is far less than the number of recovery points/pixels (N).

Multicolor MONA

To add color imaging capability to MONA platform, the calibration basis can be extended using red, green, and blue focused spots (Figure 12.11a). In this calibration process, these point sources at each color scan the surface of the nanostructured substrate, while the sensor array is recording the resulting lensfree diffraction patterns corresponding to each calibration location and illumination color/wavelength on the structured chip. To increase the modulation as a function of the illumination color, an RGB sensor chip with raw output is used.

Figure 12.11 summarizes our proof-of-concept experiment with three tightly focused spots (red, green, and blue), which are located within a single-pixel area and separated by ~1–2 μm. As expected from this experiment, the lensfree diffraction image on the sensor array is simply a white-looking blob (see the debayered raw image, Figure 12.11b), without any visible features that resemble the shape or the color of our subpixel multicolor object. The checkerboard features observed on the raw lensfree image (Figure 12.11c) are due to the arrangement of

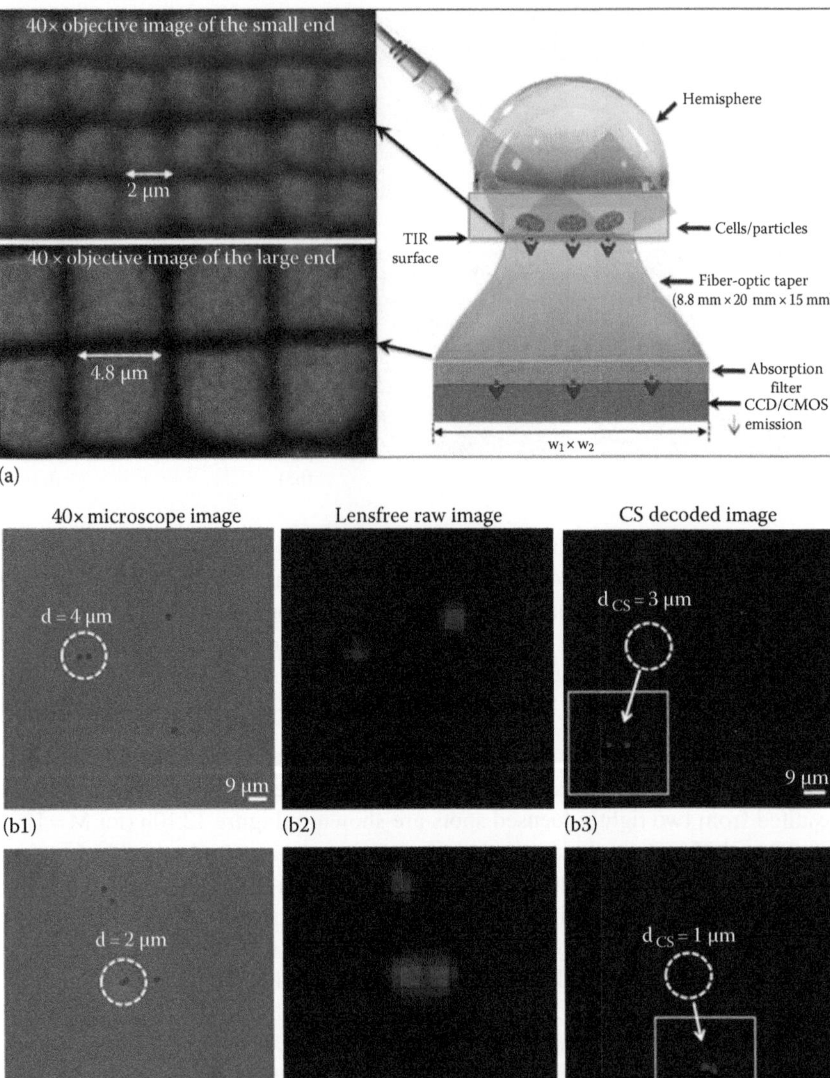

Figure 12.9 (a) FLUOCHIP with a tapered faceplate and a hemisphere is illustrated. $w_1 \times w_2 = 25$ mm × 35 mm. The lensfree images are magnified by a factor of 2.4 because of the tapering ratio between top and bottom facets of the fiber-optic taper. (a, insets) Microscope images from the top and bottom surfaces of the tapered fiber-optic faceplate. (b and c) Microbeads with 2 μm diameters are used to characterize the resolving power of the imaging system. (b2 and c2) Lensfree raw fluorescent images are used to compute (b3 and c3) compressively decoded high-resolution fluorescent images. Insets are the zoomed-in images of the closely spaced bead pairs within the dashed line. (b1 and c1) are microscope images recorded by using a 40× objective lens used for comparison purposes. (Coskun, A.F. et al., *Analyst*, 136(17), 3512, 2011a. Reproduced by permission of The Royal Society of Chemistry.)

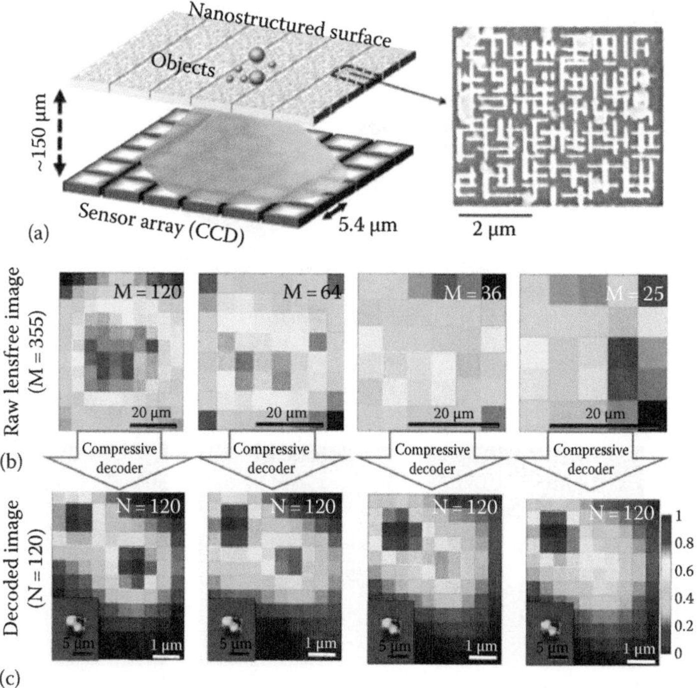

Figure 12.10 (a) Schematic diagram of MONA is illustrated. (a, inset) Scanning electron microscope image of the nanostructured chip surface is shown. (b) Raw lensfree diffraction images with different number of pixels (i.e., the number of measurements, M = 120, 64, 36, and 25). Images with lower number of pixels are synthesized by pixel-binning to show how the compressive decoder performs in an undersampled (N > M) condition. (c) Compressive decoding results for each M value are shown. Note that two subpixel spots are faithfully resolved even when the number of measurements is around 50% of the number of unknowns, N. (c, insets) Conventional, lens-based, reflection microscope images are also provided for comparison purposes. Light is transmitted only through the black area, which refers to the nanostructured chip surface, whereas the red areas show the metal-coated, nontransmissive regions. (Reproduced with permission from Khademhosseinieh, B., Sencan, I., Biener, G., Su, T.-W., Coskun, A.F., Tseng, D., and Ozcan, A., Lensfree on-chip imaging using nanostructured surfaces, *Appl. Phys. Lett.*, 96, 171106, 2010b. Copyright 2010, American Institute of Physics.)

the color filters on the sensor array chip. Decoded microscopic image is reported in Figure 12.11d, which is faithfully showing both the colors and the locations of the three tightly focused spots within a single-pixel area. This agreement between the decoded image and the 40× microscope objective-lens comparison (Figure 12.11e) suggests that MONA can perform lensfree on-chip color imaging with deeply subpixel resolution.

Spectral Demultiplexing of Sunlight Holograms

Sunlight has been the very first and the most accessible light source for initial optical microscopes. The broad spectral content of sunlight makes it a good alternative for multispectral imaging of objects. In this subsection, we will

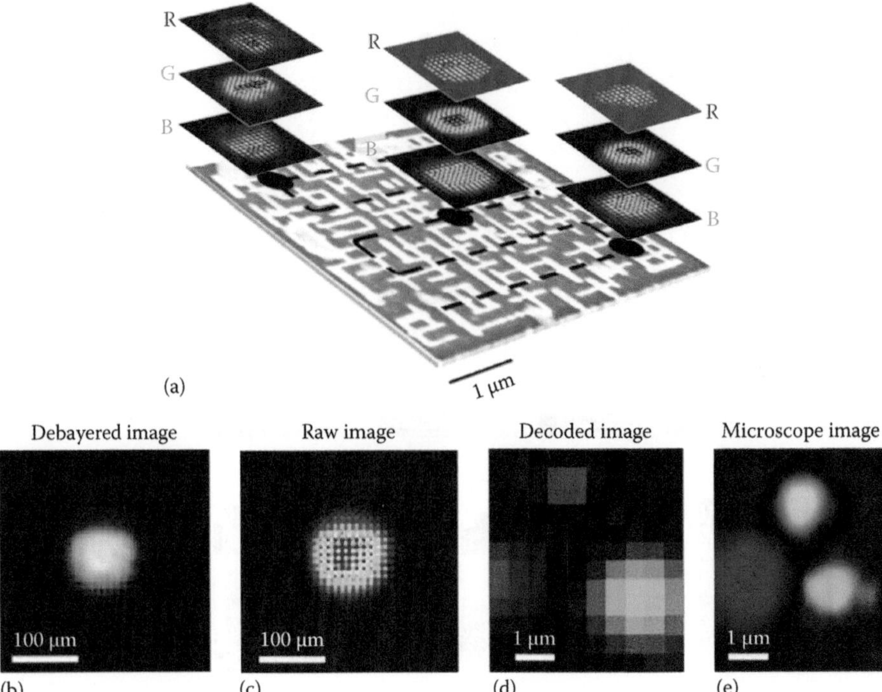

Figure 12.11 (a) Calibration process for multicolor MONA platform is illustrated. Raw lensfree diffraction images are recorded, while scanning tightly focused red, green, and blue spots, sequentially. These raw lensfree images are forming the measurement matrix/basis. For a proof-of-concept experiment, a subpixel multicolor object composed of three tightly focused (red, green, and blue) spots that are separated from each other by 1–2 μm is used. (b) RGB debayered lensfree diffraction image of this multicolor subpixel object. (c) Raw lensfree diffraction image of the same subpixel multicolor object is shown. Neither the location nor the colors of the individual spots are discernable from these lensfree images. (d) Compressive decoding results show correct location and color information for each spot within the subpixel object. Our results agree well with (e) the reflection microscope comparison image (40× objective lens, numerical aperture of 0.6). (Reproduced with permission from Khademhosseinieh, B., Biener, G., Sencan, I., and Ozcan, A., Lensfree color imaging on a nanostructured chip using compressive decoding, *Appl. Phys. Lett.*, 97, 211112, 2010a. Copyright 2010, American Institute of Physics.)

discuss a portable, on-chip holography platform performing color/multispectral imaging over a wide field of view of, for example, 15–30 mm^2, using sunlight illumination.

In the "Partially Coherent On-Chip Holographic Microscopy" section, we discussed compact, cost-effective, and lightweight microscopy platforms with ultrawide field of views, using partially coherent holographic imaging geometries. In this particular design (Figure 12.12), sunlight is used to homogeneously illuminate objects of interest, which are placed on a color sensor chip with ~500 μm gap to the active area of the sensor array (RGB CMOS, 1.4 μm pixel pitch). For this purpose, we designed a simple sunlight collection unit with a semiflexible light pipe, inspired by the current methods used in solar cell designs. After the glass-based light pipe (1 mm in diameter), a plastic diffuser is placed to prevent projecting features on the sky onto the detector, like a pinhole camera does. The

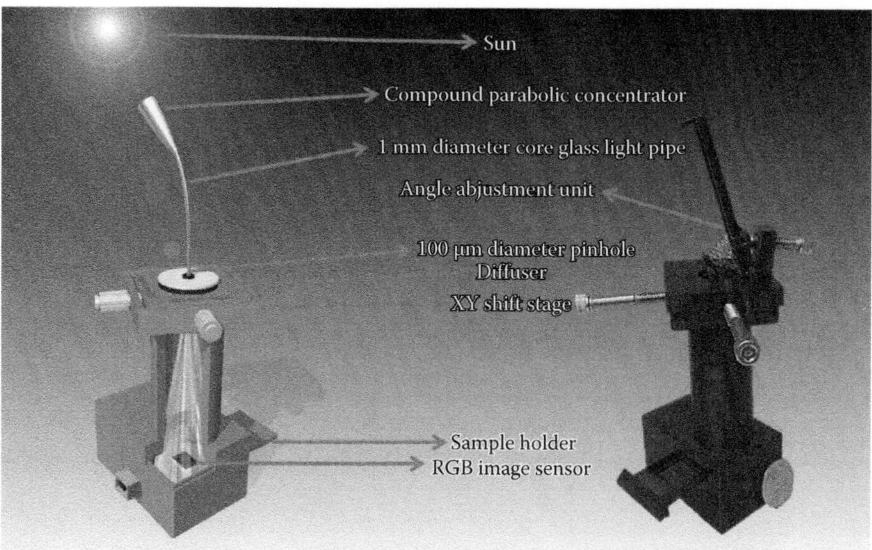

Figure 12.12 A prototype of the field-portable lensfree holographic microscope using sunlight.

sunlight, after passing through the diffuser, is then filtered by a large pinhole (with a diameter of 100 μm) and propagates 7.5 cm to acquire enough spatial coherence before it impinges on the objects of interest. Owing to the on-chip imaging geometry, we can record multiple slightly shifted holograms of the same static objects, as we move the pinhole by using a simple x–y translation stage. These laterally shifted holograms of the same area of interest are digitally combined to synthesize a finer sampled hologram for each color channel using a pixel superresolution algorithm (Hardie 2007). To obtain high-resolution lensfree microscopic images from these preprocessed holograms, we follow a two-step reconstruction method, summarized in Figure 12.13a. The first step is *spectral demultiplexing*: pixel-superresolved sunlight holograms are well sampled in spatial domain; however, as discussed in the "Partially Coherent On-Chip Holographic Microscopy" section and demonstrated in Figure 12.4, they still suffer from spectral smearing artifact due to the broadband illumination. To mitigate this problem, we decode (using Equation 12.5) these pixel-superresolved holograms to digitally remove/reduce the cross-talk among various color channels and retrieve the multispectral features of the specimen. For this purpose, we also utilize the spectral response information of the sensor array chip as well as the spectrum of sunlight and assume that multispectral spatial data are naturally redundant. Finally, these spectrally demultiplexed holograms are individually processed to obtain high-resolution lensfree microscopic images either by using iterative monochromatic phase retrieval or a holographic back-projection approach.

To experimentally demonstrate the proof of concept of this approach, we recoded a sunlight hologram of the USAF chart using our field-portable, on-chip holographic microscope shown in Figure 12.12. The green channel hologram (effective bandwidth of 70 nm under sunlight) is spectrally decoded into 21 images, each with ~14 nm effective spectral bandwidth. Holographic back-projection results using the raw green channel hologram as well as the spectrally decoded hologram, both at a center wavelength of 538 nm, are reported in Figure 12.13b and c, respectively. Part of these reconstructed images is also

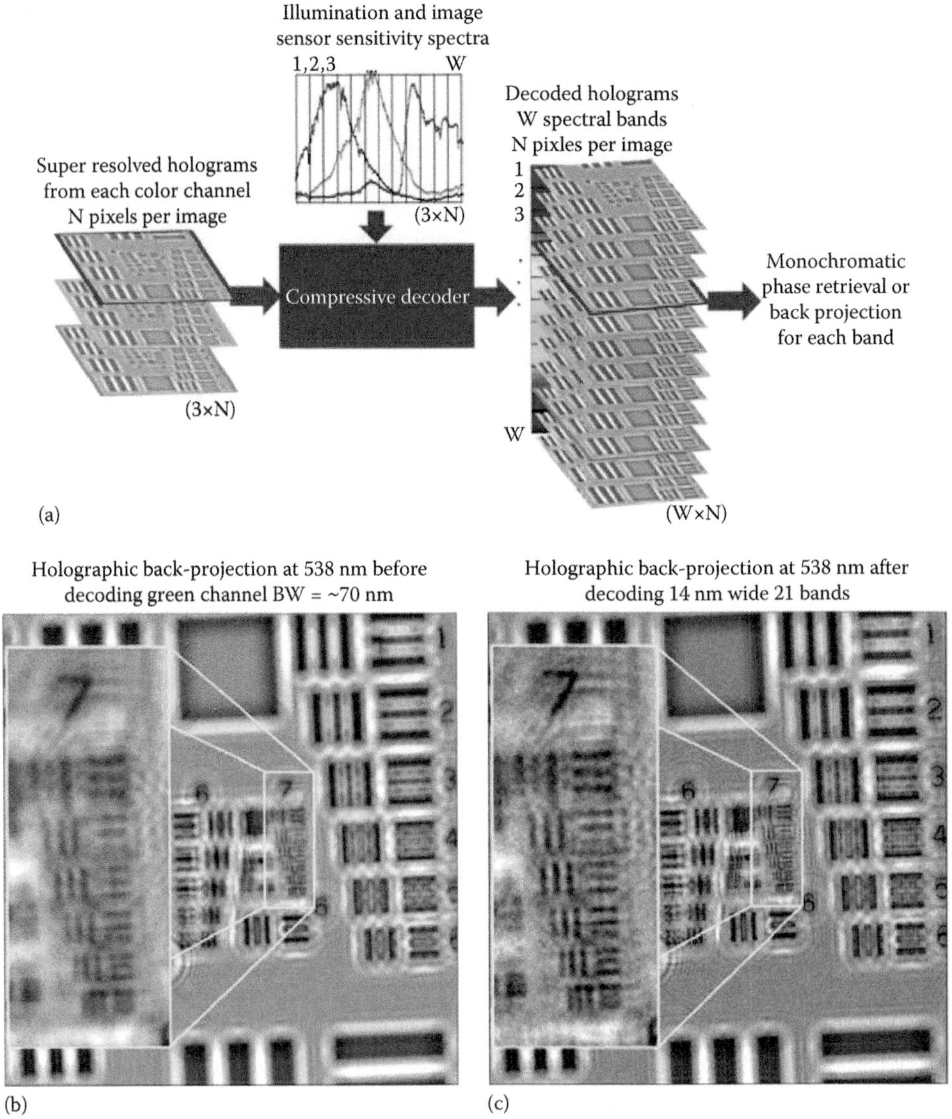

Figure 12.13 The redundant nature of the multispectral data is used to decode the sunlight hologram into narrower spectral bands. (a) Schematic diagram summarizes the spectral demultiplexing of broadband on-chip holograms. (b and c) Results of simple holographic back projection from a sunlight hologram corresponding to USAF test chart with 538 nm wavelength (b) before spectral demultiplexing, and (c) after spectral demultiplexing into 21 holograms, each with 14 nm effective illumination bandwidth. (b and c, insets) Group 7 is zoomed-in to better illustrate the resolution improvement as a result of this spectral demultiplexing step.

zoomed-in to better visualize the resolution improvement as a result of this digital spectral demultiplexing step, as illustrated in Figure 12.13b and c, insets.

Conclusions

In this chapter, we summarized some examples of how compressive decoding methods can be used for incoherent and partially coherent lensfree on-chip microscopy platforms. In fact, there are various other aspects of lensfree on-chip imaging, which can benefit from exploiting compressive decoding in spatial, spectral, and angular domains. For instance, the use of compressive sensing/sampling framework for holographic reconstruction (Brady et al. 2009) and iterative phase retrieval (Candès et al. 2015; Chan et al. 2008) can also be extended to on-chip hologram reconstruction methods. Moreover, synthetic aperture (Luo et al. 2015) and phytography (Dong et al. 2014) approaches could also get more efficient and robust in their reconstructions by exploiting sparse signal recovery methods. As another example, on-chip tomography (Isikman et al. 2011) can greatly benefit from compressive decoding approaches to more efficiently fill the angular spectrum cone with less number of illumination angles. Finally, in the temporal domain (Llull et al. 2013), compressive sensing/sampling methods can also be used to effectively boost the frame rate of the cameras used for on-chip detection/tracking so that they can be used to monitor very fast dynamics (e.g., of micro-swimmers) at extreme throughputs.

Acknowledgments

The Ozcan Research Group at UCLA gratefully acknowledges the support of the Presidential Early Career Award for Scientists and Engineers (PECASE), the Army Research Office (ARO; W911NF-13-1-0419 and W911NF-13-1-0197), the ARO Life Sciences Division, the ARO Young Investigator Award, the National Science Foundation (NSF) CAREER Award, the NSF CBET Division Biophotonics Program, the NSF Emerging Frontiers in Research and Innovation (EFRI) Award, the NSF EAGER Award, NSF INSPIRE Award, Office of Naval Research (ONR), and the Howard Hughes Medical Institute (HHMI).

References

Betzig, E., G.H. Patterson, R. Sougrat, O.W. Lindwasser, S. Olenych, J.S. Bonifacino, M.W. Davidson, J. Lippincott-Schwartz, and H.F. Hess. 2006. Imaging intracellular fluorescent proteins at nanometer resolution. *Science* 313 (5793): 1642–1645.

Bioucas-Dias, J.M. and M.A.T. Figueiredo. 2007. A new TwIST: Two-step iterative shrinkage/thresholding algorithms for image restoration. *IEEE Transactions on Image Processing* 16(12): 2992–3004.

Bishara, W., S.O. Isikman, and A. Ozcan. 2012. Lensfree optofluidic microscopy and tomography. *Annals of Biomedical Engineering* 40(2): 251–262.

Brady, D.J., K. Choi, D.L. Marks, R. Horisaki, and S. Lim. 2009. Compressive holography. *Optics Express* 17(15): 13040–13049.

Candès, E., Y. Eldar, T. Strohmer, and V. Voroninski. 2015. Phase retrieval via matrix completion. *SIAM Review* 57(2): 225–251.

Chan, W.L., M.L. Moravec, R.G. Baraniuk, and D.M. Mittleman. 2008. Terahertz imaging with compressed sensing and phase retrieval. *Optics Letters* 33(9): 974.

Chokshi, T.V., D. Bazopoulou, and N. Chronis. 2010. An automated microfluidic platform for calcium imaging of chemosensory neurons in *Caenorhabditis elegans*. *Lab on a Chip* 10(20): 2758–2763.

Coskun, A.F., I. Sencan, T.-W. Su, and A. Ozcan. 2010. Lensless wide-field fluorescent imaging on a chip using compressive decoding of sparse objects. *Optics Express* 18(10): 10510–10523.

Coskun, A.F., I. Sencan, T.-W. Su, and A. Ozcan. 2011a. Wide-field lensless fluorescent microscopy using a tapered fiber-optic faceplate on a chip. *Analyst* 136(17): 3512–3518.

Coskun, A.F., I. Sencan, T.-W. Su, and A. Ozcan. 2011b. Lensfree fluorescent on-chip imaging of transgenic *Caenorhabditis elegans* over an ultra-wide field-of-view. *PLoS ONE* 6(1): e15955.

Dong, S., Z. Bian, R. Shiradkar, and G. Zheng. 2014. Sparsely sampled Fourier ptychography. *Optics Express* 22(5): 5455.

Fienup, J.R. 1982. Phase retrieval algorithms: A comparison. *Applied Optics* 21(15): 2758–2769.

Greenbaum, A., N. Akbari, A. Feizi, W. Luo, and A. Ozcan. 2013. Field-portable pixel super-resolution colour microscope. *PLoS ONE* 8(9): e76475.

Greenbaum, A., W. Luo, T.-W. Su, Z. Göröcs, L. Xue, S.O. Isikman, A.F. Coskun, O. Mudanyali, and A. Ozcan. 2012. Imaging without lenses: Achievements and remaining challenges of wide-field on-chip microscopy. *Nature Methods* 9(9): 889–895.

Gustafsson, M.G.L. 2005. Nonlinear structured-illumination microscopy: Wide-field fluorescence imaging with theoretically unlimited resolution. *Proceedings of the National Academy of Sciences of the United States of America* 102(37): 13081–13086.

Hardie, R. 2007. A fast image super-resolution algorithm using an adaptive wiener filter. *IEEE Transactions on Image Processing* 16(12): 2953–2964.

Hell, S.W. 2003. Toward fluorescence nanoscopy. *Nature Biotechnology* 21(11): 1347–1355.

Hess, S.T., T.P.K. Girirajan, and M.D. Mason. 2006. Ultra-high resolution imaging by fluorescence photoactivation localization microscopy. *Biophysical Journal* 91(11): 4258–4272.

Isikman, S.O., W. Bishara, O. Mudanyali, I. Sencan, T.-W. Su, D.K. Tseng, O. Yaglidere, U. Sikora, and A. Ozcan. 2012. Lensfree on-chip microscopy and tomography for biomedical applications. *IEEE Journal of Selected Topics in Quantum Electronics* 18(3): 1059–1072.

Isikman, S.O., W. Bishara, U. Sikora, O. Yaglidere, J. Yeah, and A. Ozcan. 2011. Field-Portable lensfree tomographic microscope. *Lab on a Chip* 11(13): 2222–2230.

Isikman, S.O., I. Sencan, O. Mudanyali, W. Bishara, C. Oztoprak, and A. Ozcan. 2010. Color and monochrome lensless on-chip imaging of *Caenorhabditis elegans* over a wide field-of-view. *Lab on a Chip* 10(9): 1109–1112.

Khademhosseinieh, B., G. Biener, I. Sencan, and A. Ozcan. 2010a. Lensfree color imaging on a nanostructured chip using compressive decoding. *Applied Physics Letters* 97: 211112.

Khademhosseinieh, B., I. Sencan, G. Biener, T.-W. Su, A.F. Coskun, D. Tseng, and A. Ozcan. 2010b. Lensfree on-chip imaging using nanostructured surfaces. *Applied Physics Letters* 96: 171106.

Kim, S.-J., K. Koh, M. Lustig, S. Boyd, and D. Gorinevsky. 2007. An interior-point method for large-scale ℓ_1-regularized least squares. *IEEE Journal of Selected Topics in Signal Processing* 1(4): 606–617.

Lehner, B., C. Crombie, J. Tischler, A. Fortunato, and A.G. Fraser. 2006. Systematic mapping of genetic interactions in *Caenorhabditis elegans* identifies common modifiers of diverse signaling pathways. *Nature Genetics* 38(8): 896–903.

Llull, P., X. Liao, X. Yuan, J. Yang, D. Kittle, L. Carin, G. Sapiro, and D.J. Brady. 2013. Coded aperture compressive temporal imaging. *Optics Express* 21(9): 10526.

Luo, W., A. Greenbaum, Y. Zhang, and A. Ozcan. 2015. Synthetic aperture-based on-chip microscopy. *Light: Science & Applications* 4(3): e261.

McLeod, E., W. Luo, O. Mudanyali, A. Greenbaum, and A. Ozcan. 2013. Toward giga-pixel nanoscopy on a chip: A computational wide-field look at the nano-scale without the use of lenses. *Lab on a Chip* 13(11): 2028–2035.

Mellem, J.E., P.J. Brockie, D.M. Madsen, and A.V. Maricq. 2008. Action potentials contribute to neuronal signaling in *C. Elegans*. *Nature Neuroscience* 11(8): 865–867.

Mudanyali, O., E. McLeod, W. Luo, A. Greenbaum, A.F. Coskun, Y. Hennequin, C.P. Allier, and A. Ozcan. 2013. Wide-field optical detection of nanoparticles using on-chip microscopy and self-assembled nanolenses. *Nature Photonics* 7(3): 247–254.

Mudanyali, O., D. Tseng, C. Oh, S.O. Isikman, I. Sencan, W. Bishara, C. Oztoprak, S. Seo, B. Khademhosseini, and A. Ozcan. 2010. Compact, light-weight and cost-effective microscope based on lensless incoherent holography for telemedicine applications. *Lab on a Chip* 10(11): 1417–1428.

Neifeld, M.A. 1998. Information, resolution, and space-bandwidth product. *Optics Letters* 23(18): 1477.

Ozcan, A. 2014. Democratization of diagnostics and measurement tools through computational imaging and sensing, in *Imaging and Applied Optics 2014*, OSA Technical Digest (online), paper IM1C.1. doi: 10.1364/ISA.2014.IM1C.1.

Pinkston-Gosse, J. and C. Kenyon. 2007. DAF-16/FOXO targets genes that regulate tumor growth in *Caenorhabditis elegans*. *Nature Genetics* 39(11): 1403–1409.

Rohde, C.B., F. Zeng, R. Gonzalez-Rubio, M. Angel, and M.F. Yanik. 2007. Microfluidic system for on-chip high-throughput whole-animal sorting and screening at subcellular resolution. *Proceedings of the National Academy of Sciences of the United States of America* 104(35): 13891–13895.

Rust, M.J., M. Bates, and X. Zhuang. 2006. Sub-diffraction-limit imaging by stochastic optical reconstruction microscopy (STORM). *Nature Methods* 3(10): 793–796.

Sencan, I. 2013. Lensfree computational microscopy tools and their biomedical applications. Dissertation, University of California, Los Angeles, CA. http://gradworks.umi.com/35/94/3594279.html, accessed July 12, 2015.

Sencan, I., A.F. Coskun, U. Sikora, and A. Ozcan. 2014. Spectral demultiplexing in holographic and fluorescent on-chip microscopy. *Scientific Reports* (article no. 3760): 1–9.

Seo, S., S.O. Isikman, I. Sencan, O. Mudanyali, T.-W. Su, W. Bishara, A. Erlinger, and A. Ozcan. 2010. High-throughput lens-free blood analysis on a chip. *Analytical Chemistry* 82(11): 4621–4627.

Strange, K., ed. 2006. Techniques for analysis, sorting, and dispensing of *C. elegans* on the COPAS™ flow-sorting system. *Methods in Molecular Biology* 351: 275–286.

Su, T.-W., I. Choi, J. Feng, K. Huang, E. McLeod, and A. Ozcan. 2013. Sperm trajectories form chiral ribbons. *Scientific Reports* 3 (April): 1664.

Su, T.-W., A. Erlinger, D. Tseng, and A. Ozcan. 2010. Compact and light-weight automated semen analysis platform using lensfree on-chip microscopy. *Analytical Chemistry* 82(19): 8307–8312.

Tseng, D., O. Mudanyali, C. Oztoprak, S.O. Isikman, I. Sencan, O. Yaglidere, and A. Ozcan. 2010. Lensfree microscopy on a cellphone. *Lab on a Chip* 10(14): 1787.

Zhu, H., I. Sencan, J. Wong, S. Dimitrov, D. Tseng, K. Nagashima, and A. Ozcan. 2013. Cost-effective and rapid blood analysis on a cell-phone. *Lab on a Chip* 13(7): 1282–1288.

V
Phase Retrieval

13 Phase Retrieval

An Overview of Recent Developments

Kishore Jaganathan,
Yonina C. Eldar, and Babak Hassibi

Contents

Introduction

In many physical measurement systems, one can only measure the power spectral density, that is, the magnitude square of the Fourier transform of the underlying signal. For example, in an optical setting, detection devices like CCD cameras and photosensitive films cannot measure the phase of a light wave and instead measure the photon flux. In addition, at a large enough distance from the imaging plane the field is given by the Fourier transform of the image (up to a known phase factor). Thus, in the far field, optical devices essentially measure the Fourier transform magnitude. Since the phase encodes a lot of the structural content of the image, important information is lost. The problem of reconstructing a signal from its Fourier magnitude is known as phase retrieval [1,2]. This reconstruction problem is one with a rich history and arises in many areas of engineering and applied physics, including optics [3], x-ray crystallography [4], astronomical imaging [5], speech processing [6], computational biology [7], and blind deconvolution [8].

Reconstructing a signal from its Fourier magnitude alone is generally a very difficult task. It is well known that Fourier phase is quite often more important than Fourier magnitude in reconstructing a signal from its Fourier transform [9]. To demonstrate this fact, a synthetic example, courtesy of Shechtman et al. [10], is provided in Figure 13.1. The figure shows the result of the following numerical simulation: two images are Fourier transformed, their Fourier phases are swapped, and then they are inverse Fourier transformed. The result clearly demonstrates the importance of Fourier phase. Therefore, simply ignoring the phase and performing an inverse Fourier transform does not lead to satisfactory recovery. Instead, algorithmic phase retrieval can be used, offering a means for recovering the phase from the given magnitude measurements and possibly additional prior knowledge, providing an alternative to sophisticated measurement setups as in holography, which attempt to directly measure the phase by requiring interference with another known field.

To set up the phase retrieval problem mathematically, we focus on the discretized 1D setting. Let $\mathbf{x} = (x[0], x[1], \ldots, x[N-1])^T$ be a signal of length N such that it has nonzero values only within the interval $[0, N-1]$. Denote by $\mathbf{y} = (y[0], y[1], \ldots, y[N-1])^T$ its N point discrete Fourier transform (DFT) and let $\mathbf{z} = (z[0],$

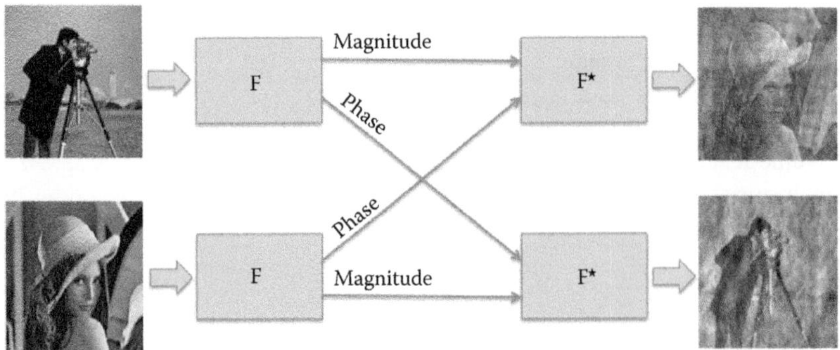

Figure 13.1 A synthetic example demonstrating the importance of Fourier phase in reconstructing a signal from its Fourier transform. (Courtesy of Shechtman, Y. et al., *IEEE Signal Process. Mag.*, 32(3), 87, 2015.)

$z[1], \ldots, z[N-1])^T$ be the Fourier magnitude-square measurements $z[m]=|y[m]|^2$. Phase retrieval can be mathematically stated as

$$\begin{aligned} &\text{find} && \mathbf{x} \\ &\text{subject to} && z[m]=\left|\langle \mathbf{f}_m, \mathbf{x}\rangle\right|^2 \quad \text{for } 0 \le m \le N-1, \end{aligned} \tag{13.1}$$

where

$\mathbf{f}_m$ is the conjugate of the mth column of the N point DFT matrix, with elements $e^{i2\pi(mn/N)}$

$\langle \cdot,\cdot \rangle$ is the inner product operator

Classic Approaches

For any given Fourier magnitude, the Fourier phase can be chosen from an N-dimensional set. Since distinct phases correspond to different signals in general, the feasible set of Equation 13.1 is an N-dimensional manifold, rendering phase retrieval a very ill-posed problem. One approach to try and overcome the ill-posedness is to employ oversampling by using an $M>N$ point DFT (see Fienup [11] for a comprehensive survey). A typical choice is $M=2N$. The term *oversampling* is used in this chapter to refer to measurements from the $M=2N$ point DFT.

Phase retrieval with oversampling can be equivalently stated as the problem of reconstructing a signal from its autocorrelation measurements $\mathbf{a}=(a[0], a[1], \ldots, a[N-1])^T$, that is,

$$\begin{aligned} &\text{find} && \mathbf{x} \\ &\text{subject to} && a[m]=\sum_{j=0}^{N-1-m} x[j]x^*[m+j] \quad \text{for } 0 \le m \le N-1. \end{aligned} \tag{13.2}$$

This is because the length $M=2N$ DFT of $\mathbf{a}$ is given by $\mathbf{z}$.

Observe that the operations of time shift, conjugate flip, and global phase change on the signal do not affect the autocorrelation, because of which there are trivial ambiguities. Signals obtained by these operations are considered equivalent, and in most applications, it is good enough if any equivalent signal is recovered. For example, in astronomy, where the underlying signal corresponds to stars in the sky, or in x-ray crystallography, where the underlying signal corresponds to atoms or molecules in a crystal, equivalent solutions are equally informative [4,5].

In the 1D setup, it has been shown, using spectral factorization, that there is no uniqueness and the feasible set of Equation 13.2 can include up to 2^N nonequivalent solutions [12]. While this is a significant improvement (when compared to the feasible set of Equation 13.1), 2^N is still a prohibitive number because of which phase retrieval with oversampling remains ill-posed. Furthermore, adding support constraints on $\mathbf{x}$ does not help to ensure uniqueness. However, for higher dimensions (2D and above), it has been shown by Hayes [13] using dimension counting that, with the exception of a set of signals of measure zero, phase retrieval with oversampling is well posed up to trivial ambiguities.

As we discuss further in the following, to guarantee unique identification in the 1D case, it is necessary to assume (or enforce) additional constraints on the unknown signal such as sparsity or to introduce specific redundancy into the measurements. Even when unique identification of the underlying signal is theoretically possible, it is unclear how to find the unique solution efficiently and robustly. Earlier approaches to phase retrieval were based on alternating projections, pioneered by the work of Gerchberg and Saxton [14]. In this framework, phase retrieval is reformulated as the following least-squares problem:

$$\min_{\mathbf{x}} \sum_{m=0}^{M-1} \left(z[m] - \left|\langle \mathbf{f}_m, \mathbf{x} \rangle\right|^2 \right)^2. \tag{13.3}$$

The Gerchberg–Saxton (GS) algorithm attempts to minimize this nonconvex objective by starting with a random initialization and iteratively imposing the time-domain (support) and Fourier magnitude constraints using projections. The details of the various steps are provided in Algorithm 13.1. The objective is shown to be monotonically decreasing as the iterations progress. However, since the projections are between a convex set (for the time-domain constraints) and a nonconvex set (for the Fourier magnitude constraints), the convergence is often to a local minimum, due to which the algorithm has limited recovery abilities even in the noiseless setting.

Algorithm 13.1 Gerchberg–Saxton (GS) Algorithm

Input: Fourier magnitude-square measurements $\mathbf{z}$

Output: Estimate $\hat{\mathbf{x}}$ of the underlying signal

Initialize: Choose a random input signal $\mathbf{x}^{(0)}$, $\ell = 0$

while halting criterion false **do**

  $\ell \leftarrow \ell + 1$

  Compute the DFT of $\mathbf{x}^{(\ell-1)}$: $\mathbf{y}^{(\ell)} = \mathbf{F}\mathbf{x}^{(\ell-1)}$

  Impose Fourier magnitude constraints: $y'^{(\ell)}[m] = \dfrac{y^{(\ell)}[m]}{\left|y^{(\ell)}[m]\right|} \sqrt{z[m]}$

  Compute the inverse DFT of $\mathbf{y}'^{(\ell)}$: $\mathbf{x}'^{(\ell)} = \mathbf{F}^{-1}\mathbf{y}'^{(\ell)}$

  Impose time-domain constraints to obtain $\mathbf{x}^{(\ell)}$

end while

return $\hat{\mathbf{x}} \leftarrow \mathbf{x}^{(\ell)}$

Fienup, in his seminal work [15], extended the alternating projection framework by adding additional time-domain restrictions and introducing several variants to the time-domain projection step. The most popular of these is the hybrid input-output (HIO) algorithm. While the basic framework of the HIO technique is similar to the GS method, the former includes an additional time-domain *correction step* in order to improve convergence (see Fienup [15] for details). The HIO algorithm is not guaranteed to converge, and when it does converge, it may be to a local minimum. Nonetheless, HIO methods and their variants are often used in optical phase retrieval. We refer the readers to Bauschke et al. [16] and Marchesini [17] for a theoretical and numerical investigation of these techniques.

Recent Approaches

Recently, phase retrieval has benefited from a surge of research in both the optics and the mathematical communities due to various new imaging techniques in optics, and advances in modern optimization tools and structured-based information processing [18–22]. For example, one imaging modality that has gained large interest in the past 15 years is coherent diffractive imaging (CDI) [23] in which an object is illuminated with a coherent wave and the far-field diffraction intensity pattern is measured. Phase retrieval algorithms are a key component in enabling CDI imaging. On the algorithmic side, as we detail further in the following, ideas of semidefinite relaxation and sparsity have played a key role in modern phase retrieval.

Recent approaches to phase retrieval can be broadly classified into two categories:

1. *Additional prior information*: Inspired by advances in the field of compressed sensing [19–22], various researchers have explored the idea of sparsity as a prior information on the signal [24–34]. A signal of length N is said to be k-sparse if it has k locations with nonzero values and $k \ll N$. The exact locations and values of the nonzero elements are not known a priori. The approach to sparse phase retrieval has been to develop conditions under which only one sparse signal satisfies the autocorrelation measurements and to suggest algorithms that exploit the sparsity to improve convergence and robustness to noise.

2. *Additional magnitude-only measurements*: Technological advances have enabled the possibility of obtaining more information about the signal via additional magnitude-only measurements. This can be done in various ways depending on the application; common approaches include the use of masks [35,36], optical gratings [37], oblique illuminations [38], and short-time Fourier transform (STFT) magnitude measurements that utilize overlap between adjacent short-time sections [39–44]. An important line of research is to identify the necessary additional magnitude-only measurements needed in order to render the recovery unique, efficient, and robust.

Another popular trend in the phase retrieval literature for analysis purposes is to replace the Fourier measurements with random measurements so that $z[m] = |\langle \mathbf{v}_m, \mathbf{x} \rangle|^2$ for random vectors $\mathbf{v}_m$. This allows to derive uniqueness and recovery guarantees more easily than for Fourier measurements [45–53]. Since this review focuses on Fourier phase retrieval that naturally arises in optics, astronomy, etc., we do not pursue this line of work here.

On the algorithmic front, one of the recent popular approaches to treat phase retrieval problems of both categories mentioned earlier is to use semidefinite programming (SDP) methods. SDP algorithms have been shown to yield robust solutions to various quadratic-constrained optimization problems (see References [53–56] and references therein). Since phase retrieval results in quadratic constraints, it is natural to use SDP techniques to try and solve such problems [25–27]. An SDP formulation of phase retrieval (Equation 13.1) can be obtained by a procedure popularly known as *lifting*—we embed $\mathbf{x}$ in a higher-dimensional space using the transformation $\mathbf{X} = \mathbf{x}\mathbf{x}^*$. The Fourier magnitude measurements are then linear in the matrix $\mathbf{X}$:

$$z[m] = \left|\langle \mathbf{f}_m, \mathbf{x} \rangle\right|^2 = \mathbf{x}^* \mathbf{f}_m \mathbf{f}_m^* \mathbf{x} = \operatorname{trace}\left(\mathbf{f}_m \mathbf{f}_m^* \mathbf{x}\mathbf{x}^*\right) = \operatorname{trace}\left(\mathbf{f}_m \mathbf{f}_m^* \mathbf{X}\right).$$

Consequently, phase retrieval reduces to finding a rank-one positive semidefinite matrix $\mathbf{X}$, which satisfies these affine measurement constraints, leading to the following reformulation:

$$\begin{aligned} &\text{Minimize} && \operatorname{rank}(\mathbf{X}) \\ &\text{Subject to} && z[m] = \operatorname{trace}\left(\mathbf{f}_m \mathbf{f}_m^* \mathbf{X}\right) \quad \text{for } 0 \le m \le N-1 \\ &&& \mathbf{X} \succeq 0. \end{aligned}$$

Unfortunately, rank($\mathbf{X}$) is a nonconvex function of $\mathbf{X}$. To obtain a convex program, one possibility is to replace it by the convex surrogate trace($\mathbf{X}$) [22,57], resulting in the convex SDP

$$\begin{aligned} &\text{Minimize} && \operatorname{trace}(\mathbf{X}) \\ &\text{Subject to} && z[m] = \operatorname{trace}\left(\mathbf{f}_m \mathbf{f}_m^* \mathbf{X}\right) \quad \text{for } 0 \le m \le N-1 \\ &&& \mathbf{X} \succeq 0. \end{aligned} \tag{13.4}$$

This approach is referred to as PhaseLift [25].

SDP algorithms are known to be robust in general. However, due to the high-dimensional transformation involved, they can be computationally demanding. A recent alternative is to use gradient-based techniques with appropriate initialization. For the random measurement case, an alternating minimization algorithm is proposed in Netrapalli et al. [52], and a nonconvex algorithm based on Wirtinger flow (WF) is suggested in Candes et al. [58]. Beyond these general-purpose techniques, efficient phase retrieval algorithms have been developed, which take advantage of sparsity [59] and specific magnitude-only measurements. In particular, clever mask designs allow for recovery using simple combinatorial algorithms [60–63].

The purpose of this chapter is to provide an overview of some of the recent developments, both theoretical and algorithmic, in phase retrieval. In particular, we focus on sparse phase retrieval, the use of masks, and recovery from the STFT magnitude. We refer the readers to the recent overview [10], which contains applications to optics as well as a more comprehensive review of the history of phase retrieval and some of the algorithmic aspects.

The rest of this chapter is organized as follows. In the "Sparse Phase Retrieval" section, we motivate sparse phase retrieval, which is the problem of reconstructing a sparse signal from its Fourier magnitude. The "Phase Retrieval Using Masks" section considers phase retrieval using masks. In the "STFT Phase Retrieval" section, we study STFT phase retrieval in which the measurements correspond to the STFT magnitude. For each of these three problems, we provide a literature survey of the available uniqueness guarantees and describe various recovery algorithms. This chapter ends with the "Conclusions" section.

Sparse Phase Retrieval

In many applications of phase retrieval, the underlying signals are naturally sparse due to the physical nature of the setup. For instance, electron microscopy deals with sparsely distributed atoms or molecules [4], while astronomical imaging tends to consider sparsely distributed stars [5]. More generally, a sparsity prior corresponds to having advance knowledge that the unknown signal has some characteristic structure, or, equivalently, that it has a small number of degrees of freedom. The simplest setting is when the basis in which the object is represented compactly is known in advance. This basis is referred to as the sparsity basis, or dictionary. When such a dictionary is not given a priori, it can often be learned from the measurements themselves [64] or from data with similar features that may be available from other sources [19]. Sparsity priors have been used extensively in many fields of engineering and statistics and are known to model well various classes of images and signals. Recently, the use of sparsity has also become popular in optical applications including holography [65], superresolution and subwavelength imaging [66–72], ankylography [73,74] and tomography [75].

If it is known a priori that the signal of interest is sparse, then one could potentially solve for the sparsest solution satisfying the Fourier magnitude measurements. Our problem can then be written as

$$\begin{aligned} &\text{Minimize} && \|\mathbf{x}\|_0 \\ &\text{Subject to} && z[m] = |\langle \mathbf{f}_m, \mathbf{x}\rangle|^2 \quad \text{for } 0 \le m \le M-1, \end{aligned} \tag{13.5}$$

where $\|\cdot\|_0$ is the ℓ_0 norm that counts the number of nonzero entries of its argument. When $M=2N$, sparse phase retrieval is equivalent to the problem of reconstructing a sparse signal from its autocorrelation.

Uniqueness

We begin by reviewing existing uniqueness results for sparse phase retrieval (Equation 13.5). These results are summarized in Table 13.1. Since the operations of time shift, conjugate flip, and global phase change on the signal do not affect its sparsity, similar to standard phase retrieval, recovery is possible only up to these trivial ambiguities.

A general uniqueness result is derived in Jaganathan et al. [26], where it is shown, using dimension counting, that most sparse signals with *aperiodic support* can be uniquely identified from their autocorrelation. A signal is said to have periodic or aperiodic support if the locations of its nonzero components are uniformly spaced or not uniformly spaced, respectively. As an example, consider the signal $x=(x[0], x[1], x[2], x[3], x[4])^T$ of length $N=5$.

1. Possible aperiodic supports: $\{n \mid x[n] \neq 0\} = \{0, 1, 3\}, \{1, 2, 4\}$.

2. Possible periodic supports: $\{n \mid x[n] \neq 0\} = \{0, 2, 4\}, \{0, 1, 2, 3, 4\}$.

Consequently, if the signal of interest is known to have aperiodic support, then sparse phase retrieval is *almost surely* well posed. Note, however, that this still does not provide an efficient robust method for finding the sparse input. An

Table 13.1 Uniqueness Results for Sparse Phase Retrieval (with $M \geq 2N$ Point DFT)

1D signals	Almost surely uniqueness for signals with aperiodic support [26]
	Uniqueness for signals with collision-free autocorrelation (and $k \neq 6$) [33], k^2-k+1 Fourier magnitude measurements are sufficient [34]
≥2D signals	Almost surely uniqueness [13]
	Uniqueness for signals with collision-free autocorrelation [33]

explanation as to why recovery of a sparse signal with periodic support is generally not possible is provided in Lu and Vetterli [31]. Specifically, it is argued that a sparse signal with periodic support can be viewed as an upsampled version of a nonsparse signal. Sparse phase retrieval in this case is then seen to be equivalent to phase retrieval, because of which most such signals cannot be uniquely identified from their autocorrelation.

In Ranieri et al. [33], it is shown that the knowledge of the autocorrelation is sufficient to uniquely identify 1D sparse signals if the autocorrelation is *collision-free* as long as the sparsity $k \neq 6$. A signal $\mathbf{x}$ is said to have a collision-free autocorrelation if for all indices $\{i_1, i_2, i_3, i_4\}$ such that $\{x[i_1], x[i_2], x[i_3], x[i_4]\} \neq 0$, we have $|i_1 - i_2| \neq |i_3 - i_4|$. In words, a signal is said to have a collision-free autocorrelation if no two pairs of locations with nonzero values in the signal are separated by the same distance. For higher dimensions, the authors show that the requirement $k \neq 6$ is not necessary. This result has been further refined in Ohlsson and Eldar [34], where it is shown that $k^2 - k + 1$ Fourier magnitude measurements are sufficient to recover the autocorrelation.

Algorithms

Phase retrieval algorithms based on alternating projections and SDP have been adapted to solve sparse phase retrieval. In this section, we first describe these two adaptations, explain their limitations, and then describe two powerful sparse phase retrieval algorithms: two-stage sparse phase retrieval (TSPR) [28] and GrEedy Sparse PhAse Retrieval (GESPAR) [30]. TSPR can *provably* recover most $O(N^{1/2-\varepsilon})$-sparse signals up to the trivial ambiguities. Further, for most $O(N^{1/4-\varepsilon})$-sparse signals, the recovery is robust in the presence of noise. GESPAR has been shown to yield fast and accurate recovery results and has been used in several phase retrieval optics applications [67,70–72,74].

1. *Alternating projections*: The Fienup HIO algorithm has been extended to solve sparse phase retrieval by adapting the step involving time-domain constraints to promote sparsity. This can be achieved in several ways. For example, the locations with absolute values less than a particular threshold may be set to zero. Alternatively, the k locations with the highest absolute values can be retained and the rest set to zero [32]. In the noiseless setting, the sparsity constraint partially alleviates the convergence issues if multiple random initializations are considered and the underlying signals are sufficiently sparse. However, in the noisy setting, convergence issues still remain.

2. *SDP-based methods*: It is well known that ℓ_1-minimization, which is a convex surrogate for ℓ_0-minimization, promotes sparse solutions [21]. Hence, a natural convex program to solve sparse phase retrieval is

$$\begin{aligned} &\text{Minimize} && \operatorname{trace}(\mathbf{X}) + \lambda \|\mathbf{X}\|_1 \\ &\text{Subject to} && z[m] = \operatorname{trace}\left(\mathbf{f}_m \mathbf{f}_m^* \mathbf{X}\right) \quad \text{for } 0 \le m \le M-1 \\ & && \mathbf{X} \succeq 0, \end{aligned} \tag{13.6}$$

for some regularizer $\lambda > 0$. While this approach has enjoyed considerable success in related problems like generalized phase retrieval for sparse signals [48,50,51], it often fails to solve sparse phase retrieval.

This does not come as a surprise as the issue of trivial ambiguities (due to time shift and conjugate flip) is still unresolved. If $\mathbf{X}_0 = \mathbf{x}_0\mathbf{x}_0^*$, is the desired sparse solution, then $\tilde{\mathbf{X}}_0 = \tilde{\mathbf{x}}_0\tilde{\mathbf{x}}_0^*$, where $\tilde{\mathbf{x}}_0$ is the conjugate-flipped version of $\mathbf{x}_0$, $\mathbf{X}_j = \mathbf{x}_j\mathbf{x}_j^*$, where $\mathbf{x}_j$ is the signal obtained by time-shifting x_0 by j units, and $\tilde{\mathbf{X}}_j = \tilde{\mathbf{x}}_j\tilde{\mathbf{x}}_j^*$, where $\tilde{\mathbf{x}}_j$ is the signal obtained by time-shifting $\tilde{\mathbf{x}}_0$ by j units are also feasible with the same objective value as $\mathbf{X}_0$. Since Equation 13.6 is a convex program, any convex combination of these solutions is feasible as well and has an objective value less than or equal to that of $\mathbf{X}_0$, because of which the optimizer is neither sparse nor rank one. Many iterative heuristics have been proposed to break this symmetry [25,27,29]. While these heuristics enjoy empirical success, they are computationally expensive due to the fact that each iteration involves the solution of a high-dimensional convex program and the number of iterations required for convergence can be high.

3. *TSPR*: The convex program (Equation 13.6) fails to solve sparse phase retrieval primarily because of issues due to trivial ambiguities, which stem from the fact that the support of the signal is unknown. In order to overcome this, TSPR is proposed in Jaganathan et al. [28], which involves (a) estimating the support of the signal using an efficient algorithm and (b) estimating the signal values *with known support* using the convex program (Equation 13.6).

   The first stage of TSPR, which involves recovery of the support of the signal (denoted by V) from the support of the autocorrelation (denoted by B), is equivalent to that of recovering an integer set from its pairwise distances (also known as the *Turnpike problem* [76–78]). For example, consider the integer set $V = \{2, 5, 13, 31, 44\}$. Its pairwise distance set is given by $B = \{0, 3, 8, 11, 13, 18, 26, 29, 31, 39, 42\}$. The Turnpike problem is to reconstruct V from B. In Jaganathan et al. [28], it is argued that due to trivial ambiguities, without loss of generality, one can always construct a solution $U = \{u_0, u_1, \ldots, u_{k-1}\}$, which is a subset of B. TSPR, in essence, eliminates all the integers in B that do not belong to U using two instances of the *intersection step* and one instance of the *graph step*. These steps are explained in Jaganathan et al. [28]. As to the second step, even though is involves lifting, the dimension of the problem is reduced from N to k due to knowledge of the support from the first stage. In the noisy setting, TSPR considers the pairwise distance set of B, along with a series of *generalized intersection steps*, to ensure stable recovery. TSPR provably recovers most $O(N^{1/2-\varepsilon})$-sparse signals efficiently and most $O(N^{1/4-\varepsilon})$-signals robustly. A brief overview of the steps involved is provided in Algorithm 13.2. We refer the interested readers to Jaganathan et al. [28] for further details.

Algorithm 13.2 TSPR (see Jaganathan et al. [28] for noisy setting)

Input: Autocorrelation measurements $\mathbf{a}$

Output: Sparse estimate $\hat{\mathbf{x}}$ of the underlying sparse signal

1. Obtain $B = \{n \mid a[n] \neq 0\}$
2. Infer $\{u_0, u_1, u_{k-1}\}$ from B
3. Intersection step using u_1: obtain $Z = 0 \cup (B \cap (B + u_1))$

4. Graph step using (Z, B): obtain $\{u_2, u_3, \ldots, u_t\}$ for $t = \sqrt[3]{\log(k)}$
5. Intersection step using $\{u_2, u_3, \ldots, u_t)$ obtain U
6. Obtain $\hat{\mathbf{X}}$ by solving

$$\begin{aligned} \text{Minimize} \quad & \text{trace}(\mathbf{X}) \\ \text{Subject to} \quad & z[m] = \text{trace}\left(\mathbf{f}_m \mathbf{f}_m^* \mathbf{X}\right) \quad \text{for } 0 \le m \le M-1 \\ & X[n_1, n_2] = 0 \quad \text{if } \{n_1, n_2\} \notin U \\ & \mathbf{X} \ge 0 \end{aligned} \tag{13.7}$$

7. Return $\hat{\mathbf{x}}$, where $\hat{\mathbf{x}}\hat{\mathbf{x}}^*$ is the best rank-one approximation of $\hat{\mathbf{X}}$

4. *GESPAR*: In Shechtman et al. [30], a sparse optimization-based greedy search method called GESPAR is proposed. Sparse phase retrieval is reformulated as the following sparsity-constrained least-squares problem:

$$\begin{aligned} \underset{\mathbf{x}}{\text{Min}} \quad & \sum_{m=0}^{M-1} (z[m] - |\langle \mathbf{f}_m, \mathbf{x} \rangle|^2)^2 \\ \text{Subject to} \quad & \|\mathbf{x}\|_0 \le k. \end{aligned} \tag{13.8}$$

GESPAR is a local search method, based on iteratively updating the signal support, and seeking a vector that corresponds to the measurements under the current support. A location-search method is repeatedly invoked, beginning with an initial random support set. Then, at each iteration, a swap is performed between a support and a nonsupport index. Only two elements are changed in the swap (one in the support and one in the nonsupport), following the so-called 2-opt method [79]. Given the support of the signal, phase retrieval is then treated as a nonconvex optimization problem and approximated using the damped Gauss–Newton method [80].

GESPAR has been used in several phase retrieval optics applications, including CDI of 1D objects [67,71], efficient CDI of sparsely temporally varying objects [70], and phase retrieval via waveguide arrays [72]. A brief overview of the steps involved is provided in Algorithm 13.3. We refer the interested readers to Shechtman et al. [30] for further details.

Algorithm 13.3 GESPAR (see Shechtman et al. [30] for details)

Input: Autocorrelation measurements $\mathbf{a}$, parameters τ and ITER

Output: Sparse estimate $\hat{\mathbf{x}}$ of the underlying sparse signal

Initialization: Set $T=0$ and $j=0$

1. Generate a random support set $S^{(0)}$ of size k.
2. Invoke the damped Gauss–Newton method with support $S^{(0)}$, and obtain $\mathbf{x}^{(0)}$

General Step ($j = 1, 2, \ldots$):

3. Update support: Let p be the index from $S^{(j-1)}$ corresponding to the component of $\mathbf{x}^{(j-1)}$ with the smallest absolute value and q be the index from the

complement of $S^{(j-1)}$ corresponding to the component of $\nabla f(\mathbf{x}^{(j-1)})$ with the highest absolute value, where $\nabla f(\mathbf{x})$ is the gradient of the least-squares objective function (Equation 13.8). Increase T by 1, and make a swap between the indices p and q, i.e., $S' = (S^{(j-1)} \backslash \{p\}) \cup \{q\}$

4. Minimize with given support: Invoke the damped Gauss–Newton method with support S', and obtain $\mathbf{x}'$

   If $f(\mathbf{x}') < f(\mathbf{x}^{(j-1)})$, then set $S^{(j)} = S'$, $\mathbf{x}^{(j)} = \mathbf{x}'$, advance j and go to Step 3. If none of the swaps resulted with a better objective function value, go to Step 1

Until $f(x) < \tau$ or $T >$ ITER.

Return $\hat{\mathbf{x}} \leftarrow \mathbf{x}^{(t)}$

Numerical Simulations

In this section, we demonstrate the performance of TSPR and GESPAR using numerical simulations.

We compare the recovery ability of the Fienup HIO algorithm, TSPR, and GESPAR in the noiseless setting. We consider sparse signals of length $N = 6400$ for varying sparsities $20 \le k \le 90$. For each sparsity, 100 trials are performed. The locations with nonzero values are chosen uniformly at random and the values in the nonzero locations are chosen from an i.i.d. standard normal distribution. Fienup HIO algorithm is run with 200 random initializations, GESPAR is run with parameters $\tau = 10^{-4}$ and $ITER = 10{,}000$. The probability of successful recovery of these algorithms is plotted versus sparsity in Figure 13.2a. It can be seen that GESPAR has the best empirical performance among the three algorithms: GESPAR recovers signals with sparsities up to 57 while TSPR recovers signals with sparsities up to 53. Both GESPAR and TSPR significantly outperform the Fienup HIO algorithm.

In Figure 13.2b, the probability of successful recovery of TSPR for $N = \{12{,}500, 25{,}000, 50{,}000\}$ is plotted versus sparsity. The $O(N^{1/2-\varepsilon})$ theoretical guarantee of TSPR is clearly verified empirically. For example, the choices $N = 12{,}500$, $k = 80$ and $N = 50{,}000$, $k = 160$ have a success probability of 0.5.

Phase Retrieval Using Masks

The use of masked magnitude-only measurements in order to resolve the Fourier phase uniquely has been explored by various researchers. The key idea, in a nutshell, is to obtain additional information about the signal using multiple masks in order to mitigate the uniqueness and algorithmic issues of phase retrieval. There are many ways in which this can be performed in practice, depending on the application. Several such methods are summarized in Candes et al. [25]: (1) *Masking*: the phase front after the sample is modified by the use of a mask or a phase plate [35,36]. A schematic representation, courtesy of Candes et al. [56], is provided in Figure 13.3. (2) *Optical grating*: the illuminating beam is modulated by the use of optical gratings [37], using a setup similar to Figure 13.3. (3) *Oblique illuminations*: the illuminating beam is modulated to hit the sample at specific angles [38].

Suppose Fourier magnitude-square measurements are collected using R masks (or modulated illuminations). For $0 \le r \le R-1$, let $\mathbf{D}_r$ be an $N \times N$ diagonal matrix, corresponding to the rth mask or modulated illumination, with diagonal entries

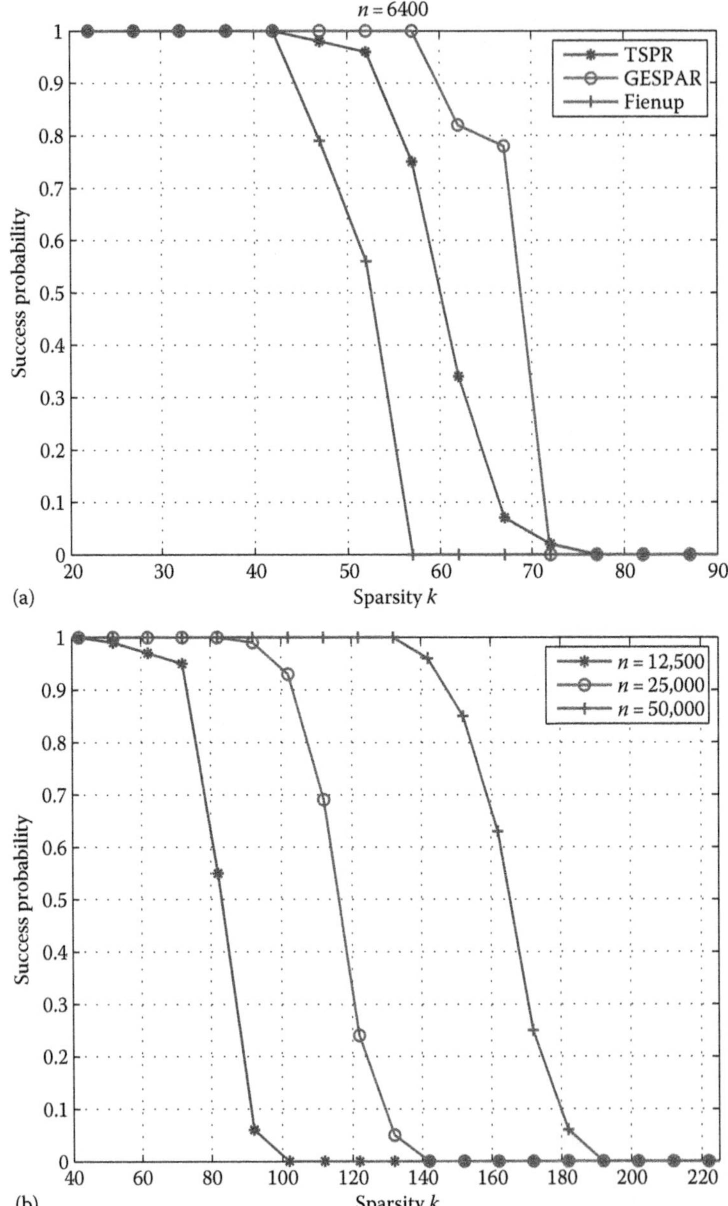

Figure 13.2 Performance of various sparse phase retrieval algorithms. (a) Probability of successful recovery for $N = 6400$ and varying choices of k. (b) Probability of successful recovery of TSPR for varying choices of N and k. (Courtesy of Jaganathan, K. et al., Sparse phase retrieval: Uniqueness guarantees and recovery algorithms, arXiv:1311.2745, 2015; Courtesy of Shechtman, Y. et al., *IEEE Trans. Signal Process.*, 62(4), 928, 2014.)

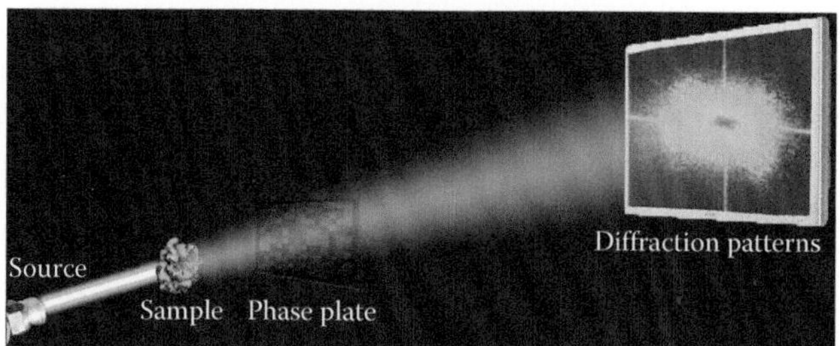

Figure 13.3 A typical setup for phase retrieval using masks or modulated illuminations. (Reprinted from *Applied and Computational Harmonic Analysis*, 39(2), Candes, E.J., Li, X., and Soltanolkotabi, M., Phase retrieval from coded diffraction patterns, 277–299, Copyright 2015, with permission from Elsevier.)

$(d_r[0], d_r[1], \ldots, d_r[N-1])$. Let $\mathbf{Z}$ denote the $N \times R$ magnitude-square measurements, such that the rth column of $\mathbf{Z}$ corresponds to the magnitude square of the N point DFT of the masked signal $\mathbf{D}_r\mathbf{x}$. Phase retrieval using masks then reduces to the following reconstruction problem:

$$\begin{aligned} &\text{find} && \mathbf{x} \\ &\text{subject to} && Z[m,r] = |\langle \mathrm{f}_m, \mathbf{D}_r\mathbf{x}\rangle|^2 \quad \text{for } 0 \le m \le N-1 \text{ and } 0 \le r \le R-1. \end{aligned} \tag{13.9}$$

A natural question to ask is how many masks are needed in order to guarantee uniqueness. While this question, in its full generality, is still open, results are available when it comes to uniquely identifying most signals. The best-known result is due to Jaganathan et al. [60], where it is shown that two generic masks are sufficient to uniquely identify most signals up to global phase if an $M \ge 2N$ point DFT is considered. While this is a strong identifiability result, it is unclear how to efficiently and robustly recover the underlying signal from these measurements. Another natural question to ask, hence, is how many and which masks allow for unique, efficient, and robust recovery of the underlying signal.

Uniqueness and Algorithms

Phase retrieval algorithms based on SDP and stochastic gradient descent (WF algorithm [58]) have been adapted to solve phase retrieval using masks uniquely, efficiently, and robustly for some choices of masks. Combinatorial algorithms have also been developed for specific mask designs, which allow for unique and efficient reconstruction in the noiseless setting. In the following, we survey the main algorithms proposed along with the corresponding choice of masks.

SDP Methods

SDP-based phase retrieval has been adapted to account for masks in Candes et al. [56] and Jaganathan et al. [60] by solving

$$\begin{aligned} &\text{Minimize} && \operatorname{trace}(\mathbf{X}) \\ &\text{Subject to} && Z[m,r] = \operatorname{trace}\left(\mathbf{D}_r^*\mathbf{f}_m\mathbf{f}_m^*,\mathbf{D}_r\mathbf{X}\right) \quad \text{for } 0 \le m \le N-1 \text{ and } 0 \le r \le R-1 \\ &&& \mathbf{X} \succeq 0. \end{aligned} \tag{13.10}$$

In order to provide recovery guarantees, the masks in Candes et al. [56] are chosen from a random model. In particular, the diagonal matrices $\mathbf{D}_r$ are assumed to be i.i.d. copies of a matrix $\mathbf{D}$, whose entries consist of i.i.d. copies of a random variable d satisfying the following properties:

$$\mathbb{E}[d]=0 \quad \mathbb{E}[d^2]=0 \quad \mathbb{E}|d|^4=2\mathbb{E}|d|^2$$

An example of an admissible random variable is given by $d=b_1b_2$, where b_1 and b_2 are independent and distributed as

$$b_1=\begin{cases}1 & \text{with prob.} & \frac{1}{4}\\ -1 & \text{with prob.} & \frac{1}{4}\\ i & \text{with prob.} & \frac{1}{4}\\ -i & \text{with prob.} & \frac{1}{4}\end{cases} \qquad b_2=\begin{cases}1 & \text{with prob.} & \frac{4}{5}\\ \sqrt{6} & \text{with prob.} & \frac{1}{5}\end{cases}. \tag{13.11}$$

Under this model, it is shown that $R \geq c \log^4 N$ masks, for some numerical constant c, are sufficient for the convex program (Equation 13.10) to uniquely recover the underlying signal up to global phase with high probability in the noiseless setting. This result has been further refined to $R \geq c \log^2 N$ in Gross et al. [61].

In Jaganathan et al. [60], specific deterministic masks are considered instead of random masks. In the noiseless setting, it is shown that two masks are sufficient for the convex program (Equation 13.10) to uniquely recover nonvanishing signals up to global phase if an $M \geq 2N$ point DFT is used. A signal $\mathbf{x}$ of length N is said to be nonvanishing if $x[n] \neq 0$ for each $0 \leq n \leq N-1$. In particular, the two masks $\{\mathbf{I}, \mathbf{D}_1\}$, where $\mathbf{I}$ is the $N \times N$ identity matrix and $\mathbf{D}_1$ is a diagonal matrix with diagonal entries given by

$$d_1[n]=\begin{cases}0 & \text{for } n=0\\ 1 & \text{for } 1 \leq n \leq N-1,\end{cases} \tag{13.12}$$

are proposed. This result has also been extended to the N point DFT setup by using five masks $\{\mathbf{I}, \mathbf{D}_2, \mathbf{D}_3, \mathbf{D}_4, \mathbf{D}_5\}$, where $\{\mathbf{D}_2, \mathbf{D}_3, \mathbf{D}_4, \mathbf{D}_5\}$ are diagonal matrices with diagonal entries

$$d_2[n]=\begin{cases}1 & \text{for } 0 \leq n \leq \left\lfloor \frac{N}{2} \right\rfloor\\ 0 & \text{otherwise}\end{cases} \qquad d_3[n]=\begin{cases}1 & \text{for } 1 \leq n \leq \left\lfloor \frac{N}{2} \right\rfloor\\ 0 & \text{otherwise}\end{cases}$$

$$d_4[n]=\begin{cases}1 & \text{for } \left\lfloor \frac{N}{2} \right\rfloor \leq n \leq N-1\\ 0 & \text{otherwise}\end{cases} \qquad d_5[n]=\begin{cases}1 & \text{for } \left\lfloor \frac{N}{2} \right\rfloor+1 \leq n \leq N-1\\ 0 & \text{otherwise.}\end{cases}$$

In the noisy setting, for both the aforementioned random masks and deterministic mask setup, empirical evidence strongly suggests that the recovery is stable. In the deterministic mask setup, stability guarantees are provided in Jaganathan et al. [60].

Wirtinger Flow Algorithm

An alternative recovery approach for masked signals is based on the WF method [58], which applies gradient descent to the least-squares problem:

$$\min_{\mathbf{x}} \sum_{r=0}^{R-1}\sum_{m=0}^{N-1}(Z[m,r]-|\langle \mathbf{f}_m,\mathbf{D}_r\mathbf{x}\rangle|^2)^2. \tag{13.13}$$

Minimizing such nonconvex objectives is known to be NP-hard in general. Gradient descent-type methods have shown promise in solving such problems; however, their performance is very dependent on the initialization and update rules due to the fact that different initialization and update strategies lead to convergence to different (possibly local) minima.

WF is a gradient descent-type algorithm that starts with a careful initialization obtained by means of a spectral method. We refer the readers to Candes et al. [58] for a discussion on various spectral method-based initialization strategies. The initial estimate is then iteratively refined using particular update rules. It is argued that the average WF update is the same as the average stochastic gradient scheme update. Consequently, WF can be viewed as a stochastic gradient descent algorithm, in which only an unbiased estimate of the true gradient is observed. The authors recommend the use of smaller step sizes in the early iterations and larger step sizes in later iterations. When the masks are chosen from a random model with a distribution satisfying properties similar to Equation 13.11, it is shown that $R \geq c \log^4 N$ masks, for some numerical constant c, are sufficient for the WF algorithm to uniquely recover the underlying signal up to global phase with high probability in the noiseless setting. A brief overview of WF is provided in Algorithm 13.4.

Algorithm 13.4 WF Algorithm

Input: Magnitude-square measurements $\mathbf{Z}$ and modulations $\{\mathbf{D}_0, \mathbf{D}_1, \ldots, \mathbf{D}_{R-1}\}$, parameters $\mu_{\max}$ and t_0

Output: Estimate $\hat{\mathbf{x}}$ of the underlying signal

Initialize $\mathbf{x}^{(0)}$ via a spectral method (see Candes et al. [58] for variations): The eigenvector corresponding to the largest eigenvalue of

$$\frac{1}{RN}\left(\sum_{r=0}^{R-1}\sum_{m=0}^{N-1} Z[m,r]\left(\mathbf{D}_r^*\mathbf{f}_m\mathbf{f}_m^*\mathbf{D}_r\right)\right)$$

while halting criterion false **do**

Update the estimate $\mathbf{x}^{(t+1)}$ using the rule

$$\mathbf{x}^{(t+1)} = \mathbf{x}^{(t)} + \frac{\mu}{\left\|\mathbf{x}^{(0)}\right\|^2}\left(\frac{1}{RN}\sum_{r=0}^{R-1}\sum_{m=0}^{N-1}\left(Z[m,r]-\left|\left\langle \mathbf{f}_m,\mathbf{D}_r\mathbf{x}^{(t)}\right\rangle\right|^2\right)\mathbf{D}_r^*\mathbf{f}_m\mathbf{f}_m^*\mathbf{D}_r\mathbf{x}^{(t)}\right)$$

$\mu = \min(1-e^{-t/t_0}, \mu_{\max})$

$t \leftarrow t+1$

end while

Return $\hat{\mathbf{x}} \leftarrow \mathbf{x}^{(t)}$

Combinatorial Methods (for the Noiseless Setting)

For some choices of masks, efficient combinatorial algorithms can be employed to recover the underlying signal, where the choice of masks is closely tied to the reconstruction technique. However, these methods are typically unstable in the presence of noise.

In Jaganathan et al. [60], a combinatorial algorithm is proposed for the two masks $\{\mathbf{I}, \mathbf{D}_1\}$, which is shown to recover signals with $x[0] \neq 0$ up to global phase. The algorithm is derived by expressing the measurements in terms of the autocorrelation as

$$a_0[m] = \sum_{j=0}^{N-1-m} x[j]x^*[j+m] \qquad a_1[m] = \sum_{j=1}^{N-1-m} x[j]x^*[j+m] \quad \text{for } 0 \le m \le N-1.$$

Noting that $a_0[0] - a_1[0] = |x[0]|^2$, the value of $|x[0]|$ can be immediately inferred. Since $a_0[m] - a_1[m] = x[0]x^*[m]$, $x[m]$ for $1 \le m \le N-1$ is next determined up to a phase ambiguity.

Another combinatorial algorithm is proposed in Candes et al. [25] for the three masks $\{\mathbf{I}, \mathbf{I}+\mathbf{D}^s, \mathbf{I}-i\mathbf{D}^s\}$, where s is any integer coprime with N and $\mathbf{D}$ is a diagonal matrix with diagonal entries

$$d[n] = e^{i2\pi\frac{n}{N}} \quad \text{for } 0 \le n \le N-1.$$

It is shown that signals with nonvanishing N point DFT can be uniquely recovered using these masks up to global phase. Indeed, the measurements obtained in this case provide the knowledge of $|y[n]|^2$, $|y[n]+y[n-s]|^2$ and $|y[n]-iy[n-s]|^2$ for $0 \le n \le N-1$ ($n-s$ is understood modulo N). Writing $y[n] = |y[n]|e^{i\phi[n]}$ for $0 \le n \le N-1$, we have

$$|y[n]+y[n-s]|^2 = |y[n]|^2 + |y[n-s]|^2 + 2|y[n]||y[n-s]| \operatorname{Re}(e^{i(\phi[n-s]-\phi[n])}),$$

$$|y[n]+\mathrm{i}y[n-s]|^2 = |y[n]|^2 + |y[n-s]|^2 + 2|y[n]||y[n-s]| \operatorname{Im}(e^{i(\phi[n-s]-\phi[n])}).$$

Consequently, if $y[n] \neq 0$ for $0 \le n \le N-1$, then the measurements provide the relative phases $\phi[n-s] - \phi[n]$ for $0 \le n \le N-1$. Since s is coprime with N, by setting $\phi[0] = 0$ without loss of generality, $\phi[n]$ can be inferred for $1 \le n \le N-1$. Since most signals have a nonvanishing N point DFT, these three masks may be used to recover most signals efficiently.

In order to be able to recover all signals (as opposed to most signals), a polarization-based technique is proposed in Bandeira et al. [62] and Alexeev et al. [77]. It is shown that $O(\log N)$ masks (see Bandeira et al. [62] for design details) are sufficient for this technique.

In Pedarsani et al. [63], the authors consider a combinatorial algorithm, based on coding theoretic tools, for the three masks $\left\{\mathbf{I}, \mathbf{I}+\mathbf{e}_0\mathbf{e}_0^*, \mathbf{I}+i\mathbf{e}_0\mathbf{e}_0^*\right\}$, where $\mathbf{e}_0$ is the $N \times 1$ column vector $(1, 0, \ldots, 0)^T$. For signals with $x[0] \neq 0$, it is shown that the value of $|x[0]|$ can be uniquely found with high probability. The phase of $x[0]$ is set to 0 without loss of generality, and the phase of $x[n]$ relative to $x[0]$ for $0 \le n \le N-1$ is inferred by solving a set of algebraic equations.

The results on masked phase retrieval reviewed in this section are summarized in Table 13.2.

Table 13.2 Uniqueness and Recovery Algorithms for Phase Retrieval Using Masks

Robust methods	$O(\log^2 N)$ random masks, whose diagonal entries are i.i.d. copies of a random variable satisfying some properties, are sufficient for the SDP algorithm with high probability [58,61]
	For nonvanishing signals, two specific masks (with oversampling) or five specific masks are sufficient for the SDP algorithm [28]
	$O(\log^4 N)$ random masks, whose diagonal entries are i.i.d. copies of a random variable satisfying some properties, are sufficient for the Wirtinger flow algorithm with high probability [58]
Combinatorial methods	For signals with nonvanishing DFT, three specific masks are sufficient for a particular combinatorial algorithm [25]
	$O(\log N)$ masks are sufficient for a polarization-based algorithm [49]
	For nonvanishing signals, two specific masks (with oversampling) or five specific masks are sufficient for a particular combinatorial algorithm [28]
	For signals satisfying $x[0] \neq 0.3$ specific masks are sufficient for a particular combinatorial algorithm [63]

Numerical Simulations

We now demonstrate the performance of the various algorithms discussed using numerical simulations.

In the first set of simulations, we consider the performance of SDP and WF, for the random mask setup proposed in Candes et al. [56,58], respectively. A total of 50 trials are performed by generating random signals of length $N = 128$, such that the values in each location are chosen from an i.i.d. standard normal distribution. For the WF algorithm, the parameters μ_{max} and t_0 were chosen to be 0.2 and 330, respectively. We refer the readers to Candes et al. [56,58] for details about the distributions used. The probability of successful recovery as a function of R is plotted in Figure 13.4 in the noiseless setting. It can be observed that $R \approx 6$ is sufficient for successful recovery with high probability for both methods.

In the second set of simulations, the performance of SDP, for the random mask setup proposed in Candes et al. [56] and the specific two-mask setup proposed in Jaganathan et al. [60], is evaluated in the noisy setting. In the case of random masks, eight masks are used and the rest of the parameters are the same as in Figure 13.4. For the two-mask setup, $N = 32$ is chosen and the 64 point DFT is considered. The normalized mean-squared error is plotted as a function of SNR for these two settings in Figure 13.5. While a direct comparison of the two methods is meaningless since the first one uses four times as many measurements, the results clearly show that the recovery is stable in the presence of noise.

STFT Phase Retrieval

In this section, we consider introducing redundancy into the Fourier measurements by using the STFT. The key idea is to introduce redundancy in the magnitude-only measurements by maintaining a substantial overlap between adjacent short-time sections. As we will see, the redundancy offered by the STFT enables unique and robust recovery in many cases. Furthermore, using the STFT leads to improved performance over recovery from the oversampled Fourier magnitude with the same number of measurements.

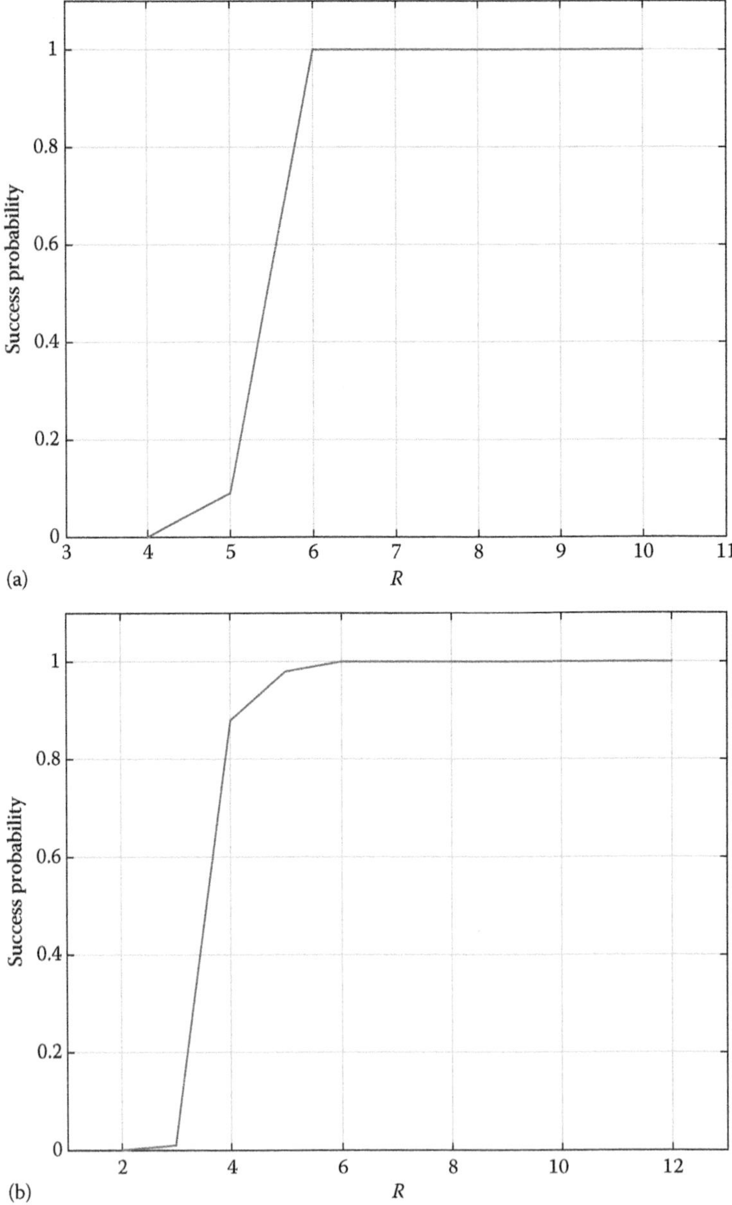

Figure 13.4 Performance of the SDP and WF algorithms in the noiseless setting for $N=128$. (a) Probability of successful recovery vs. R using SDP. (b) Probability of successful recovery vs. R using WF. (Reprinted from *Applied and Computational Harmonic Analysis*, 39(2), Candes, E.J., Li, X., and Soltanolkotabi, M., Phase retrieval from coded diffraction patterns, 277–299, Copyright 2015, with permission from Elsevier; Courtesy of Candes, E.J. et al., *IEEE Trans. Inform. Theory*, 61(4), 1985, 2015.)

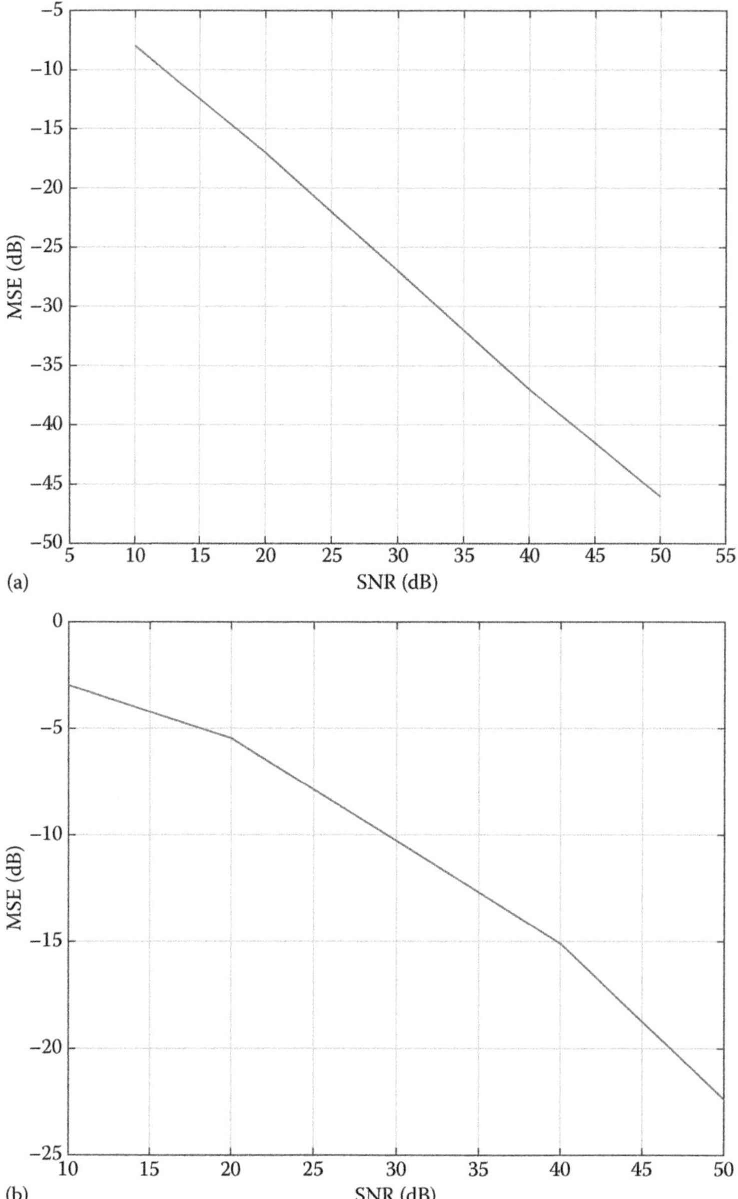

Figure 13.5 Performance of the SDP algorithm in a noisy setting. (a) MSE (dB) vs. SNR (dB) for the random mask setup proposed in (N = 128; R = 8). (b) MSE (dB) vs. SNR (dB) for the specific two-mask setup proposed in (N = 32; M = 64; R = 2). ([a] Reprinted from *Applied and Computational Harmonic Analysis*, 39(2), Candes, E.J., Li, X., and Soltanolkotabi, M., Phase retrieval from coded diffraction patterns, 277–299, Copyright 2015, with permission from Elsevier; [b] Courtesy of Jaganathan, K. et al., Phase retrieval with masks using convex optimization, *IEEE International Symposium on Information Theory Proceedings*, 2015, pp. 1655–1659.)

Phase retrieval from the STFT magnitude has been used in several signal processing applications, for example, in speech and audio processing [81–84], where the spectral content of speech changes over time [81,83]. It has also been applied extensively in optics. One example is in frequency resolved optical gating (FROG) or XFROG, which are used for characterizing ultrashort laser pulses by optically producing the STFT magnitude of the measured pulse [39,85]. In FROG, the pulse itself (or a function of the pulse) is used to gate the measured signal, while in XFROG gating is performed by a fixed known window. Another example is ptychographical CDI [86] or Fourier ptychography [40,42,43], a technology that has enabled x-ray, optical, and electron microscopy with increased spatial resolution without the need for advanced lenses.

Let $\mathrm{w} = (w[0], w[1], \ldots, w[W-1])$ be a window of length W such that it has nonzero values only within the interval $[0, W-1]$. The STFT of $\mathbf{x}$ with respect to w, denoted by $\mathbf{Y}_w$, is defined as

$$Y_w[m,r] = \sum_{n=0}^{N-1} x[n]w[rL-n]e^{-i2\pi\frac{mn}{N}} \quad \text{for } 0 \le m \le N-1 \text{ and } 0 \le r \le R-1, \tag{13.14}$$

where the parameter L denotes the separation in time between adjacent short-time sections and the parameter $R = [(N+W-1)/L]$ denotes the number of short-time sections considered.

The STFT can be interpreted as follows: suppose $\mathbf{w}_r$ denotes the signal obtained by shifting the flipped window $\mathbf{w}$ by rL time units (i.e., $w_r[n] = w[rL-n]$) and let o denote the Hadamard (element-wise) product operator. The rth column of $\mathbf{Y}_w$, for $0 \le r \le R-1$, corresponds to the N point DFT of $\mathbf{x} \circ \mathbf{w}_r$. In essence, the window is flipped and slid across the signal (see Figure 13.6 for a pictorial representation), and $\mathbf{Y}_w$ corresponds to the Fourier transform of the windowed signal recorded at regular intervals. This interpretation is known as the *sliding window* interpretation [87].

The problem of reconstructing a signal from its STFT magnitude is known as STFT phase retrieval. In fact, STFT phase retrieval can be viewed as a special instance of phase retrieval using masks, where the different masks considered are time-shifted copies of the window. This can be seen as follows: let $\mathbf{Z}_w$ be the $N \times R$ measurements corresponding to the magnitude square of the STFT of $\mathbf{x}$ with respect to w so that $Z_w[m, r] = |Y_w[m, r]|^2$. Let $\mathbf{W}_r$, for $0 \le r \le R-1$, be the $N \times N$ diagonal matrix with diagonal elements $(w_r[0], w_r[1], \ldots, w_r[N-1])$. STFT phase retrieval can then be mathematically stated as

$$\begin{aligned} &\text{find} && \mathbf{x} \\ &\text{subject to} && Z_w[m,r] = |\langle \mathbf{f}_m, \mathbf{W}_r\mathbf{x}\rangle|^2 \quad \text{for } 0 \le m \le N-1 \text{ and } 0 \le r \le R-1, \end{aligned} \tag{13.15}$$

which is equivalent to Equation 13.9.

Uniqueness

In this section, we review the main results regarding uniqueness of STFT phase retrieval. As before, a signal $\mathbf{x}$ is nonvanishing if $x[n] \neq 0$ for each $0 \le n \le N-1$. Similarly, a window w is called nonvanishing if $w[n] \neq 0$ for all $0 \le n \le W-1$. These results are summarized in Table 13.3.

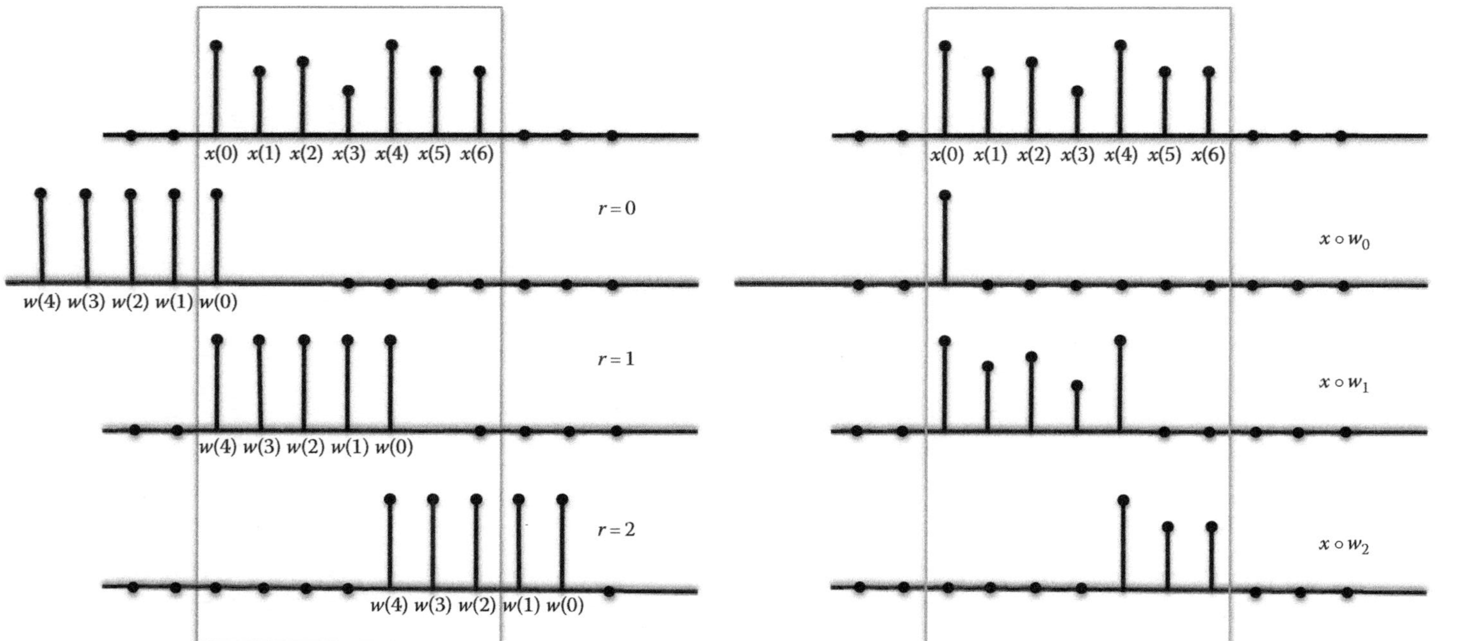

Figure 13.6 Sliding window interpretation of the STFT for $N=7$, $W=5$, and $L=4$. The shifted window overlaps with the signal for three shifts, and hence $R=3$ short-time sections are considered. (Courtesy of Jaganathan, K. et al., STFT phase retrieval: Uniqueness guarantees and recovery algorithms, arXiv:1508.02820, 2015.)

Table 13.3 Uniqueness of STFT Phase Retrieval

Nonvanishing signals $\{x[n] \neq 0 \mid 0 \leq n \leq N-1\}$	Uniqueness up to global phase if $L=1$, $2 \leq W \leq (N+1)/2$, $W-1$ coprime with N and mild conditions on $\mathbf{w}$ [44]
	Uniqueness if the first L samples are known a priori, $2L \leq W \leq N/2$ and $\mathbf{w}$ is nonvanishing [81]
	Uniqueness up to global phase for most signals if $L \leq W \leq N/2$ and $\mathbf{w}$ is nonvanishing [28]
Sparse signals $\{x[n]=0$ for at least one $0 \leq n \leq N-1\}$	No uniqueness for most signals with W consecutive zeros [44]
	Uniqueness for signals with at most $W-2L$ consecutive zeros if the first L samples, starting from the first nonzero sample, are known a priori, $2L \leq W \leq N/2$ and $\mathbf{w}$ is nonvanishing [81]
	Uniqueness up to global phase and time shift for most signals with less than $\min\{W-L, L\}$ consecutive zeros if $L < W \leq N/2$ and $\mathbf{w}$ is nonvanishing [28]

First, we argue that, for $W<N$ (which is typically the case), $L<W$ is a necessary condition in order to be able to uniquely identify most signals: If $L>W$, then the STFT magnitude does not contain any information from some locations of the signal, because of which most signals cannot be uniquely identified. If $L=W$, then the adjacent short-time sections do not overlap and hence STFT phase retrieval is equivalent to a series of nonoverlapping phase retrieval problems. Consequently, as in the case of phase retrieval, most 1D signals are not uniquely identifiable. For higher dimensions (2D and above), almost all windowed signals corresponding to each of the short-time sections are uniquely identified up to trivial ambiguities if $\mathbf{w}$ is nonvanishing. However, since there is no way of establishing relative phase, time shift, or conjugate flip between the windowed signals corresponding to the various short-time sections, most signals cannot be uniquely identified. For example, suppose we choose $L=W=2$ and $w[n]=1$ for all $0 \leq n \leq W-1$. Consider the signal $\mathbf{x}_1=(1, 2, 3)^T$ of length $N=3$. Signals $\mathbf{x}_1$ and $\mathbf{x}_2=(1, -2, -3)^T$ have the same STFT magnitude. In fact, more generally, signals $\mathbf{x}_1$ and $(1, e^{i\phi}2, e^{i\phi}3)^T$, for any ϕ, have the same STFT magnitude.

Nonvanishing Signals

In Jaganathan et al. [41], it is shown that most nonvanishing signals are uniquely identifiable from their STFT magnitude up to global phase if $1 \leq L \leq W$ and $\mathbf{w}$ is chosen such that it is nonvanishing with $W \leq \frac{N}{2}$. In other words, this result states that, if adjacent short-time sections overlap (the extent of overlap does not matter), then STFT phase retrieval is almost surely well posed under mild conditions on the window. Observe that, like in phase retrieval, the global phase ambiguity cannot be resolved due to the fact that signals $\mathbf{x}$ and $e^{i\phi}\mathbf{x}$, for any ϕ, have the same STFT magnitude regardless of the choice of $\{\mathbf{w}, L\}$. However, unlike in phase retrieval, the time-shift and conjugate-flip ambiguities are resolvable for most nonvanishing signals.

For some specific choices of $\{\mathbf{w}, L\}$, all nonvanishing signals are uniquely identified from their STFT magnitude up to global phase. In Eldar et al. [44], it is shown that the STFT magnitude can uniquely identify nonvanishing signals up to global phase for $L=1$ if the window $\mathbf{w}$ is chosen such that the N point DFT of $(|w[0]|^2, w[1]^2, \ldots, |w[N-1]|^2)$ is nonvanishing, $2 \leq W \leq \frac{N+1}{2}$ and $W-1$ is coprime with N. An alternative result is that if the first L samples are known a priori, then the STFT magnitude can uniquely identify nonvanishing signals for any L as long as the window **w** is chosen such that it is nonvanishing and $2L \le W \le \frac{N}{2}$ [81].

Sparse Signals

While the aforementioned results provide guarantees for nonvanishing signals, they do not say anything about sparse signals. In this section, a signal **x** of length N is said to be sparse if $x[n]=0$ for at least one $0 \le n \le N-1$. Based on intuitions from compressed sensing, one might expect to recover sparse signals easier than nonvanishing signals. However, this is actually not the case for STFT phase retrieval.

The following example is provided in Eldar et al. [44] to demonstrate that the time-shift ambiguity cannot be resolved for some classes of sparse signals and some choices of $\{\mathbf{w}, L\}$: Suppose $L \ge 2$, W is a multiple of L and $w[n]=1$ for all $0 \le n \le W-1$. Consider a signal $\mathbf{x}_1$ of length $N \ge L+1$ such that it has nonzero values only within an interval of the form $[(t-1)L+1, (t-1)L+L-p] \subset [0, N-1]$ for some integers $1 \le p \le L-1$ and $t \ge 1$. The signal $\mathbf{x}_2$ obtained by time-shifting $\mathbf{x}_1$ by $q \le p$ units (i.e., $x_2[i]=x_1[i-q]$) has the same STFT magnitude. The issue with this class of sparse signals is that the STFT magnitude is identical to the Fourier magnitude because of which time-shift and conjugate-flip ambiguities cannot be resolved.

It is further shown that there are sparse signals that are not recoverable even up to trivial ambiguities for some choices of $\{\mathbf{w}, L\}$. Consider two nonoverlapping intervals $[u_1, \upsilon_1]$, $[u_2, \upsilon_2] \subset [0, N-1]$ such that $u_2 - \upsilon_1 > W$, and choose signals $\mathbf{x}_1$ supported on $[u_1, \upsilon_1]$ and $\mathbf{x}_2$ supported on $[u_2, \upsilon_2]$. The magnitude square of the STFT of $\mathbf{x}_1+\mathbf{x}_2$ and of $\mathbf{x}_1-\mathbf{x}_2$ are equal for any choice of L. For such examples of sparse signals, the two intervals with nonzero values are separated by a distance greater than W because of which there is no way of establishing relative phase using a window of length W.

The aforementioned examples establish the fact that sparse signals are harder to recover than nonvanishing signals from their STFT magnitude. Since the issues are primarily due to a large number of consecutive zeros, the uniqueness guarantees for nonvanishing signals have been extended to incorporate sparse signals with limits on the number of consecutive zeros. In Jaganathan et al. [41], it is shown that most sparse signals with less than $\min\{W-L, L\}$ consecutive zeros are uniquely identified from their STFT magnitude up to global phase and time-shift ambiguities if adjacent short-time sections overlap (i.e., $L<W$) and w is chosen such that it is nonvanishing and $W \le \frac{N}{2}$. The work in Nawab et al. [81] establishes that if L consecutive samples are known a priori, starting from the first nonzero sample, then the STFT magnitude can uniquely identify signals with at most $W-2L$ consecutive zeros for any L if the window **w** is chosen such that it is nonvanishing and $2L \le W \le \frac{N}{2}$.

Algorithms

Phase retrieval techniques based on alternating projections, SDP, and greedy methods (e.g., GESPAR for sparse signals) have been modified to solve STFT phase retrieval efficiently and robustly. In this section, we provide a survey of the existing STFT phase retrieval algorithms.

1. *Alternating projections*: The classic alternating projection method has been adapted to solve STFT phase retrieval by Griffin and Lim [82].

To this end, STFT phase retrieval is formulated as the following least-squares problem:

$$\min_{\mathbf{x}} \sum_{r=0}^{R-1} \sum_{m=0}^{N-1} \left(Z_w[m,r] - \left| \langle \mathbf{f}_m, \mathbf{W}_r \mathbf{x} \rangle \right|^2 \right)^2. \tag{13.16}$$

The Griffin–Lim (GL) algorithm attempts to minimize this objective by starting with a random initialization and imposing the time-domain and STFT magnitude constraints alternately using projections. The details of the various steps are provided in Algorithm 13.5. The objective is shown to be monotonically decreasing as the iterations progress. In the noiseless setting, an important feature of the GL method is its empirical ability to converge to the global minimum when there is substantial overlap between adjacent short-time sections. However, no theoretical recovery guarantees are available. In the noisy setting, the algorithm has the same limitations as the GS and HIO techniques for standard phase retrieval.

In optics, a slight variation of the GL iterations is used, referred to as principal components generalized projections (PCGP). We refer the readers to Kane [85] for details.

Algorithm 13.5 Griffin–Lim (GL) Algorithm

Input: STFT magnitude-square measurements $\mathbf{Z}_w$ and window $\mathbf{w}$

Output: Estimate $\hat{\mathbf{x}}$ of the underlying signal

Initialize: Choose a random input signal $\mathbf{x}^{(0)}$, $\ell = 0$

while halting criterion false **do**

$\ell \leftarrow \ell + 1$

Compute the STFT of $\mathbf{x}^{(\ell-1)}$: $Y_w^{(\ell)}[m,r] = \sum_{n=0}^{N-1} x^{(\ell-1)}[n] w[rL-n] e^{-i2\pi \frac{mn}{N}}$

Impose STFT magnitude constraints: $\mathbf{Y}_w'^{(\ell)}[m,r] = \frac{Y_w^{(\ell)}[m,r]}{|Y_w^{(\ell)}[m,r]|} \sqrt{Z_w[m,r]}$

Compute the inverse DFT of $\mathrm{Y}_w'^{(\ell)}$ for each short-time section to obtain windowed signals $\mathbf{x}_r'^{(\ell)}$

Impose time-domain constraints to obtain $\mathbf{x}^{(\ell)} : x^{(\ell)}[n] = \frac{\sum_r x_r'^{(\ell)}[n] w^*[rL-n]}{\sum_r |w[rL-n]|^2}$

end while

return $\hat{\mathbf{x}} \leftarrow \mathbf{x}^{(\ell)}$

---

2. *SDP method*: In Jaganathan et al. [41] and Sun and Smith [83], the SDP-based phase retrieval approach has been applied to STFT phase retrieval, leading to the STliFT algorithm detailed in Algorithm 13.6.

   Many recent results in related problems like *generalized phase retrieval* [53] and *phase retrieval using random masks* [56,61] suggest that one can provide conditions on $\{\mathbf{w}, L\}$, which when satisfied ensure that the SDP-based algorithm correctly recovers the underlying signal. In Jaganathan et al. [41], it is shown that STliFT uniquely recovers nonvanishing signals from their STFT magnitude up to global phase if $L=1, 2 \le W \le (N/2)$ and $\mathbf{w}$ is nonvanishing. It is further proved that STliFT uniquely recovers nonvanishing signals from their STFT magnitude for any $2 \le 2L \le W \le (N/2)$ and nonvanishing $\mathbf{w}$ if the first $L/2$ samples are known a priori. These guarantees only partially explain the excellent empirical performance of STliFT. In the noiseless setting, STliFT is observed empirically to recover most signals for any $2 \le 2L \le W \le (N/2)$. In the presence of noise, like many SDP methods, STliFT appears to exhibit stable recovery. This behavior is demonstrated in the numerical simulation section.

Algorithm 13.6 STliFT

Input: STFT magnitude-square measurements $\mathbf{Z}_w$ and window $\mathbf{w}$

Output: Estimate $\hat{\mathbf{x}}$ of the underlying signal

Obtain $\hat{\mathbf{X}}$ by solving

$$\begin{aligned} \text{minimize} \quad & \text{trace}(\mathbf{X}) \\ \text{subject to} \quad & Z_w[m,r] = \text{trace}(\mathbf{f}_m \mathbf{f}_m^* (\mathbf{X} \circ \mathbf{w}_r \mathbf{w}_r^*)) \quad \text{for} \\ & 0 \le m \le N-1 \text{ and } 0 \le r \le R-1, \\ & \mathbf{X} \succeq 0. \end{aligned} \tag{13.17}$$

Return $\hat{\mathbf{x}}$, where $\hat{\mathbf{x}}\hat{\mathbf{x}}^*$ is the best rank-one approximation of $\hat{\mathbf{X}}$

3. *GESPAR for sparse signals*: If it is known a priori that the underlying signal is sparse, then the recovery performance can potentially be improved by exploiting this sparsity. One method for taking sparsity into account is by adapting GESPAR [30] to STFT phase retrieval [44]. Simulations demonstrate that GESPAR is able to exploit both the redundancy in the measurements and the sparsity of the input, leading to high probability of successful reconstruction and stable recovery in the presence of noise as long as sufficient redundancy is introduced into the measurement process. We demonstrate some of these results via simulations later.

4. *Combinatorial methods (for the noiseless setting)*: For some choices of $\{\mathbf{w}, L\}$, researchers have developed efficient combinatorial methods specific to the STFT phase retrieval setup. In Nawab et al. [81], a sequential reconstruction technique is proposed for nonvanishing signals when $L=1$, $W \ge 2$. The algorithm works as follows: $x[0]$ is reconstructed up to a phase from the measurement corresponding to the short-time section $r=0$. Using the knowledge of $x[0]$, $x[1]$ is reconstructed up to a relative phase from the measurement corresponding to the short-time section $r=1$. Continuing sequentially, the entire signal can be estimated up to global phase (see Nawab et al. [81] for details). In Eldar et al. [44], a reconstruction algorithm is proposed where $|x[n]|$

for $0 \leq n \leq N-1$ is first reconstructed by solving a linear system of equations and the phases of $x[n]$ are then sequentially reconstructed using a combinatorial method.

Like many sequential reconstruction approaches, these methods are typically unstable in the noisy setting due to error propagation.

Numerical Simulations

We now demonstrate the performance of various STFT phase retrieval algorithms using numerical simulations.

In the first set of simulations, the nonvanishing signal recovery ability of STliFT is evaluated for $N=32$ in the noiseless and noisy settings. The window $\mathbf{w}$ is chosen such that $w[n]=1$ for all $0 \leq n \leq W-1$ and the parameters L and W are varied between 1 and $N/2$. For each choice of $\{W, L\}$, 100 simulations are performed by randomly choosing nonvanishing signals, such that the values in each location are drawn from an i.i.d. standard normal distribution. The probability of successful recovery as a function of $\{W, L\}$ in the noiseless setting is shown in Figure 13.7a. Observe that STliFT successfully recovers the underlying signal with very high probability if $2 \leq W \leq \frac{N}{2}$. The case $L=\frac{N}{4}$, $W=\frac{N}{2}$ uses only *six* windowed measurements and the underlying signal is recovered with very high probability by STliFT. Given the limited success of SDP-based methods in the Fourier phase retrieval setup, this is very encouraging. In the noisy setting, the normalized mean square error is plotted in Figure 13.7b for various choices of SNR and $\{W, L\}$. The linear relationship between log(MSE) and SNR indicates that STliFT stably recovers the underlying signal. It can also be seen that when there is significant overlap between adjacent short-time sections, the signals are recovered more stably, which is not surprising.

In the second set of simulations, the sparse signal recovery abilities of STFT-GESPAR, GL algorithm, and PCGP are evaluated for $N=64$ in the noiseless and noisy settings. The window satisfies $w[n]=1$ for $0 \leq n \leq W-1$ and the parameters L and W are chosen to be $\{2, 4, 8, 16\}$ and 16, respectively. The sparsity dictionary is a random basis with i.i.d. standard normal variables, followed by normalization of the columns. To generate the sparse inputs, for each sparsity level k, the k locations for the nonzero values are chosen uniformly at random. The signal values over the selected support are then drawn from an i.i.d. standard normal distribution.

STFT-GESPAR is used with a threshold $\tau=10^{-4}$ and maximum number of swaps 50,000, PCGP and GL are run using 50 random initial points with 1000 maximal iterations. For comparison with the Fourier transform approach, the performance of GESPAR with the same parameters and oversampling factors of $\{8, 4, 2, 1\}$ is shown. In Figure 13.8a, the probability of successful recovery as a function of sparsity is plotted. Note that when $L=16$ there is no redundancy in the STFT and therefore it is not surprising that there is no advantage to the STFT method. In all other cases for which $L<16$, the STFT introduces redundancy, which leads to improved performance over simply oversampling the DFT. It is also evident that GESPAR outperforms both the GL and PCGP algorithms.

In Figure 13.8b, the effect of noise and the DFT length on the normalized mean-squared error is considered. All parameters are the same as in Figure 13.8a besides L, which is set to $L=1$ and the signal length chosen as $N=32$. The DFT length is set to $K=\{2, 4, 8, 16, 32\}$. When K is smaller than the window length

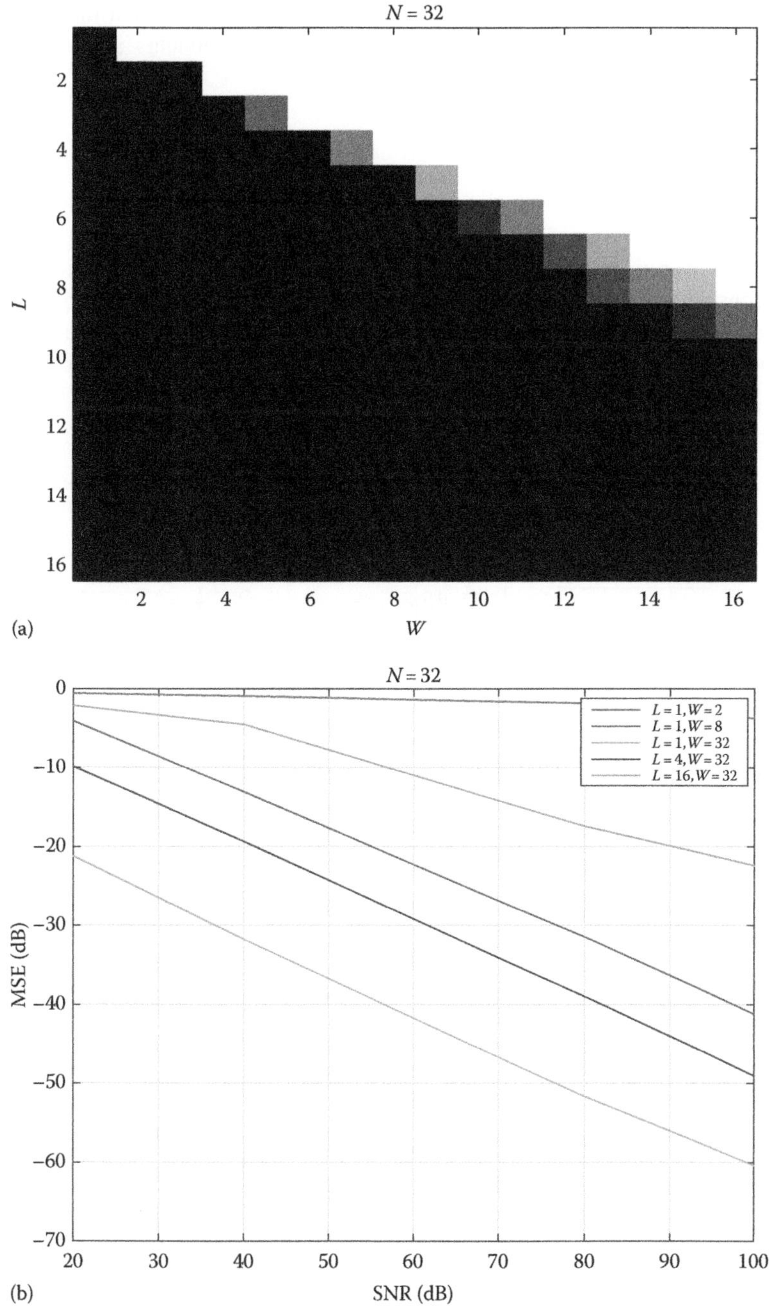

Figure 13.7 Performance of STliFT for $N=32$ and various choices of $\{W, L\}$. (a) Probability of successful recovery in the noiseless setting (white region: success with probability 1, black region: success with probability 0). (b) MSE (dB) vs. SNR (dB) in the noisy setting. (Courtesy of Jaganathan, K. et al., STFT phase retrieval: Uniqueness guarantees and recovery algorithms, arXiv:1508.02820, 2015.)

$W = 16$, a DFT of length W is used and only the first K measurements are chosen (i.e., only the K low-frequency measurements are used). As expected, increasing the DFT length improves the recovery ability. It is also evident that the performance improves significantly when all Fourier components are measured, namely, when $K \geq 16$.

Conclusions

In this chapter, we reviewed some of the recent progresses on phase retrieval. In many cases, we demonstrated that an unknown signal can be robustly and efficiently recovered from a set of phaseless Fourier measurements. In particular, we first considered the problem of sparse phase retrieval. We noted that most sparse signals can be uniquely identified from their autocorrelation and presented two efficient and robust algorithms (TSPR and GESPAR). Then, we considered the

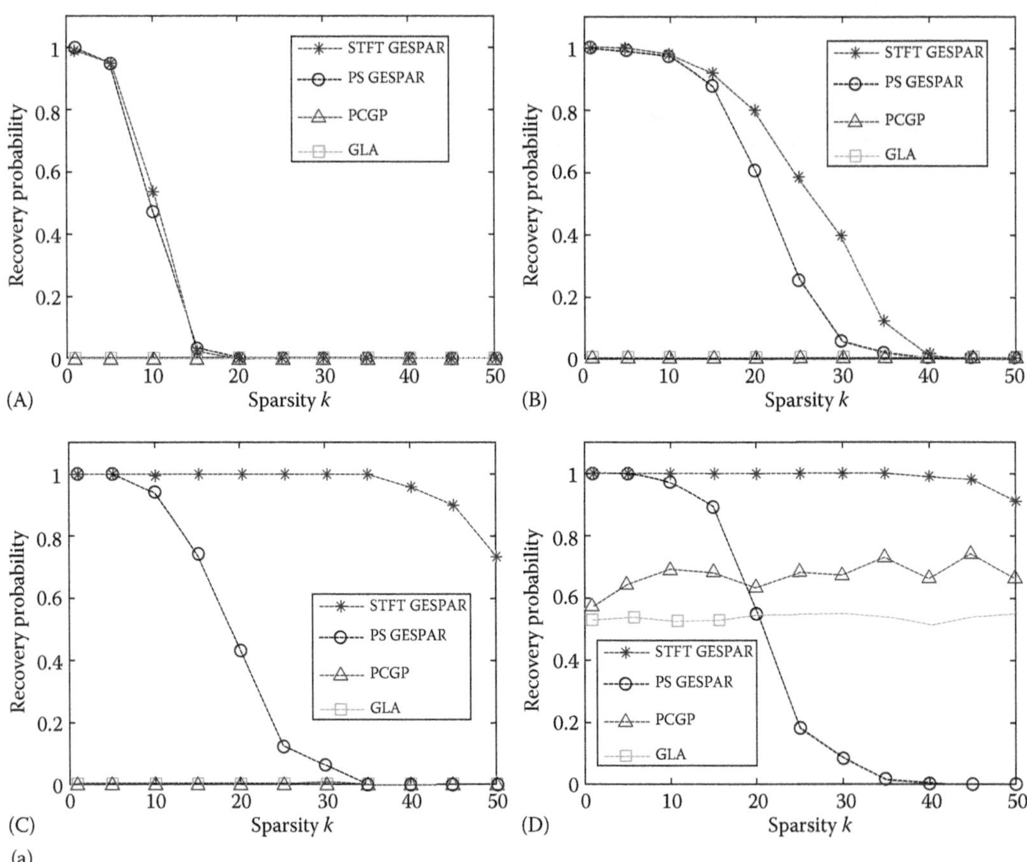

Figure 13.8 Performance of STFT-GESPAR for $N = \{32, 64\}$ and various choices of $\{K, W, L\}$. (a) Probability of successful recovery vs. sparsity in the noiseless setting for varying measurements (A) 64, (B) 128, (C) 256, and (D) 512. (From Eldar, Y.C., Sidorenko, P., Mixon, D.G., Barel, S., and Cohen, O., Sparse phase retrieval from short-time Fourier measurements, *IEEE Signal Processing Letters*, 22(5), 638–642. © 2015 IEEE.) *(Continued)*

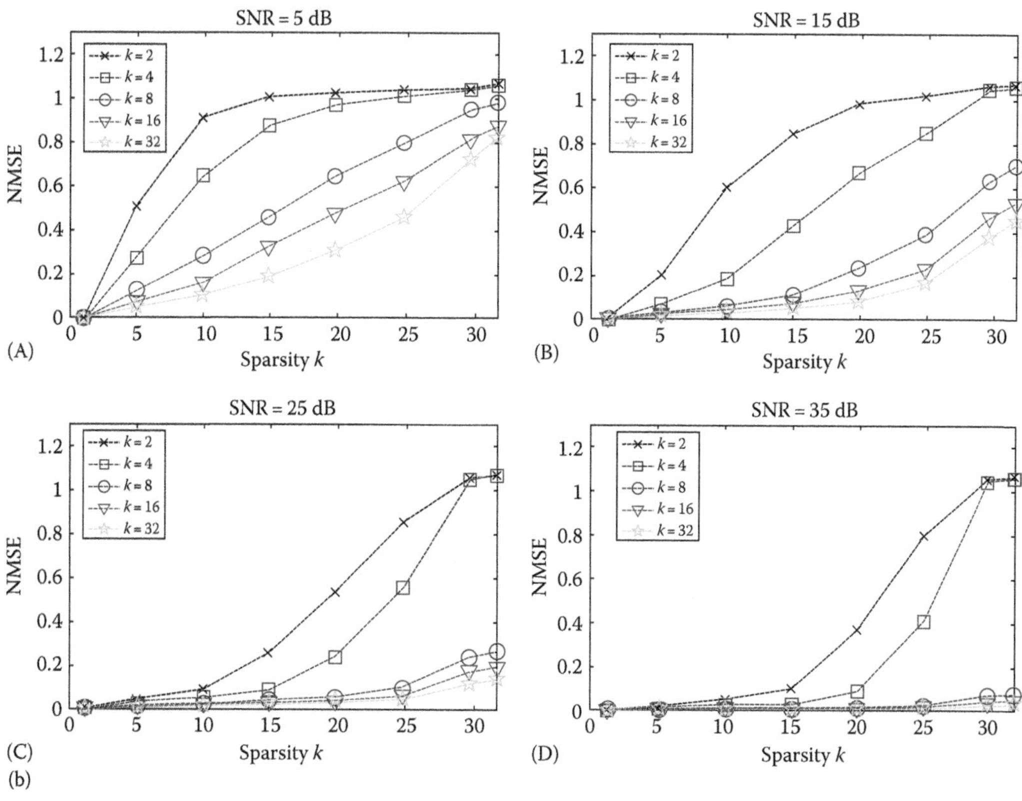

Figure 13.8 (*Continued*) Performance of STFT-GESPAR for $N = \{32, 64\}$ and various choices of $\{K, W, L\}$. (b) MSE vs. sparsity in the noisy setting for varying SNR (in [dB]) (A) 5, (B) 15, (C) 25, and (D) 35. (From Eldar, Y.C., Sidorenko, P., Mixon, D.G., Barel, S., and Cohen, O., Sparse phase retrieval from short-time Fourier measurements, *IEEE Signal Processing Letters*, 22(5), 638–642. © 2015 IEEE.)

problem of phase retrieval using masks. We presented various masks that allow for unique and robust recovery of the unknown signal. Finally, we treated the problem of STFT phase retrieval. We noted that most signals can be uniquely identified from their STFT magnitude and suggested an efficient and robust algorithm (STliFT) for signal recovery. These results clearly have many practical applications in optics and imaging systems where measuring the phase is often challenging. They may also have ramifications to other areas of engineering, such as radar and wireless communications, where the possibility of avoiding having to measure the phase could result in simpler and cost-effective practical systems.

Acknowledgments

The authors would like to thank Prof. Mordechai Segev and Oren Cohen for introducing them to the STFT phase retrieval problem and for many insightful discussions regarding phase retrieval and optics.

References

1. A. L. Patterson, A Fourier series method for the determination of the components of interatomic distances in crystals, *Physical Review* 46(5) (1934): 372.
2. A. L. Patterson, Ambiguities in the x-ray analysis of crystal structures, *Physical Review* 65(5–6) (1944): 195.
3. A. Walther, The question of phase retrieval in optics, *Journal of Modern Optics* 10(1) (1963): 41–49.
4. R. P. Millane, Phase retrieval in crystallography and optics, *Journal of the Optical Society of America A* 7(3) (1990): 394–411.
5. J. C. Dainty and J. R. Fienup, Phase retrieval and image reconstruction for astronomy, *Image Recovery: Theory and Application* (1987): 231–275.
6. L. Rabiner and B. H. Juang, *Fundamentals of Speech Recognition*, Prentice Hall (1993).
7. M. Stefik, Inferring DNA structures from segmentation data, *Artificial Intelligence* 11(1) (1978): 85–114.
8. B. Baykal, Blind channel estimation via combining autocorrelation and blind phase estimation, *IEEE Transactions on Circuits and Systems* 51(6) (2004): 1125–1131.
9. A. V. Oppenheim and J. S. Lim, The importance of phase in signals, *Proceedings of the IEEE* 69(5) (1981): 529–541.
10. Y. Shechtman, Y. C. Eldar, O. Cohen, H. N. Chapman, J. Miao, and M. Segev, Phase retrieval with application to optical imaging, *IEEE Signal Processing Magazine* 32(3) (2015): 87–109.
11. J. R. Fienup, Phase retrieval algorithms: A personal tour [invited], *Applied Optics* 52(1) (2013): 45–56.
12. E. M. Hofstetter, Construction of time-limited functions with specified autocorrelation functions, *IEEE Transactions on Information Theory* 10(2) (1964): 119–126.
13. M. H. Hayes, The reconstruction of a multidimensional sequence from the phase or magnitude of its Fourier transform, *IEEE Transactions on Acoustics, Speech and Signal Processing* 30(2) (1982): 140–154.
14. R. W. Gerchberg and W. O. Saxton, A practical algorithm for the determination of the phase from image and diffraction plane pictures, *Optik* 35 (1972): 237.
15. J. R. Fienup, Phase retrieval algorithms: A comparison, *Applied Optics* 21(15) (1982): 2758–2769.
16. H. H. Bauschke, P. L. Combettes, and D. R. Luke, Phase retrieval, error reduction algorithm, and Fienup variants: A view from convex optimization, *Journal of the Optical Society of America A* 19(7) (2002): 1334–1345.
17. S. Marchesini, Invited article: A unified evaluation of iterative projection algorithms for phase retrieval, *Review of Scientific Instruments* 78(1) (2007): 011301.
18. D. P. Palomar and Y. C. Eldar, *Convex Optimization in Signal Processing and Communications*, Cambridge University Press (2010).
19. Y. C. Eldar and G. Kutyniok, *Compressed Sensing: Theory and Applications*, Cambridge University Press (2012).
20. Y. C. Eldar, *Sampling Theory: Beyond Bandlimited Systems*, Cambridge University Press (2015).

21. E. J. Candes and T. Tao, Decoding by linear programming, *IEEE Transactions on Information Theory* 51(12) (2005): 4203–4215.
22. E. J. Candes and B. Recht, Exact matrix completion via convex optimization, *Foundations of Computational Mathematics* 9(6) (2009): 717–772.
23. J. Miao, P. Charalambous, J. Kirz, and D. Sayre, Extending the methodology of x-ray crystallography to allow imaging of micrometre-sized non-crystalline specimens, *Nature* 400(6742) (1999): 342–344.
24. M. L. Moravec, J. K. Romberg, and R. G. Baraniuk, Compressive phase retrieval, *International Society for Optics and Photonics* (2007): 670120.
25. E. J. Candes, Y. C. Eldar, T. Strohmer, and V. Voroninski, Phase retrieval via matrix completion, *SIAM Journal on Imaging Sciences* 6(1) (2013): 199–225.
26. K. Jaganathan, S. Oymak, and B. Hassibi, Recovery of sparse 1-D signals from the magnitudes of their Fourier transform, *IEEE International Symposium on Information Theory Proceedings* (2012), pp. 1473–1477.
27. Y. Shechtman, Y. C. Eldar, A. Szameit, and M. Segev, Sparsity based sub-wavelength imaging with partially incoherent light via quadratic compressed sensing, *Optics Express* 19 (2011): 14807–14822.
28. K. Jaganathan, S. Oymak, and B. Hassibi, Sparse phase retrieval: Uniqueness guarantees and recovery algorithms, arXiv:1311.2745 (2015).
29. K. Jaganathan, S. Oymak, and B. Hassibi, Sparse phase retrieval: Convex algorithms and limitations, *IEEE International Symposium on Information Theory Proceedings* (2013), pp. 1022–1026.
30. Y. Shechtman, A. Beck, and Y. C. Eldar, GESPAR: Efficient phase retrieval of sparse signals, *IEEE Transactions on Signal Processing* 62(4) (2014): 928–938.
31. Y. M. Lu and M. Vetterli, Sparse spectral factorization: Unicity and reconstruction algorithms, *IEEE International Conference on Acoustics, Speech and Signal Processing* (2011), pp. 5976–5979.
32. S. Mukherjee and C. Seelamantula, An iterative algorithm for phase retrieval with sparsity constraints: Application to frequency domain optical coherence tomography, *IEEE International Conference on Acoustics, Speech and Signal Processing* (2012), pp. 553–556.
33. J. Ranieri, A. Chebira, Y. M. Lu, and M. Vetterli, Phase retrieval for sparse signals: Uniqueness conditions, arXiv:1308.3058 (2013).
34. H. Ohlsson and Y. C. Eldar, On conditions for uniqueness in sparse phase retrieval, *IEEE International Conference on Acoustics, Speech and Signal Processing* (2014), pp. 1841–1845.
35. I. Johnson, K. Jefimovs, O. Bunk, C. David, M. Dierolf, J. Gray, D. Renker, and F. Pfeiffer, Coherent diffractive imaging using phase front modifications, *Physical Review Letters* 100(15) (2008): 155503.
36. Y. J. Liu et al. Phase retrieval in x-ray imaging based on using structured illumination, *Physical Review A* 78(2) (2008): 023817.
37. E. G. Loewen and E. Popov, *Diffraction Gratings and Applications*, CRC Press (1997).
38. A. Faridian, D. Hopp, G. Pedrini, U. Eigenthaler, M. Hirscher, and W. Osten, Nanoscale imaging using deep ultraviolet digital holographic microscopy, *Optics Express* 18(13) (2010): 14159–14164.
39. R. Trebino, *Frequency-Resolved Optical Gating: The Measurement of Ultrashort Laser Pulses*, Springer (2002).

40. J. M. Rodenburg, Ptychography and related diffractive imaging methods, *Advances in Imaging and Electron Physics* 150 (2008): 87–184.
41. K. Jaganathan, Y. C. Eldar, and B. Hassibi, STFT phase retrieval: Uniqueness guarantees and recovery algorithms, arXiv:1508.02820 (2015).
42. M. J. Humphry, B. Kraus, A. C. Hurst, A. M. Maiden, and J. M. Rodenburg, Ptychographic electron microscopy using high-angle dark-field scattering for sub-nanometre resolution imaging, *Nature Communications* 3 (2012): 730.
43. G. Zheng, R. Horstmeyer, and C. Yang, Wide-field, high-resolution Fourier ptychographic microscopy, *Nature Photonics* 7(9) (2013): 739–745.
44. Y. C. Eldar, P. Sidorenko, D. G. Mixon, S. Barel, and O. Cohen, Sparse phase retrieval from short-time Fourier measurements, *IEEE Signal Processing Letters* 22(5) (2015): 638–642.
45. Y. C. Eldar and S. Mendelson, Phase retrieval: Stability and recovery guarantees, *Applied and Computational Harmonic Analysis* 36(3) (2014): 473–494.
46. R. Balan, P. Casazza, and D. Edidin, On signal reconstruction without phase, *Applied and Computational Harmonic Analysis* 20(3) (2006): 345–356.
47. R. Balan, B. G. Bodmann, P. G. Casazza, and D. Edidin, Painless reconstruction from magnitudes of frame coefficients, *Journal of Fourier Analysis and Applications* 15(4) (2009): 488–501.
48. H. Ohlsson, A. Yang, R. Dong, and S. Sastry, Compressive phase retrieval from squared output measurements via semidefinite programming, arXiv:1111.6323 (2011).
49. A. S. Bandeira, J. Cahill, D. G. Mixon, and A. A. Nelson, Saving phase: Injectivity and stability for phase retrieval, *Applied and Computational Harmonic Analysis* 37(1) (2014): 106–125.
50. X. Li and V. Voroninski, Sparse signal recovery from quadratic measurements via convex programming, *SIAM Journal on Mathematical Analysis* 45(5) (2013): 3019–3033.
51. S. Oymak, A. Jalali, M. Fazel, Y. C. Eldar, and B. Hassibi, Simultaneously structured models with application to sparse and low-rank matrices, *IEEE Transactions on Information Theory* 61(5) (2015): 2886–2908.
52. P. Netrapalli, P. Jain, and S. Sanghavi, Phase retrieval using alternating minimization, *Advances in Neural Information Processing Systems* (2013): 2796–2804.
53. E. J. Candes, T. Strohmer, and V. Voroninski, Phaselift: Exact and stable signal recovery from magnitude measurements via convex programming, *Communications on Pure and Applied Mathematics* 66(8) (2013): 1241–1274.
54. M. X. Goemans and D. P. Williamson, Improved approximation algorithms for maximum cut and satisfiability problems using semidefinite programming, *Journal of the ACM* 42(6) (1995): 1115–1145.
55. I. Waldspurger, A. d'Aspremont, and S. Mallat, Phase recovery, maxcut and complex semidefinite programming, *Mathematical Programming* 149(1–2) (2015): 47–81.
56. E. J. Candes, X. Li, and M. Soltanolkotabi, Phase retrieval from coded diffraction patterns, *Applied and Computational Harmonic Analysis* 39(2) (2015): 277–299.
57. B. Recht, M. Fazel, and P. Parrilo, Guaranteed minimum-rank solutions of linear matrix equations via nuclear norm minimization, *SIAM Review* 52(3) (2010): 471–501.

58. E. J. Candes, X. Li, and M. Soltanolkotabi, Phase retrieval via Wirtinger flow: Theory and algorithms, *IEEE Transactions on Information Theory* 61(4) (2015): 1985–2007.
59. A. Beck and Y. C. Eldar, Sparsity constrained nonlinear optimization: Optimality conditions and algorithms, *SIAM Journal on Optimization* 23(3) (2013): 1480–1509.
60. K. Jaganathan, Y. C. Eldar, and B. Hassibi, Phase retrieval with masks using convex optimization, *IEEE International Symposium on Information Theory Proceedings* (2015), pp. 1655–1659.
61. D. Gross, F. Krahmer, and R. Kueng, Improved recovery guarantees for phase retrieval from coded diffraction patterns, arXiv:1402.6286 (2014).
62. A. S. Bandeira, Y. Chen, and D. G. Mixon, Phase retrieval from power spectra of masked signals, *Information and Inference* (2014): iau002.
63. R. Pedarsani, K. Lee, and K. Ramchandran, PhaseCode: Fast and efficient compressive phase retrieval based on sparse-graph codes, *Annual Allerton Conference in Communication, Control, and Computing* (2014), pp. 842–849.
64. S. Gleichman and Y. C. Eldar, Blind compressed sensing, *IEEE Transactions on Information Theory* 57(10) (2011): 6958–6975.
65. Y. Rivenson, A. Stern, and B. Javidi, Compressive Fresnel holography, *Journal of Display Technology* 6(10) (2010): 506–509.
66. S. Gazit, A. Szameit, Y. C. Eldar, and M. Segev, Super-resolution and reconstruction of sparse sub-wavelength images, *Optics Express* 17(26) (2009): 23920–23946.
67. A. Szameit et al., Sparsity-based single-shot subwavelength coherent diffractive imaging, *Nature Materials*, Supplementary Info 11(5) (2012): 455–459.
68. C. Luo, S. G. Johnson, J. D. Joannopoulos, and J. B. Pendry, Subwavelength imaging in photonic crystals, *Physical Review B* 68(4) (2003): 045115.
69. E. A. Ash and G. Nicholls, Super-resolution aperture scanning microscope, *Nature* (1972): 510–512.
70. Y. Shechtman, Y. C. Eldar, O. Cohen, and M. Segev, Efficient coherent diffractive imaging for sparsely varying dynamics, *Optics Express* 21(5) (2013): 6327–6338.
71. P. Sidorenko, A. Fleischer, Y. Shechtman, Y. C. Eldar, M. Segev, and O. Cohen, Sparsity-based super-resolved coherent diffraction imaging of one-dimensional objects, *Nature Communications* 6 (2015).
72. Y. Shechtman, E. Small, Y. Lahini, M. Verbin, Y. C. Eldar, Y. Silberberg, and M. Segev, Sparsity-based superresolution and phase-retrieval in waveguide arrays, *Optics Express* 21(20) (2013): 24015–24024.
73. J. Miao, C. Chen, Y. Mao, L. S. Martin, and H. C. Kapteyn, Potential and challenge of ankylography, arXiv:1112.4459 (2011).
74. M. Mutzafi, Y. Shechtman, Y. C. Eldar, O. Cohen, and M. Segev, Sparsity-based ankylography for recovering 3D molecular structures from single-shot 2D scattered light intensity, *Nature Communications* 6 (2015).
75. T. Heinosaari, L. Mazzarella, and M. M. Wolf, Quantum tomography under prior information, *Communications in Mathematical Physics* 318(2) (2013): 355–374.

76. S. S. Skiena, W. D. Smith, and P. Lemke, Reconstructing sets from interpoint distances (extended abstract), *Annual Symposium on Computational Geometry* (1990), pp. 332–339.
77. T. Dakic, On the turnpike problem, PhD thesis, Simon Fraser University, Burnaby, British Columbia, Canada (2000).
78. K. Jaganathan and B. Hassibi, Reconstruction of integers from pairwise distances, arXiv:1212.2386 (2012).
79. C. H. Papadimitriou and K. Steiglitz, *Combinatorial Optimization: Algorithms and Complexity*, Courier Corporation (1998).
80. D. P. Bertsekas, *Nonlinear Programming*, Athena Scientific, Belmont, MA (1999).
81. S. H. Nawab, T. F. Quatieri, and J. S. Lim, Signal reconstruction from short-time Fourier transform magnitude, *IEEE Transactions on Acoustics, Speech and Signal Processing* 31(4) (1983): 986–998.
82. D. Griffin and J. S. Lim, Signal estimation from modified short-time Fourier transform, *IEEE Transactions on Acoustics, Speech and Signal Processing* 32(2) (1984): 236–243.
83. D. L. Sun and J. O. Smith, Estimating a signal from a magnitude spectrogram via convex optimization, arXiv:1209.2076 (2012).
84. J. S. Lim and A. V. Oppenheim, Enhancement and bandwidth compression of noisy speech, *Proceedings of the IEEE* 67(12) (1979): 1586–1604.
85. D. J. Kane, Principal components generalized projections: A review [invited], *Journal of the Optical Society of America B* 25(6) (2008): A120–A132.
86. M. Guizar-Sicairos and J. R. Fienup, Phase retrieval with transverse translation diversity: A nonlinear optimization approach, *Optics Express* 16(10) (2008): 7264–7278.
87. L. R. Rabiner and R. W. Schafer, *Digital Processing of Speech Signals*, Prentice Hall (1978).

Index

R

S

T

W

Z

Printed and bound by CPI Group (UK) Ltd, Croydon, CR0 4YY

01/11/2024

01782604-0009